T0074898

Dedicated to all students who attended our lectures on Computational Physics.
— Lucas Böttcher and Hans J. Herrmann

I also dedicate this book to my great-grandparents Doris and Gerhard.
— Lucas Böttcher

Dedicated to all students who attended our lectures on Computational Physics.

— Lucas Böttcher and Hans J. Herrmann

I also dedicate this book to my great-grandparents Doris and Gerhard.

— Lucas Böttcher

Computational Statistical Physics

Providing a detailed and pedagogical account of the rapidly growing field of computational statistical physics, this book covers both the theoretical foundations of equilibrium and nonequilibrium statistical physics and modern, computational applications such as percolation, random walks, magnetic systems, machine learning dynamics, and spreading processes on complex networks. A detailed discussion of molecular dynamics simulations, a topic of great importance in biophysics and physical chemistry, is also included. The accessible and self-contained approach adopted by the authors makes this book suitable for teaching courses at the graduate level, and numerous worked examples and end-of-chapter problems allow students to test their progress and understanding.

Lucas Böttcher is Assistant Professor of Computational Social Science at Frankfurt School of Finance and Management and Research Scientist at UCLA's Department of Computational Medicine. His research areas include statistical physics, applied mathematics, complex systems science, and computational physics. He is interested in the application of concepts and models from statistical physics to other disciplines, including biology, ecology, and sociology.

Hans J. Herrmann is Directeur de Recherche at CNRS in Paris, visiting Professor at the Federal University of Ceará in Brazil and Emeritus of ETH Zürich. He is a Guggenheim Fellow (1986), a member of the Brazilian Academy of Science, and a Fellow of the American Physical Society. He has been the recipient of numerous prestigious awards, including the Max-Planck Prize (2002), the Gentner-Kastler Prize (2005), the Aneesur Rahman Prize (2018), and an IBM Faculty Award (2009), and he has received an ERC Advanced Grant (2012). He is Managing Editor of *International Journal of Modern Physics* C and *Granular Matter*. He has coauthored more than 700 publications and coedited 13 books.

Computational Statistical Physics

LUCAS BÖTTCHER
Frankfurt School of Finance and Management, University of California, Los Angeles

HANS J. HERRMANN
Centre National de la Recherche Scientifique (CNRS), Federal University of Ceará, ETH Zürich

CAMBRIDGE
UNIVERSITY PRESS

University Printing House, Cambridge CB2 8BS, United Kingdom

One Liberty Plaza, 20th Floor, New York, NY 10006, USA

477 Williamstown Road, Port Melbourne, VIC 3207, Australia

314–321, 3rd Floor, Plot 3, Splendor Forum, Jasola District Centre, New Delhi – 110025, India

103 Penang Road, #05–06/07, Visioncrest Commercial, Singapore 238467

Cambridge University Press is part of the University of Cambridge.

It furthers the University's mission by disseminating knowledge in the pursuit of education, learning, and research at the highest international levels of excellence.

www.cambridge.org
Information on this title: www.cambridge.org/9781108841429
DOI: 10.1017/9781108882316

First published 2021

Printed in the United Kingdom by TJ Books Limited, Padstow Cornwall

A catalogue record for this publication is available from the British Library.

ISBN 978-1-108-84142-9 Hardback

Contents

Preface

This textbook was born from the lecture notes of a course given first at the University of Stuttgart (1999–2006) as a core subject of the bachelor curriculum in computational physics and then at ETH Zurich as part of the bachelor and master curriculum in physics and computational science and engineering. Over the years, the course was continuously modified according to current scientific findings and breakthroughs in the fields of statistical physics and complex systems science.

The book is divided into two parts: Stochastic Methods (Part I) and Molecular Dynamics (Part II). In Part I, we cover computational approaches to studying percolation, random walks, spin systems, and complex networks. We include examples and exercises to illustrate that computational methods are invaluable in obtaining further insights into certain systems when analytical approaches are not feasible. In the context of Boltzmann machines, we also describe recent developments in connecting statistical physics concepts to machine learning. We conclude Part I with a discussion of computational methods to study nonequilibrium systems and highlight applications in modeling epidemic spreading, opinion dynamics, and irreversible growth.

In Part II, we focus on molecular dynamics and establish a connection between the study of microscopic particle interactions and their emerging macroscopic properties that can be analyzed statistically. We provide an overview of different simulation techniques that enable the reader to computationally study a broad range of interaction potentials and particle shapes. We discuss thermostat and barostat methods to simulate molecular interactions in canonical temperature and pressure ensembles. For the simulation of rigid particles, we cover the method of contact dynamics. To account for quantum-mechanical effects in molecular-dynamics simulations, we conclude with outlining the basic concepts of density functional theory and the Car–Parrinello method.

Throughout the book, we point to applications of the discussed mathematical and computational methods in different fields besides physics, including engineering, social sciences, and biology. In addition to providing a solid theoretical background in computational statistical physics, the aim of the book is to inspire the reader to develop their own simulation codes and computational experiments. We therefore include a carefully chosen selection of detailed exercises, which will enable the reader to implement and better understand the discussed theory and algorithms. For the interested reader, we also include side notes and information boxes that explore certain topics in greater depth.

We assume that the reader is familiar with at least one programming language and has some basic knowledge in classical mechanics, electrodynamics, statistical physics, and numerical mathematics.

This book benefited from the comments of many colleagues. In particular, we thank Stefan Luding, Malte Henkel, Dirk Kadau, Lorenz Müller, Marco-Andrea Buchmann, Nicola Offeddu, Marcel Thielmann, Madis Ollikainen, Alisha Föry, Josh LeClair, Fernando Alonso-Marroquin, Yupeng Jiang, and Thomas Asikis for their feedback. We also thank Giovanni Balduzzi for his support with performing different simulations. Moreover, we are grateful to the numerous scientists in the fields of computational and statistical physics for contributing photos that we included in biographical panels to highlight their work. Historically, a gender imbalance has been very much ingrained in computational physics, but fortunately the situation is starting to change.

What is Computational Physics?

Computational physics is the study and implementation of numerical algorithms to solve problems in physics by means of computers. As we will see throughout the book, finding numerical solutions to a given problem is useful because there are only very few systems that can be solved analytically. In particular, computational physics methods are used to simulate many-body particle systems. The simulated "virtual reality" is sometimes referred to as the third branch of physics (between experiment and theory) [1].

The analysis and visualization of large data sets that are generated numerically and experimentally is also part of computational physics but will not be treated in the present book.

Computational physics plays an important role in the following fields:

- Computational fluid dynamics (CFD): solving and analyzing problems that involve fluid flows
- Classical phase transitions: percolation, critical phenomena
- Solid state physics (quantum mechanics)
- High-energy physics/particle physics: in particular, lattice quantum chromodynamics ("lattice QCD")
- Astrophysics: many-body simulations of stars, galaxies, etc.
- Geophysics and solid mechanics: earthquake simulations, fracture, rupture, crack propagation, etc.
- Agent-based modeling (and interdisciplinary applications): complex networks in biology, economy, social sciences, and other disciplines

PART I

STOCHASTIC METHODS

1 Random Numbers

Random numbers (RNs) are important for scientific simulations. We will see that they are used in many different applications, including the following:

- To simulate random events in experimental data (e. g., radioactive decay)
- To simulate thermal fluctuations (e. g., Brownian motion)
- To incorporate a lack of detailed knowledge (e. g., traffic or stock market simulations)
- To test the stability of a system with respect to perturbations
- To perform random sampling

1.1 Definition of Random Numbers

Random numbers are a sequence of numbers in random order. The probability that a given number occurs next in such a sequence is always the same. Physical systems can produce random events, for example, in electronic circuits (*electronic flicker noise*) and in systems where quantum effects play an important role (e. g., radioactive decay or photon emissions in semiconductors). However, physical random numbers are usually *bad* in the sense that they are often correlated and not reproducible.

Generating random numbers with algorithms is also a bit problematic because computers are completely deterministic, while randomness should be nondeterministic. We therefore must content ourselves with generating so-called pseudo-random numbers, which are computed with deterministic algorithms based on strongly non-linear functions. These numbers should follow a well-defined distribution and should have long periods. Furthermore, we should be able to compute them quickly and in a reproducible way.

A very important function for generating pseudo-random numbers is the modulo operator **mod** (% in C++). It determines the remainder of a division of one integer number by another one.

Given two numbers a (dividend) and n (divisor), we write a **modulo** n or a **mod** n, which represents the remainder of dividing a by n. For the mathematical definition of the modulo operator, we consider the integers a, q, and r. We then express a as

$$a = nq + r, \tag{1.1}$$

where r ($0 \leq r < |n|$) is the remainder (i.e., the result of a **mod** n). One useful property of the **mod** operator for generating RNs is that it is a strongly nonlinear function.

We distinguish between two classes of pseudo-random number generators (RNGs): multiplicative and additive RNGs.

- Multiplicative RNGs are simpler and faster to program and are based on integers.
- Additive RNGs are more difficult to implement and are based on binary variables.

In the following sections, we describe these RNGs in detail and outline different methods that allow us to examine the quality of random sequences.

1.2 Congruential RNG (Multiplicative)

The simplest form of a congruential RNG [2, 3] was proposed by Lehmer (see Figure 1.1). It is based on the **mod** operator.

Congruential RNG

To define the congruential RNG, we choose two integer numbers, c and p, and a seed value x_0 such that $c, p, x_0 \in \mathbb{Z}$. We then generate the sequence $x_i \in \mathbb{Z},\ i \in \mathbb{N}$ iteratively according to

$$x_i = (c x_{i-1}) \,\mathbf{mod}\, p\,. \tag{1.2}$$

Figure 1.1 Derrick H. Lehmer (1905–1991) was an American mathematician who worked on number theory and theory of computation.

This iteration generates random numbers in the interval $[0, p-1]$[1]. To transform a random number $x_i \in [0, p-1]$ into a normalized random number $z_i \in [0, 1)$, we simply divide x_i by p and obtain

$$0 \leq z_i = \frac{x_i}{p} < 1\,, \tag{1.3}$$

where z_i is a rational number (i.e., $z_i \in \mathbb{Q}$). The random numbers z_i are homogeneously distributed, which means that every number between 0 and 1 is equally probable.

Since all integers are smaller than p, the sequence must repeat after maximally $(p-1)$ iterations. Thus, the *maximal period* of the RNG defined by eq. (1.2) is $(p-1)$. If we pick the seed value $x_0 = 0$, the sequence remains at this value. Therefore, $x_0 = 0$ cannot be used as a seed of the described congruential RNG.

In 1910, the American mathematician Robert D. Carmichael [4] proved that the maximal period of a congruential RNG can be obtained if p is a Mersenne prime number and if it is the smallest integer number that satisfies

$$c^{p-1} \,\mathbf{mod}\, p = 1\,. \tag{1.4}$$

[1] Throughout the book, we adopt the notation that closed square brackets [] in intervals are equivalent to "$\leq$" and "$\geq$" and parentheses () correspond to "<" and ">," respectively. Thus, the interval $[0, 1]$ corresponds to $0 \leq x \leq 1$, $x \in \mathbb{R}$ and $(0, 1)$ means $0 < x < 1$, $x \in \mathbb{R}$.

A *Mersenne number* is defined as $M_n = 2^n - 1$ with $n \in \mathbb{N}$. If the number M is prime, it is referred to as *Mersenne prime*. As of May 2021, only 51 Mersenne primes were known. The largest has 24,862,048 digits and was discovered by Patrick Laroche in December 2018 within the Great Internet Mersenne Prime Search (GIMPS) [5]. In 1988, Park and Miller [6] proposed the following numbers to generate the maximal period of congruential RNGs, here, in pseudo-code:

```
const int p=2147483647;

const int c=16807;

int rnd=42; // seed

rnd=(c*rnd)%p;

print rnd;
```

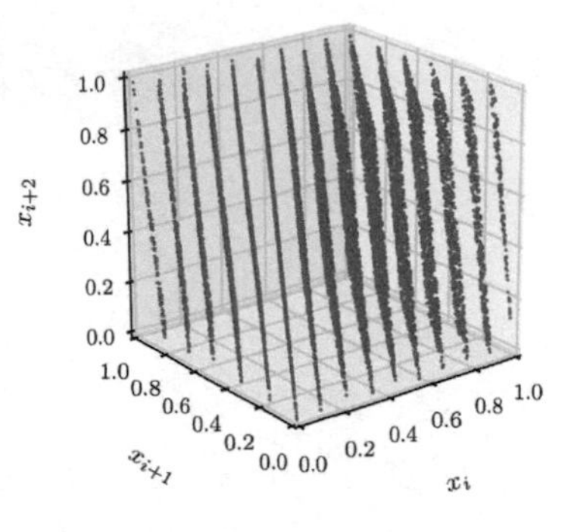

Figure 1.2 Cube test: plot of consecutive random numbers x_i, x_{i+1}, x_{i+2} with 15 clearly visible planes ("RANDU" algorithm with $c = 65539$, $p = 2^{31}$, $x_0 = 1$).

The number $p = 2147483648$ is a Mersenne prime and corresponds to the maximal integer of 32 bits: $2^{31} - 1$.

To assess the homogeneity of numbers generated by an RNG, we can either plot two consecutive pseudo-random numbers (x_i, x_{i+1}) (the so-called square test) or employ a "cube test" for three consecutive numbers (x_i, x_{i+1}, x_{i+2}). In Figure 1.2, we show an example of an RNG that clearly fails the cube test. All numbers that are generated by this particular RNG lie on just 15 parallel planes, indicating correlation effects that may lead to undesired effects in simulations of stochastic processes. In 1968, George Marsaglia (see Figure 1.3) showed that all congruential RNGs as defined in eq. (1.2) produce pseudo-random numbers that lie on equally spaced hyperplanes [8]. The distance between these planes decreases with the length of the period. For the interested reader, we briefly summarize the corresponding theorem on the following page.

Figure 1.3 George Marsaglia (1924–2011) made many contributions to improving and testing RNGs [7].

Additional Information: Marsaglia's Theorem

Let $\{z_i\} = \{x_i/p\}$ ($i \in \mathbb{N}$) denote normalized numbers of an RNG sequence, and let $\pi_1 = (z_1, \ldots, z_n)$, $\pi_2 = (z_2, \ldots, z_{n+1})$, $\pi_3 = (z_3, \ldots, z_{n+2})$, ... be the points in the unit n-cube formed by n successive numbers z_i. Marsaglia showed that all points $\pi_1, \pi_2, \ldots$ lie on parallel hyperplanes [8]. Formally, if $a_1, a_2, \ldots, a_n \in \mathbb{Z}$ is any choice of integers such that

$$a_1 + a_2 c + a_3 c^2 + \cdots + a_n c^{n-1} \equiv 0 \text{ mod } p,$$

then all the points $\pi_1, \pi_2, \ldots$ will lie in the set of parallel hyperplanes defined by the equations

$$a_1 y_1 + a_2 y_2 + \cdots + a_n y_n = 0, \pm 1, \pm 2, \ldots, \quad y_i \in \mathbb{R}, \; 1 \le i \le n.$$

Additional Information: Marsaglia's Theorem (*cont.*)

There are at most

$$|a_1| + |a_2| + \cdots + |a_n|$$

of these hyperplanes that intersect the unit n-cube. Note that there is always a choice of $a_1, \ldots, a_n$ such that all of the points fall in fewer than $(n!p)^{1/n}$ hyperplanes. The theorem can be proven in four steps:
Step 1: If

$$a_1 + a_2c + a_3c^2 + \cdots + a_nc^{n-1} \equiv 0 \,\mathbf{mod}\, \mathbf{p},$$

one can prove that

$$a_1z_i + a_2z_{i+1} + \cdots + a_nz_{i+n-1}$$

is an integer for every i.
Step 2: The point $\pi_i = (z_i, z_{i+1}, \ldots, z_{i+n-1})$ must lie in one of the hyperplanes

$$a_1y_1 + a_2y_2 + \cdots + a_ny_n = 0, \pm1, \pm2, \ldots, \quad y_i \in \mathbb{R},\ 1 \leq i \leq n.$$

Step 3: The number of hyperplanes of the above type, which intersect the unit n-cube, is at most

$$|a_1| + |a_2| + \cdots + |a_n|.$$

Step 4: For every multiplier c and modulus p, there is a set of integers $a_1, \ldots, a_n$ (not all zero) such that

$$a_1 + a_2c + a_3c^2 + \cdots + a_nc^{n-1} \equiv 0 \,\mathbf{mod}\, \mathbf{p}$$

and

$$|a_1| + |a_2| + \cdots + |a_n| \leq (n!p)^{1/n}.$$

This is of course only the outline of the proof. The exact details are described in Marsaglia (1968) [8]. In a similar way, it is possible to show that for congruential RNGs the distance between the planes must be larger than $\sqrt{\frac{p}{n}}$.

Exercise: Congruential RNG

Task 1
Write a program that generates random numbers using a congruential random number generator. First, use small values for c (< 50) and p (< 50), for example, $c = 3$ and $p = 31$.

- Check for correlations using the square test. That is, create a plot of two consecutive random numbers (x_i, x_{i+1}). (What is the maximum number of random numbers that have to be created until you can see all possible lines/planes for this specific random number generator?)
- Create a corresponding 3D plot for the cube test.
- Do the same for other random number generators (at least one!), for example, by changing c and p. You may also compare your results to those obtained with C++ built-in generators, such as `rand()` and `drand48()`.

Exercise: Congruential RNG (*cont.*)

Task 2

Generate a homogeneous distribution of random points inside a circle. How should the coordinates r and ϕ be chosen using uniformly distributed random numbers?

Task 3

Test your RNG for different c and p using the χ^2 test:

- Divide the range of random numbers into k bins. That is, divide the range into discrete intervals of the same size, so that the probability of a random number lying in interval i is $p_i = 1/k$.
- Using each RNG, generate at least one sequence of n numbers. For each sequence, determine N_i, the number of random numbers in interval i (choose n such that all $np_i \geq 5$).
- Compute the χ^2 value for one specific sequence s of random numbers

$$\chi_s^2 = \sum_{i=1}^{k} \frac{(N_i - np_i)^2}{np_i} .$$

  Use the table from Knuth [9] to check if the random numbers are uniformly distributed and compare the quality of different RNGs.
- Calculate χ_s^2 for different sequences (i. e., different seeds of the RNG). You can then plot the cumulative probability for the χ^2 in comparison to the theoretically expected values (values again from the table of Ref. [9]).

1.3 Lagged Fibonacci RNG (Additive)

A slightly more complicated RNG is the lagged Fibonacci algorithm proposed by Robert C. Tausworthe (see Figure 1.4) in 1965 [10]. Lagged Fibonacci-type generators can achieve extremely long periods and allow us to make some predictions about their underlying correlation effects.

To define lagged Fibonacci RNGs, we start with a sequence of binary numbers $x_i \in \{0, 1\}$ $(1 \leq i \leq b)$. The next bit in our sequence x_{b+1} is

$$x_{b+1} = \left(\sum_{j \in \mathcal{J}} x_{b+1-j} \right) \mathbf{mod}\, 2 , \tag{1.5}$$

Figure 1.4 Robert C. Tausworthe is a retired senior research engineer at the Jet Propulsion Laboratory, California Institute of Technology.

where $\mathcal{J} \subset [1, \ldots, b]$. The sum $\sum_{j \in \mathcal{J}} x_{b+1-j}$ includes only a subset of all the other bits, so the new bit could, for instance, simply be based on the first and third bits, $x_{b+1} = (x_1 + x_3)\mathbf{mod}\, 2$ (or any other subset). We now focus on the properties of RNGs that are defined by eq. (1.5) and consider a two-element lagged Fibonacci generator.

Two-Element Lagged Fibonacci RNG

Let $c, d \in \{1, \ldots, b\}$ with $d \leq c$. The RNG sequence elements x_{b+1} are recursively generated according to

$$x_{b+1} = (x_{b+1-c} + x_{b+1-d}) \,\mathbf{mod}\, \mathbf{2}\,.$$

We immediately see that we need some initial sequence of at least c bits to start from (a so-called *seed sequence*). Except for all bits equal to zero, any other seed configuration can be used. One possibility is to use a seed sequence that was generated by a congruential RNG.

The maximum period of sequences that are generated by the outlined two-element lagged Fibonacci RNG is 2^c-1. As for congruential RNGs, there are conditions for the choice of the parameters c and d to obtain the maximum period. For lagged Fibonacci generators, c and d must satisfy the Zierler–Trinomial condition, which states that the trinomial

$$T_{c,d}(z) = 1 + z^c + z^d\,, \tag{1.6}$$

where z is a binary number, cannot be factorized into subpolynomials (i.e., is primitive). A possible choice of numbers that satisfy the Zierler condition is $(c,d) = (250, 103)$. The generator is named after Kirkpatrick (see Figure 9.10) and Stoll, who proposed these numbers in 1981 [11].

Some examples of known pairs (c,d) follow:

$$\begin{array}{cl} (c,d) & \\ (250, 103) & \text{Kirkpatrick–Stoll (1981) [11]} \\ (132049, 54454) & \text{J. R. Heringa et al. (1992) [12]} \\ (6972593, 3037958) & \text{R. P. Brent et al. (2003) [13].} \end{array} \tag{1.7}$$

We may use one of the following methods to convert the obtained binary sequences to natural numbers (e.g., 32 bit unsigned integers):

- Running 32 lagged Fibonacci generators in parallel (this can be done very efficiently): The problem with this method is the initialization, because all 32 initial sequences have to be uncorrelated. The quality of the initial sequences has a major impact on the quality of the produced random numbers.
- Extracting a 32-bit-long part from the sequence: This method is relatively slow because for each random number we have to generate 32 new elements in the binary sequence.

1.4 Available Libraries

In general, we should always make sure that we use a high-quality RNG. For some purposes it might be sufficient to use `drand48` (in C/C++). If your compiler supports the C++11 standard (or above), there are different implementations already available in the "random" library. For example, linear congruential RNGs are available by calling `minstd_rand` from the `linear_congruential_engine` class. A useful general-purpose RNG is the so-called *Mersenne twister* [14], which was developed in 1997 by Makoto Matsumoto and Takuji Nishimura. The Mersenne twister has some structural similarities to lagged Fibonacci RNGs and belongs to the class of generalized feedback shift register algorithms. The name "Mersenne twister" was chosen because its period length is a Mersenne prime. In C++, we can use the Mersenne twister to generate uniformly distributed random numbers as follows:

```
#include <iostream>
#include <random>

using namespace std;
int main()
{

    random_device rd;
    mt19937 mt_rng(rd());
    uniform_real_distribution<double> u(0.0, 1.0);

    double rnd = u(mt_rng);

    cout << rnd << endl;

    return 0;
}
```

In the above code listing, we use the `random_device` as seed for the `mt19937` generator. This generator has a very long period of $2^{19937} - 1$. For the sake of reproducibility of numerical results, it is recommended to store all seeds in a file. In PYTHON, no further efforts are required to use the Mersenne twister because the `mt19937` is the core RNG in all PYTHON distributions.

1.5 How Good is an RNG?

There are many possibilities to test how "random" a certain RNG sequence is. Possible tests for a given sequence $\{s_i\}$, $i \in \mathbb{N}$ include the following:

1. Square test (see Section 1.2 for details)
2. Cube test (see Section 1.2 for details)

3. Average value: the arithmetic mean of all numbers in the sequence $\{s_i\}$ should correspond to the analytical mean value. Let us assume here that the numbers s_i are rescaled to lie in the interval $s_i \in [0, 1)$. The arithmetic mean should then be

$$\bar{s} = \lim_{N\to\infty} \frac{1}{N} \sum_{i=1}^{N} s_i = \frac{1}{2}. \tag{1.8}$$

The more numbers that are averaged, the better $\frac{1}{2}$ will be approximated.
4. Fluctuation of the mean value (χ^2 test): the distribution around the mean value should behave like a Gaussian distribution.
5. Spectral analysis (Fourier analysis): Let $\{s_i\}$ denote values of a function. It is possible to perform a Fourier transform of such a function by means of the fast Fourier transform (FFT; see details in Section 15.2.1). If the frequency distribution corresponds to white noise (uniform distribution), the randomness is good; otherwise, peaks will show up (resonances).
6. Correlation test: Analysis of correlations such as

$$\langle s_i s_{i+d} \rangle - \langle s_i^2 \rangle, \tag{1.9}$$

for different values of d.

Of course, this list is not complete. Many other tests can be used to check the quality of RNG sequences.

Probably the most famous set of RNG tests is the Marsaglia's "Diehard" set. These Diehard tests are a battery of statistical tests for measuring the quality of a set of random numbers. They were developed over many years and published for the first time by Marsaglia on a CD-ROM with random numbers in 1995 [15]. Marsaglia's tests were inspired by different applications, and each can measure different types of correlations.

Additional Information: Marsaglia's "Diehard" Tests

- Birthday spacings: If random points are chosen in a large interval, the spacings between points should be asymptotically Poisson distributed. The name stems from the birthday paradox.[1]
- Overlapping permutations: When analyzing five consecutive random numbers, the 120 possible orderings should occur with equal probability.
- Ranks of matrices: A number of bits of some number of random numbers is formed into a matrix over {0,1}. The rank of this matrix is then determined, and the ranks are counted.
- Monkey test: Sequences of some number of bits are taken as words, and the number of overlapping words in a stream is counted. The number of words not appearing should follow a known distribution. The name is based on the infinite monkey theorem.[2]
- Parking lot test: Randomly place unit circles in a 100×100 square. If the circle overlaps an existing one, try again. After 12,000 tries, the number of successfully "parked" circles should follow a certain normal distribution.

Additional Information: Marsaglia's "Diehard" Tests (*cont.*)

- Minimum distance test: Find the minimum distance of 8,000 uniformly randomly placed points in a $10{,}000 \times 10{,}000$ square. The square of this distance should be exponentially distributed with a certain mean.
- Random spheres test: Put 4,000 randomly chosen points in a cube of side length 1,000. Now a sphere is placed on every point with a radius corresponding to the minimum distance to another point. The smallest sphere's volume should then be exponentially distributed.
- Squeeze test: 2^{31} is multiplied by random floats in $[0, 1)$ until 1 is reached. After 100,000 repetitions, the number of floats needed to reach 1 should follow a certain distribution.
- Overlapping sums test: Sequences of 100 consecutive floats are summed up in a very long sequence of random floats in $[0, 1)$. The sums should be normally distributed.
- Runs test: Ascending and descending runs in a long sequence of random floats in $[0, 1)$ are counted. The counts should follow a certain distribution.
- Craps test: 200,000 games of craps[3] are played. The number of wins and the number of throws per game should follow a certain distribution.

[1] The birthday paradox states that the probability of two randomly chosen persons having the same birthday in a group of 23 (or more) people is more than 50 percent. For 57 or more people, the probability is already larger than 99 percent. Finally, for at least 366 people, the probability is exactly 100 percent. This is not paradoxical in a logical sense; it is called a paradox nevertheless since intuition would suggest probabilities much lower than 50 percent.

[2] The infinite monkey theorem states that a monkey hitting keys at random on a typewriter keyboard for an infinite amount of time will almost surely (i.e., with probability 1) write a certain text, such as the complete works of William Shakespeare.

[3] A dice game.

1.6 Nonuniform Distributions

Thus far, we have only considered uniform distributions of pseudo-random numbers. The congruential and lagged Fibonacci RNGs produce numbers that can easily be mapped to the interval [0, 1) or any other interval by simple shifts and multiplications. However, if we want to generate random numbers that are distributed according to a certain distribution (e. g., a Gaussian distribution), the algorithms presented so far are not able to do so. However, we may employ techniques that permit us to transform uniform pseudo-random numbers to other distributions. There are essentially two different ways to perform this transformation:

- We can apply *transformation methods* if the target distribution is known analytically, is integrable, and the resulting expression is invertible.
- However, if the target distribution is not known analytically or if it cannot be analytically integrated and inverted, we have to use the so-called *rejection method*.

These methods are explained in the following sections.

1.6.1 Transformation Method

For a certain class of distributions, we can generate pseudo-random numbers from uniformly distributed random numbers by applying a mathematical transformation. The transformation method works particularly nicely for the most common distributions (exponential and normal distributions). While the transformation is rather straightforward, it is not always feasible – depending on the analytical form of the distribution.

For a positive continuous random variable z and probability density function (PDF) $f(z)$, the probability that z lies between z and $z+\mathrm{d}z$ is $f(z)\,\mathrm{d}z$. Thus, the probability that z takes on values between a and b is

$$\Pr(a \le z \le b) = \int_a^b f(z)\,\mathrm{d}z. \tag{1.10}$$

The corresponding cumulative distribution function (CDF) is

$$F(z) = \int_0^z f(z')\,\mathrm{d}z'. \tag{1.11}$$

The idea behind the transformation method is to find the equivalence between the CDFs of a uniform distribution f_u and the distribution of interest. The PDF of the uniform distribution with support $[0,1]$ is

$$f_\mathrm{u}(z) = \begin{cases} 1, & \text{for } z \in [0,1], \\ 0, & \text{otherwise}. \end{cases} \tag{1.12}$$

Now we want to obtain random numbers that are distributed according to a second PDF $f(y)$. If we compare the areas of integration (i.e., CDFs), we find

$$z = \int_0^z f_\mathrm{u}(z')\,\mathrm{d}z' = \int_0^y f(y')\,\mathrm{d}y', \tag{1.13}$$

where z is a uniformly distributed random variable and y a random variable distributed according to the desired distribution $f(y)$. The first equality in eq. (1.13) follows from the definition of the uniform distribution in eq. (1.12) and we impose the second equality to find a suitable change of variables such that the CDFs of f_u and f are the same. Inverting $F(y)$ leads to

$$y = F^{-1}(z). \tag{1.14}$$

This shows that a transformation between the two distributions can be found if and only if

1. the integral $F(y) = \int_0^y f(y')\,\mathrm{d}y'$ can be solved analytically in a closed form (i.e., $f(y)$ must be integrable),
2. there exists a closed-form analytical expression of the inverse of $z = F(y)$ such that $y = F^{-1}(z)$ (i.e., $F(y)$ must be invertible).

Of course, these conditions can be overcome to a certain extent by precomputing/tabulating and inverting $F(y)$ numerically if the integral is well behaved (i. e., nonsingular). Then, it is possible to transform the uniform numbers numerically.

We are now going to demonstrate this method for the two most commonly used distributions: the exponential distribution and the Gaussian distribution. We will see that already in the case of a Gaussian distribution, quite some work is required to perform such a transformation.

Figure 1.5 George E. P. Box (1919–2013) was a statistician working at different research institutions, including Princeton University and the University of Wisconsin–Madison.

The Exponential Distribution

The exponential distribution is

$$f(y) = ke^{-yk}. \tag{1.15}$$

By applying the area equality of eq. (1.13), we find

$$z = \int_0^y ke^{-y'k}\,\mathrm{d}y', \tag{1.16}$$

and thus

$$z = -e^{-y'k}|_0^y = 1 - e^{-yk}. \tag{1.17}$$

Solving for y yields

$$y = -\frac{1}{k}\ln(1-z), \tag{1.18}$$

so that for each homogeneously distributed random number z we get an exponentially distributed random number y through eq. (1.18).

Figure 1.6 Mervin E. Muller (1928–2018) was a computer scientist working at the Ohio State University. Photograph courtesy The Ohio State University.

The Gaussian Distribution

Methods for generating normally distributed random numbers are very useful because there are many applications and examples where such numbers are needed.

The Gaussian or normal distribution is

$$f(y) = \frac{1}{\sqrt{2\pi\sigma^2}}e^{-\frac{y^2}{2\sigma^2}}, \tag{1.19}$$

where σ represents the standard deviation of the distribution. Unfortunately, we can only solve its integral between $-\infty$ and ∞ analytically:

$$\frac{1}{\sqrt{2\pi\sigma^2}}\int_{-\infty}^{\infty} e^{-\frac{y'^2}{2\sigma^2}}\,\mathrm{d}y' = 1. \tag{1.20}$$

However, Box and Muller [16] (see Figures 1.5 and 1.6) have introduced the following elegant trick to circumvent this restriction. We take two (uncorrelated) uniform random variables z_1 and z_2. We apply the area equality of eq. (1.13) again, but this time we write it as a product of the two random variables

$$z_1 \cdot z_2 = \int_{-\infty}^{y_1} \frac{1}{\sqrt{2\pi\sigma^2}} e^{-\frac{y_1'^2}{2\sigma}} \mathrm{d}y_1' \cdot \int_{-\infty}^{y_2} \frac{1}{\sqrt{2\pi\sigma^2}} e^{-\frac{y_2'^2}{2\sigma}} \mathrm{d}y_2'$$
$$= \int_{-\infty}^{y_2} \int_{-\infty}^{y_1} \frac{1}{2\pi\sigma^2} e^{-\frac{y_1'^2+y_2'^2}{2\sigma}} \mathrm{d}y_1' \mathrm{d}y_2' \,. \tag{1.21}$$

We now solve the integral by transforming the variables y_1' and y_2' into polar coordinates

$$r^2 = y_1^2 + y_2^2 \,, \tag{1.22}$$

$$\tan\phi = \frac{y_1}{y_2} \,, \tag{1.23}$$

with

$$\mathrm{d}y_1' \mathrm{d}y_2' = r' \mathrm{d}r' \mathrm{d}\phi' \,. \tag{1.24}$$

We apply these coordinate transformations to eq. (1.21) to obtain

$$z_1 \cdot z_2 = \frac{1}{2\pi\sigma^2} \int_0^{\phi} \int_0^{r} e^{-\frac{r'^2}{2\sigma^2}} r' \, \mathrm{d}r' \mathrm{d}\phi' \tag{1.25}$$

$$= \frac{\phi}{2\pi\sigma^2} \int_0^{r} e^{-\frac{r'^2}{2\sigma^2}} r' \, \mathrm{d}r' \tag{1.26}$$

$$= \frac{\phi}{2\pi\sigma^2} \cdot \sigma^2 \left(1 - e^{-\frac{r^2}{2\sigma^2}}\right), \tag{1.27}$$

giving

$$z_1 \cdot z_2 = \underbrace{\frac{1}{2\pi} \arctan\left(\frac{y_1}{y_2}\right)}_{\equiv z_1} \cdot \underbrace{\left(1 - e^{-\frac{y_1^2+y_2^2}{2\sigma^2}}\right)}_{\equiv z_2} . \tag{1.28}$$

By separating these two terms (and associating them to z_1 and z_2, respectively), we can invert this function and find

$$y_1^2 + y_2^2 = -2\sigma^2 \ln(1 - z_2) \,, \tag{1.29}$$

$$\frac{y_1}{y_2} = \tan(2\pi z_1) = \frac{\sin(2\pi z_1)}{\cos(2\pi z_1)} \,. \tag{1.30}$$

One solution of these two coupled equations is

$$y_1 = \sqrt{-2\sigma^2 \ln(1 - z_2)} \sin(2\pi z_1) \,, \tag{1.31}$$

$$y_2 = \sqrt{-2\sigma^2 \ln(1 - z_2)} \cos(2\pi z_1) \,. \tag{1.32}$$

Thus, by using two uniformly distributed random numbers z_1 and z_2, we obtain (through Eqs. (1.31) and (1.32)) two normally distributed random numbers y_1 and y_2.

1.6.2 Random Points on a Sphere

Another application of the transformation method is to generate points that are distributed uniformly at random over the surface of a sphere with radius R. We denote the coordinates of the points by coordinates (ϕ, θ) where $\phi \in [0, 2\pi)$ and $\theta \in [0, \pi]$.

An overly simplified attempt would be to generate two uniformly distributed random numbers $u, v \in [0, 1]$ and then set $\phi = 2\pi u$ and $\theta = v\pi$. We show the resulting distribution of points in the top panel of Figure 1.7. It is evident that there is a larger concentration of points in the pole regions when compared to the remaining parts of the sphere. The reason is that we are not properly accounting for the surface element $dA = R^2 \sin(\theta) d\phi d\theta$.

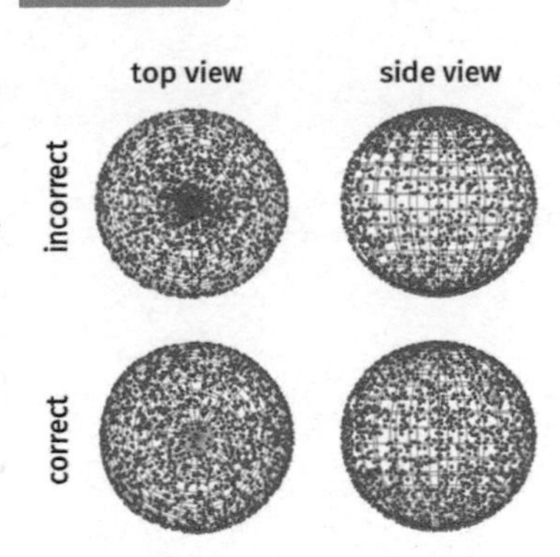

Figure 1.7 Distribution of random points on a sphere.

To obtain points that are distributed uniformly at random over a sphere surface, we have to apply the transformation method to the distribution of θ:

$$v = \frac{1}{2} \int_0^{\theta} \sin(\theta')\, d\theta' . \quad (1.33)$$

The prefactor 1/2 guarantees that the cumulative distribution function is normalized to unity for $\theta = \pi$. Based on eq. (1.33), we find $\theta = \arccos(2v - 1)$ and as in the first attempt we use $\phi = 2\pi u$. The points are now uniformly distributed over the sphere's surface as we show in the bottom panel of Figure 1.7.

1.6.3 The Rejection Method

We have seen in Section 1.6.1 that there are two conditions that a function has to satisfy to apply the transformation method: integrability and invertability. If either of these conditions is not satisfied, there exists no analytical method to obtain random numbers for such a distribution. It is important to note that this is particularly relevant for experimentally obtained data, where no analytical description is available. In such a case, we have to resort to a numerical method (rejection method) to obtain arbitrarily distributed random numbers.

Let $f(y)$ be the distribution from which we would like to obtain random numbers. A necessary condition for the rejection method to work is that $f(y)$ is bounded and only exists over a finite domain. That is, $f(y) < A$ for $y \in [0, B]$, with $A, B \in \mathbb{R}$ and $A, B < \infty$. We then define a box with edge lengths B and A (see Figure 1.8).

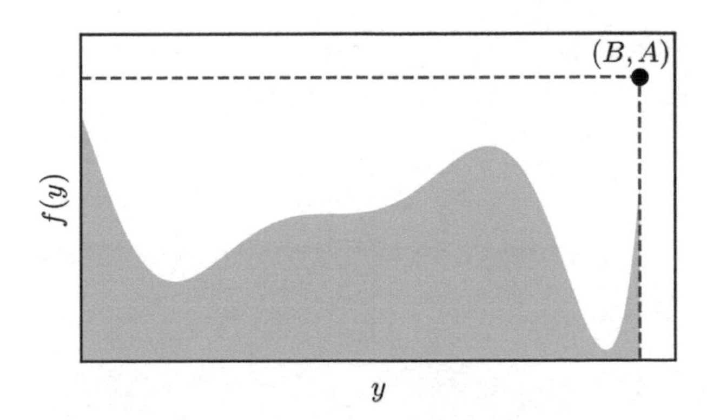

Figure 1.8 Illustration of the rejection method. Sample points are placed within the box and rejected if they lie above the curve (i.e, inside the blue-shaded area), and accepted otherwise.

We now generate two pseudo-random variables z_1 and z_2 with $z_1, z_2 \in [0, 1)$. If we consider the point with coordinates (Bz_1, Az_2), we see that it surely lies within the defined box. If this point lies above the curve describing the distribution $f(y)$ (i.e., $Az_2 > f(Bz_1)$), the point is rejected (hence the name of the method). Otherwise $y = Bz_1$ is retained as a random number that is distributed according to $f(y)$.

In principle, the method works quite well. It can be improved considering the following points:

- It is desirable to have a good guess for the upper bound. Obviously, the better the guess, the less points are rejected. In the above description of the algorithm, we have assumed a rectangular box. This is, however, not a necessary condition. In principle, one can take any domain on which homogeneously distributed random numbers are easily generated.
- There is a method to make the rejection method faster (but also more complicated). We may use N boxes to cover $f(y)$ and define the individual box with side length A_i and $b_i = B_{i+1} - B_i$ for $l \leq i \leq N$. Then, the approximation of $f(y)$ is much better (this is related to the definition of the Riemann integral).

2 Random-Geometrical Models

2.1 Percolation

In materials science and chemistry, *percolation* refers to the movement or filtering of fluids through porous media. The term stems from a Latin word and is still common in Italian.[1]

A very simple and basic model of such a process was first introduced by Broadbent and Hammersley (see Figure 2.1) in 1957 [17].

While the original idea was to model fluid motion through a porous material (e. g., a container filled with glass beads), it was found that the model has many other applications. Furthermore, it was observed that the model possesses some interesting universal features characteristic of so-called critical phenomena.[2]

Applications include:

- Porous media: for example, used in the oil industry and as a model for the pollution of soils
- Sol-gel transitions
- "Mixtures" of conductors and insulators: finding the point at which a conducting material becomes insulating
- Spreading of (forest) fires [18]
- Spreading of epidemics and computer viruses [19, 20]
- Crashing of stock markets (D. Sornette [21])
- Landslide election victories (S. Galam [22])
- Recognition of pathogenic antigens by T-cells (A. Perelson [23])

Figure 2.1 John Hammersley (1920–2004) was a British mathematician who made substantial contributions to probability theory.

2.2 The Sol–Gel Transition

The formation of gelatine is quite special from a chemical and physical point of view. Initially, it is a fluid containing many small monomers (emulsion) which is referred to as sol. If we place the sol in a fridge, it becomes a gel. The process taking place

[1] *ita*: percolare: 1 Passare attraverso. ~ filtrare. 2 Far filtrare. (1 pass through ~ filter, 2 make sth. filter).

[2] The notion of *critical phenomena* describes the physics associated with critical points and phase transitions and is discussed later in this book.

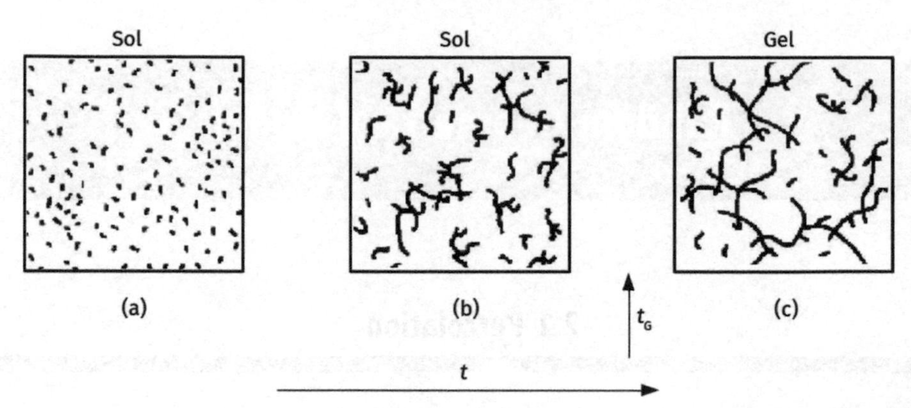

Figure 2.2 Sol–Gel transition. (a) Monomers are dispersed in the liquid (sol). (b) When the liquid is cooled, the monomers start polymerizing and growing. (c) Once a huge macromolecule (which stretches across the whole container) has formed, the sol–gel transition occurs at the gel time t_G.

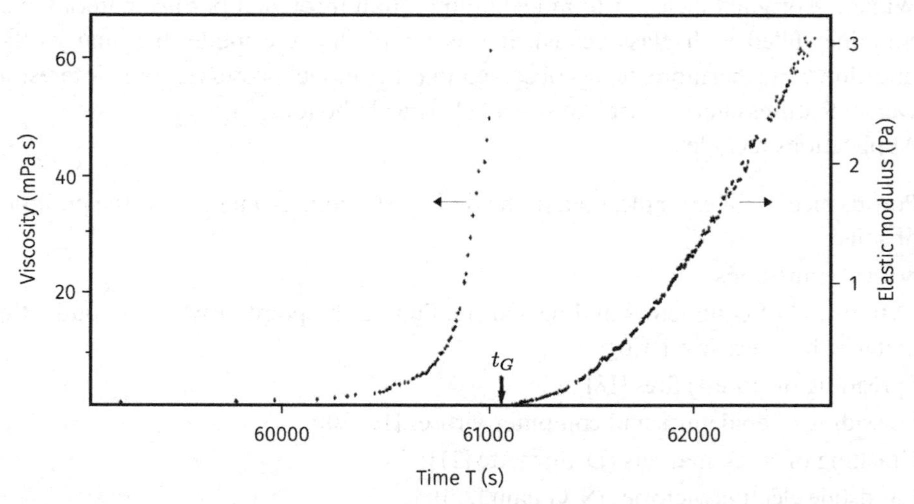

Figure 2.3 Viscosity and elastic modulus for the system acrylamide-bis-acrylamide as a function of time. At the gelation time t_G, the viscosity diverges to infinity, and the elastic modulus, previously at zero, increases to a finite value. The figure is taken from Ref. [24].

is schematically illustrated in Figure 2.2. Upon cooling, monomers start polymerizing and polymers start growing. At some point in time, one molecule becomes sufficiently long to span from one side of the container to the other, which is when the so-called percolation transition occurs. The polymerization and growth process is experimentally accessible; one can, for instance, measure the elastic modulus or the viscosity as a function of time. After a characteristic time ("gel time") t_G, the elastic modulus suddenly increases from zero to a finite value. Similarly, the viscosity increases and diverges at t_G; this is reflected in experimental findings (see Figure 2.3).

2.3 The Percolation Model

Modeling the process of gelation is surprisingly simple [25]. Let us assume that we have a square lattice of side length L (e. g., $L = 40$) such that every cell can either be occupied or empty. We fill each cell with probability p. That is, for each cell, we create a random number $z \in [0, 1)$ and compare it to p. If $z < p$, the cell will be marked as occupied; otherwise, the cell remains empty. This can be done for different values of p (see Figure 2.4). For small values of p, most cells are empty, whereas for big values of p, most cells are occupied. If two nearest neighbors are occupied, one says that they are connected and a group of connected sites is called a cluster. There exists a critical probability $p = p_c$ ($p_c = 0.592\ldots$ for an infinite lattice) where, for the first time, a fully connected cluster of cells will span the box (such a configuration is called a percolating cluster). Although p_c is uniquely defined for infinite lattice sizes, we can also find it for finite sizes; in that case, p_c is the average probability at which a percolating cluster occurs. The critical probability is referred to as the percolation threshold.

2.3.1 The Burning Method

In order to calculate the critical probability associated with a lattice, we need a method to detect if a spanning cluster exists in a given system (such as that shown in right panel of Figure 2.4).

The burning method [26] not only provides us with a Boolean feedback (yes/no), but also calculates the minimal path length (the minimal distance between opposite sides, following only connected sites). The name of the method stems from its implementation. Imagine a grid with occupied and unoccupied sites; an occupied site represents a tree while an unoccupied site is simply empty space. If we start a fire at the very top of our grid, all trees in the first row will start to light up as they are in the immediate

Figure 2.4 Percolation on a square lattice with side length $L = 40$ for different values of p. For $p < p_c$, there are many small clusters, while for $p > p_c$, one big spanning cluster is present.

Figure 2.5 The burning method. A fire is started in the first line. At every iteration, the fire from a burning tree lights up occupied neighboring cells. The algorithm ends if all reachable trees are burned or if the burning method has reached the bottom line.

vicinity of the fire. Obviously, this forest fire will spread and nearest neighbors of the trees in the first row will soon catch on fire as well. Thus, in the second iterative step all trees immediately adjacent to burning trees get torched.

Clearly, the iterative method only comes to an end if the fire has reached the bottom or if the fire runs out of fuel (i.e., no unburned trees neighboring a burning site) and consequently dies out. If the inferno has reached the bottom, we can determine the length of the shortest path from the iteration counter; that is, the number of iterations needed to reach the other side defines the shortest-path length of the percolating cluster.

The algorithm is as follows:

Burning Method

1. Label occupied sites by a one and unoccupied ones by zero.
2. Label all occupied cells in the top line with the marker $t = 2$.
3. Iteration step $t + 1$:
   (a) Go through all the cells which have label t.
   (b) For each t-labeled cell:
      i. Check if any nearest neighbor (on the square lattice: North, East, South, West) is occupied and not burning (label is 1).
      ii. Assign any such neighbors the label $t + 1$.

Burning Method (*cont.*)

4. Repeat step 2 (with $t \to t+1$) until either there are no neighbors to burn anymore or the bottom line has been reached. In the latter case, the latest label minus 1 defines the shortest-path length.

A graphical representation of the burning algorithm on the square lattice is given in Figure 2.5. In the shown example, the shortest-path length is $22 - 1 = 21$ units.

Exercise: Percolation and Forest Fire

Task 1
Write a program to simulate percolation with occupation probability p on a square lattice of size $L \times L$. Consider different lattice sizes and illustrate the resulting lattice with occupied and unoccupied sites.

Task 2
Write a program for a simple model for forest fires. This can be used to check if a sample contains a spanning cluster (i. e., if it percolates):

- For a set of parameters L and p, create a square lattice, like in the previous task (representing the forest). Color the occupied sites with green (representing trees) or assign a number to them that represents unburned forest.
- In the first time step burn all the trees in the first row (change the color to red or assign a number that represents burning trees).
- In the next time step, burn all the trees neighboring the burning trees (red ones) and burn out the previously burning trees (change their color to black or assign a number that represents burned trees).
- Repeat until the fire stops.
- Could you find a spanning cluster?
- You can now also measure the following quantities:
  - the shortest-path length (i.e., the number of time steps that the fire needs to reach the other side).
  - the lifetime of the fire defined as the total number of time steps needed before the fire stops.

2.3.2 The Percolation Threshold

When we carry out simulations of the percolation model, we observe that the probability of obtaining a spanning cluster or "wrapping probability" $W(p)$ depends on the occupation probability p (see Figure 2.6). When we also vary the lattice size, we see that the transition from low- to high-wrapping probabilities $W(p)$ becomes more abrupt with increasing lattice size. If we choose very large lattices, the wrapping probability will start to resemble a step function. The occupation probability at which we

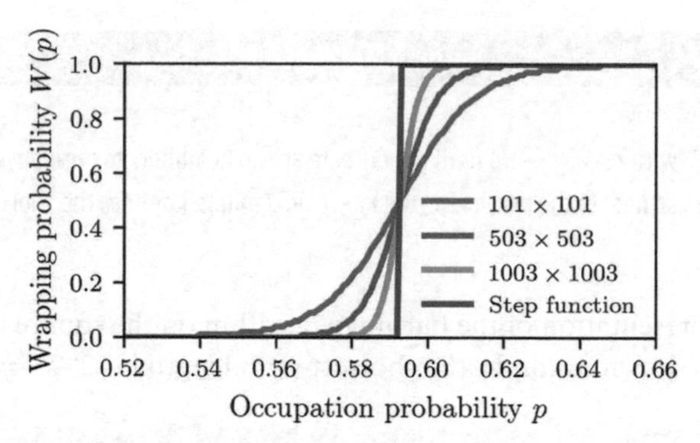

Figure 2.6 We show the wrapping probability $W(p)$ for site percolation on the square lattice as a function of the occupation probability for different lattice sizes along with a step function for comparison. A more detailed analysis reveals that the percolation threshold is $p_c \approx 0.592746$ [27].

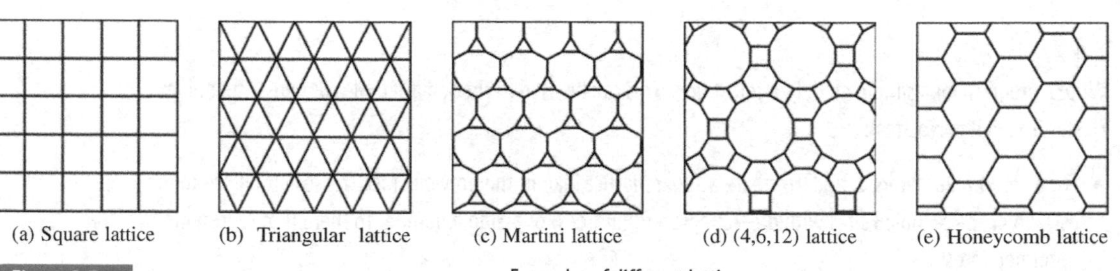

Figure 2.7 Examples of different lattices.

observe the increasingly abrupt transition is the *percolation threshold* p_c. When we carry out a sequence of simulations with increasing lattice sizes, we can determine the percolation threshold on the square lattice to be approximately $p_c = 0.59274\ldots$ [27]. The percolation threshold is a characteristic value for a given type of lattice. We find different percolation thresholds depending on the type of lattice under consideration. We show values for some lattice types in Table 2.1.

Percolation can be performed not only for lattice sites (site percolation) but also for the corresponding bonds by randomly occupying them (bond percolation). For a given lattice, the number of neighbors may be different for bond percolation when compared to site percolation. This difference will be also reflected in the corresponding bond percolation threshold.[3] Looking at the different lattices in Figure 2.7 and Table 2.1, it becomes clear that thresholds decrease with the number of neighbors (also called "coordination number" or "degree"). For example, the honeycomb lattice (see Figure 2.7e) has the highest threshold in two dimensions. Furthermore, the threshold of some lattices such as the triangular lattice (see Figure 2.7b) can be calculated analytically [28] (denoted by an asterisk * in Table 2.1). Intuitively, it is clear that

[3] Please note that unless otherwise stated, we subsequently use the term *percolation* instead of *site percolation* for the sake of brevity. The general behavior of bond percolation is qualitatively similar to that of site percolation.

Table 2.1 Percolation thresholds for different lattices for site percolation and bond percolation

Lattice	Site	Bond
Square	0.59275 [29]	1/2* [30]
Triangular	1/2* [30]	2 sin (π/18)* [30]
Honeycomb	0.696 [29]	1-2 sin (π/18)* [31]
Cubic (body-centered)	0.246 [32]	0.180 [33]
Cubic (face-centered)	0.198 [32]	0.120 [33]
Cubic (simple)	0.312 [32]	0.249 [33]
Diamond	0.430 [34]	0.390 [34]
4-Hypercubic	0.197 [35]	0.160 [35]
5-Hypercubic	0.141 [35]	0.118 [35]
6-Hypercubic	0.109 [35]	0.094 [35]
7-Hypercubic	0.089 [35]	0.079 [35]

Note: Numbers with an asterisk (*) can be calculated analytically.

the threshold should depend on the geometry of the lattice; thinking of the burning algorithm, the speed at which the "fire" spreads depends on the spatial configuration of "trees."

2.3.3 The Order Parameter

We have determined the percolation threshold in the previous section and outlined that the transition between spanning and nonspanning configurations occurs exactly at p_c for infinitely large systems. Let us now consider probabilities $p > p_c$, where we will always find a spanning cluster (for sufficiently large systems). Naturally, we may ask ourselves how many of the sites belong to the spanning cluster (which is also the biggest one). More precisely, we can define the fraction of sites which belong to the largest cluster $P(p)$ as a function of the occupation probability p. We call this quantity the order parameter (for its dependence on p, see Figure 2.8).

Obviously, for $p < p_c$ the order parameter is zero, as there are no spanning clusters. For $p > p_c$ the order parameter approaches a finite value and increases with p. When analyzing the behavior of the order parameter, we find that $P(p)$ behaves like a power law in the region close to the percolation threshold

$$P(p) \propto (p - p_c)^{\beta}, \tag{2.1}$$

with a dimension-dependent exponent β. This exponent neither depends on the type of lattice nor on the detailed connectivity rule (bond or site percolation) and is therefore

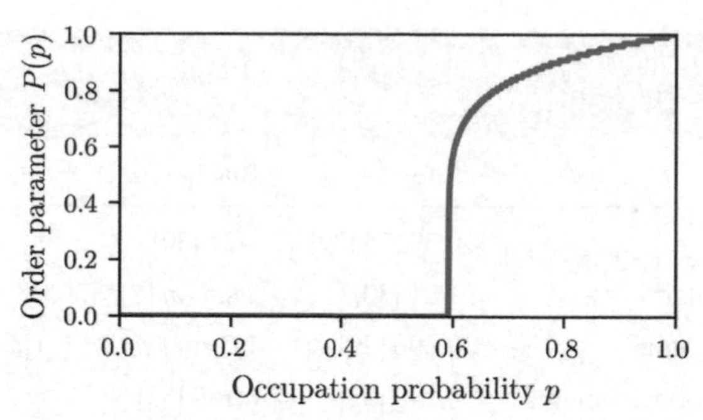

Figure 2.8 Order parameter $P(p)$ as a function of the occupation probability p for site percolation on the square lattice. The system exhibits a "critical" behavior at p_c. Starting at p_c, the order parameter grows as $P(p) \propto (p - p_c)^{\beta}$, where $\beta = \frac{5}{36}$ in 2D and $\beta \approx 0.41$ in 3D.

considered "universal." We will discuss this concept in more detail in the following sections.

Exercise: Percolation Threshold and Order Parameter

Consider the percolation model and determine the probability to find a spanning cluster, shortest-path length, and the lifetime of the fire. You may proceed as follows:

- For a given number of lattice sites N and occupation probability p make several measurements (e.g., 1,000) using different seeds for the random number generator, and compute the fraction of samples with a spanning cluster, the average shortest-path length (when percolation occurs), and the lifetime of the fire.
- Repeat the same for different values of p. Plot the results against p.
- Change N and compare the results. (For each N you should obtain a curve.)
- Find the value of the threshold probability p_c for which the system changes its behavior.

Please consider the following point:

- Choosing a site to be occupied with probability p can be simply done in the following way: generate a random number between 0 and 1. If it is smaller than p set the site as occupied.

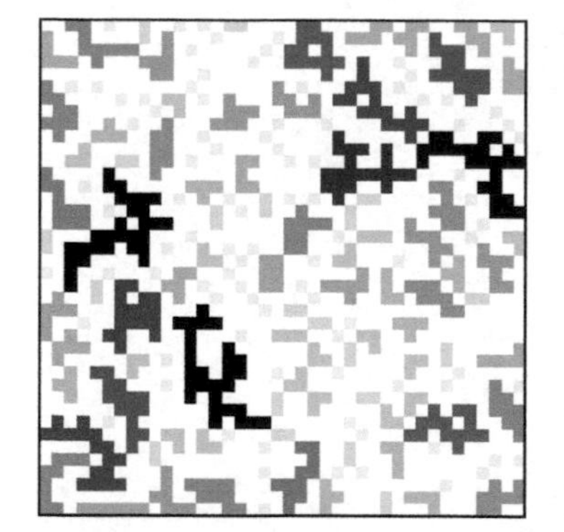

Figure 2.9 Percolation on a square lattice with $L = 40$ for $p = 0.4$. The different shades of gray indicate lattice sites that belong to different clusters.

2.3.4 The Cluster-Size Distribution

After having investigated the percolation threshold and the fraction of sites in the spanning cluster, the natural extension would be to identify all the clusters (see Figure 2.9). An efficient algorithm is necessary to perform such a task. In fact, there are several such algorithms; the most popular (and for this purpose most efficient) algorithm is the Hoshen–Kopelman algorithm [36], which was developed in 1976 by Joseph Hoshen and Raoul Kopelman (see Figure 2.10).

We represent the percolation configuration as a matrix N_{ij} that can have values of 0 (site is unoccupied) and 1 (site is occupied). Furthermore, let k be a running index labeling the clusters in N_{ij}. We also introduce an array M_k to keep track of the mass of cluster k (the number of sites belonging to the cluster k). We start the algorithm by setting $k = 2$ (since 0 and 1 are already taken) and search for the first occupied site in N_{ij} beginning in the upper left corner of the lattice. We then add this site to the array $M_{k=2} = 1$ and set the entry in N_{ij} to k (so it is branded as pertaining to the cluster k).

Figure 2.10 Raoul Kopelman is a professor at the University of Michigan.

We then go over all lattice sites N_{ij}, line by line as a type writer, and try to detect whether an occupied site belongs to an already known cluster or a new one. We comb through the lattice from top-left to bottom-right; the criteria are rather simple:

- If a site is occupied and the top and left neighbors are empty, we have found a new cluster and we set k to $k+1$, $N_{ij} = k$ and $M_k = 1$.
- If only one of the sites (top or left) has the value k_0 (i.e., is part of a cluster), we increase the corresponding value in the array, M_{k_0}, by one (setting M_{k_0} to $M_{k_0} + 1$). We label the new site accordingly. That is, we set $N_{ij} = k_0$.
- If both neighboring sites are occupied with k_1 and k_2, respectively (assuming $k_1 \neq k_2$) – meaning that they are already part of a cluster – we choose one of them (e.g., k_1). We set the matrix entry to the chosen value, $N_{ij} = k_1$ and increase the array value not only by one but also by the whole number of sites already in the second cluster (here k_2), $M_{k_1} \to M_{k_1} + M_{k_2} + 1$. Of course we have to mark the second array M_{k_2} in some way so that we know that its cluster size has been transferred over to M_{k_1} which we do by setting it to $-k_1$. We have thus branded M_{k_2} in such a way that we immediately recognize that it does not serve as a counter anymore (as a cluster cannot consist of a negative number of sites). Furthermore, should we encounter an occupied site neighboring a k_2 site, we can have a look at M_{k_2} to see that we are actually dealing with cluster k_1 (revealing the "true" cluster number).

The last point is crucial, as we usually deal with a number of sites that are marked in such a way that we have to first recursively detect which cluster they pertain to before carrying out the algorithm any further. The recursive detection stops as soon as we have found a k_0 with $M_{k_0} \geq 0$ (i.e., a "true" cluster number).

Once all the sites N_{ij} have been visited, the algorithm ends up with a number l of clusters, where l is smaller than the largest obtained cluster index $k_{\max}$. The only thing left to do is to construct a histogram of the different cluster sizes. This is done by looping through all the clusters $k \in \{2, \ldots, k_{\max}\}$ while skipping negative M_k.

We visualize the Hoshen–Kopelman algorithm in Figure 2.11. The algorithm is very efficient because it visits every site only once and it scales linearly with the number of sites.

Once we have run the algorithm for a given lattice (or collection of lattices and taken the average), we can calculate the cluster-size distribution n_s (i.e., the number of clusters of size s per site) as a function of the occupation probability p. These results are illustrated in Figure 2.12, where the first graph (Figure 2.12a) represents the behavior

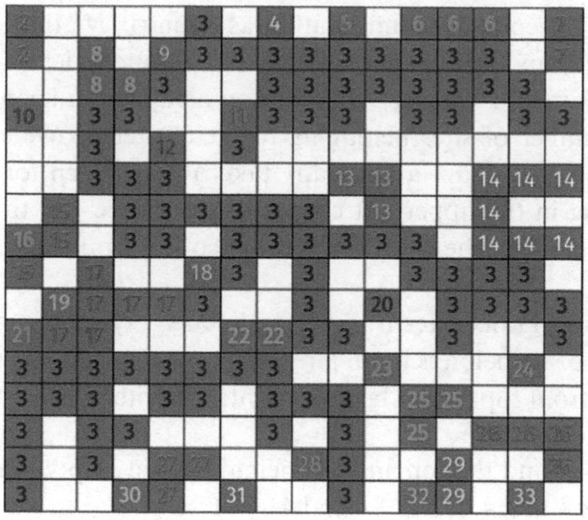

Figure 2.11 The Hoshen–Kopelman algorithm applied to a percolation configuration on the square lattice. The numbers denote the running cluster variable k.

Hoshen–Kopelman Algorithm

1. $k = 2, M_k = 1$
2. For all i, j of N_{ij}
   (a) If top and left are empty (or nonexistent) $k \to k+1, N_{ij} = k, M_k = 1$.
   (b) If one is occupied with k_0 then $N_{ij} = k_0, M_{k_0} \to M_{k_0} + 1$.
   (c) If both are occupied with k_1 and k_2 (and $k_1 \neq k_2$) then choose one, for example, k_1 and $N_{ij} \to k_1$, $M_{k_1} \to M_{k_1} + M_{k_2} + 1, M_{k_2} = -k_1$.
   (d) If both are occupied with $k_1, N_{ij} = k_1, M_{k_1} \to M_{k_1} + 1$.
   (e) If any of the ks considered has a negative mass M_k, find the original cluster they reference to and use its cluster number and weight instead by using while($M_k < 0$) $k = -M_k$.
3. For $k \in \{2, \ldots, k_{\max}\}$, if $M_k > 0$ then $N(M_k) \to N(M_k) + 1$.
4. For all cluster sizes s, determine $n_s = N(s)/N$ where n_s is the desired cluster-size distribution.

for subcritical occupation probabilities ($p < p_c$), the second graph (Figure 2.12b) shows the behavior at the critical occupation probability ($p = p_c$), and the third graph (Figure 2.12c) depicts the behavior for overcritical occupation probabilities ($p > p_c$). In the subcritical regime ($p < p_c$), we find that $n_s(p)$ obeys an exponential function multiplied with a power law. In contrast, in the overcritical region ($p > p_c$), we observe a distribution exhibiting an exponential decay but with an argument that is stretched with a power $s^{1-1/d}$.

We observe in Figure 2.12b that $n_s(p = p_c)$ follows as a straight line in a double logarithmic plot, implying a power law behavior (i.e., $n_s(p = p_c) \propto s^{-\tau}$). The exponent

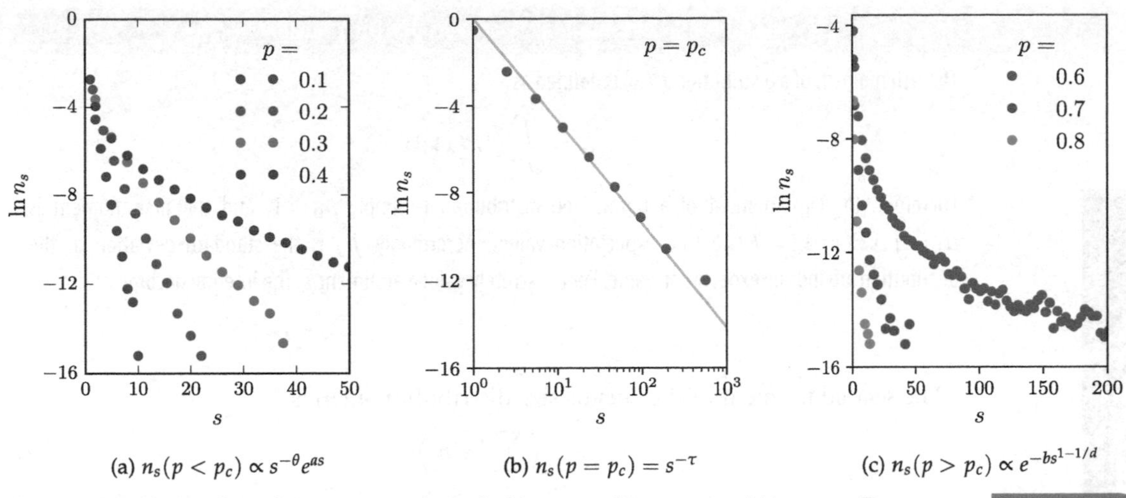

(a) $n_s(p < p_c) \propto s^{-\theta} e^{as}$ (b) $n_s(p = p_c) = s^{-\tau}$ (c) $n_s(p > p_c) \propto e^{-bs^{1-1/d}}$

Figure 2.12 Cluster-size distribution of percolation on a square lattice for (a) $p < p_c$, (b) $p = p_c$, and (c) $p > p_c$. The gray solid line in the center panel has slope $-\tau = -187/91 \approx -2.05$.

τ depends on the system dimension and is $\frac{187}{91}$ in 2D while numerical results indicate that it is 2.1892(1) in 3D [35]). The bounds for τ are $2 \leq \tau \leq 5/2$.

We summarize the behavior of the cluster-size distribution n_s in the three different regions:

$$n_s(p) \propto \begin{cases} s^{\theta} e^{as}, & p < p_c\,, \\ s^{-\tau}, & p = p_c\,, \\ e^{-bs^{1-1/d}}, & p > p_c\,. \end{cases} \tag{2.2}$$

We now compare the distributions to the cluster-size distribution at p_c and may plot a rescaled distribution

$$\tilde{n}_s(p) = n_s(p)/n_s(p = p_c)\,,$$

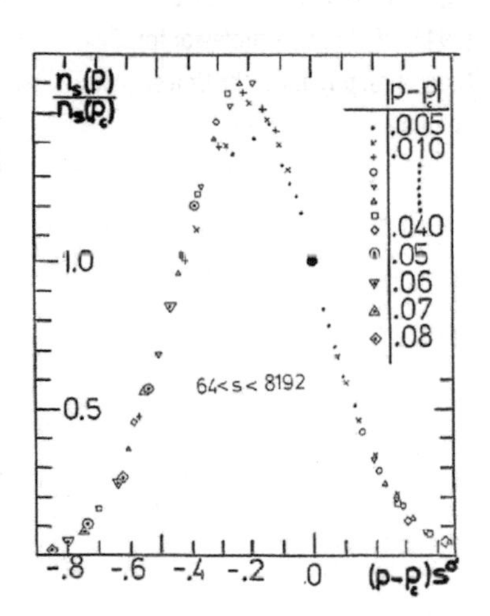

Figure 2.13 Scaling behavior of the cluster-size distribution for percolation on the square lattice. Original plot made by Prof. Dietrich Stauffer.

which is defined in all three regions of eq. (2.2). A plot of $\tilde{n}_s(p)$ as a function of $(p - p_c)s^\sigma$ is shown in Figure 2.13. Choosing the right value of σ, we see that all data for different s and p fall on one single curve. This is called data collapse. In 2D, $\sigma = 36/91$ and in 3D the exponent $\sigma = 0.445(10)$ [33].

We notice that the cluster-size distributions are all described by the following scaling relation

$$n_s(p) = s^{-\tau}\Re_\pm[(p - p_c)s^\sigma], \tag{2.3}$$

with the scaling functions[4] $\Re_\pm$, where the subscript $\pm$ stands for $p > p_c$ (+) and $p < p_c$ (−), respectively. This scaling law was first described by Dietrich Stauffer (see Figure 2.14).

[4] Scaling function: A function of two variables that can be expressed as a function of one single variable.

Additional Information: Moments of a Distribution

The nth moment of a distribution $P(x)$ is defined as

$$\mu_n = \int x^n P(x)\,\mathrm{d}x\,.$$

Therefore, the 0th moment of a normalized distribution is simply $\mu_0 = 1$, and the first moment is $\mu_1 = \int xP(x)\,\mathrm{d}x = E(x)$ (the expectation value). Accordingly, μ_2 is the standard deviation of the distribution around the expectation value. For a discretely valued argument s, the integral becomes a sum.

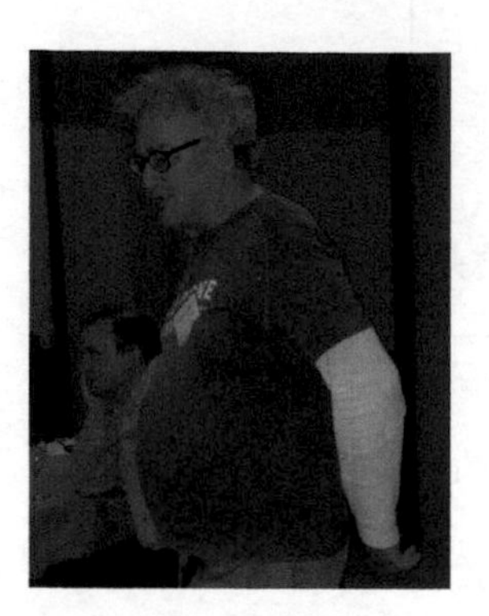

Figure 2.14 Dietrich Stauffer (1943–2019) was a professor for theoretical physics at the University of Cologne.

The second moment of the cluster-size distribution $n_s(p)$ is

$$\chi = \left\langle \sum_s{}' s^2 n_s \right\rangle. \tag{2.4}$$

The prime (′) indicates the exclusion of the largest cluster in the sum (as the largest cluster would make χ infinite at $p > p_c$ for $N \to \infty$). We find

$$\chi \propto |p - p_c|^{-\gamma}, \tag{2.5}$$

with $\gamma = 43/18$ in 2D and $\gamma \approx 1.8357$ in 3D [37]. The second moment is a very strong indicator of p_c, as we can see a very clear divergence around p_c.

In the gray box below, we show that the scaling exponents we have seen so far are related by [38]

$$\gamma = \frac{3 - \tau}{\sigma}. \tag{2.6}$$

An overview of the different exponents is given in Ref. [18].

Additional Information: Scaling Relation for the Moments of the Cluster-Size Distribution

For the second moment of the cluster-size distribution, we have $\langle \sum_s{}' s^2 n_s \rangle \propto |p - p_c|^{-\gamma}$ with $\gamma = (3 - \tau)/\sigma$ (see eq. (2.6)). For general moments of the cluster-size distribution, we find $\langle \sum_s{}' s^k n_s \rangle \propto |p - p_c|^{(\tau-1-k)/\sigma}$ using eq. (2.3) [38]:

$$\begin{aligned}
\sum_s s^k n_s &= \int_0^{\infty} s^k n_s\,\mathrm{d}s = \int_0^{\infty} s^{k-\tau} \mathfrak{R}_{\pm}[(p - p_c)s^{\sigma}]\,\mathrm{d}s \\
&= \sigma^{-1} \int_0^{\pm\infty} s^{k-\tau+1} z^{-1} \mathfrak{R}_{\pm}(z)\,\mathrm{d}z \\
&= \sigma^{-1} |p - p_c|^{(\tau-1-k)/\sigma} \int |z|^{(1+k-\tau)/\sigma} z^{-1} \mathfrak{R}_{\pm}(z)\,\mathrm{d}z\,,
\end{aligned} \tag{2.7}$$

where we substituted $z = (p - p_c)s^{\sigma}$ in the third step. Integrals above p_c run from 0 to $+\infty$ and below p_c from 0 to $-\infty$. For large $\pm z$, the scaling function $\mathfrak{R}_{\pm}(z)$ decays exponentially. We have thus confirmed $\langle \sum_s{}' s^k n_s \rangle \propto |p - p_c|^{(\tau-1-k)/\sigma}$.

Exercise: Cluster-Size Distribution

Implement the Hoshen–Kopelman algorithm for a square lattice of size $L \times L$ and occupation probability p.

- The algorithm determines which sites belong to each cluster. Each cluster gets a number k (starting from $k = 2$ to distinguish from "untouched" sites). The size (mass) $M(k)$ of a cluster is stored.
- Here, we will go through the system from left to right and top to bottom.
- There are three possible situations for the top and left neighboring sites of an occupied site:
  - both are unoccupied: new cluster ($k \rightarrow k + 1, M(k) = 1$).
  - one is occupied or both are occupied with same cluster (Cluster k_0): site belongs to cluster k_0, $M(k_0)$ increased by 1.
  - both are occupied with two different clusters (k_1 and k_2): clusters meet each other and have to be united to one cluster (e. g., $M(k_1) \rightarrow M(k_1) + M(k_2) + 1$). Make an annotation that, for example, cluster k_2 belongs to cluster k_1 ($M(k_2) = -k_1$). (Note: Typically the larger k-value is chosen to be added to the smaller one.)
- For determining the "true" cluster number in cases when $M(k)$ is negative, it is useful to implement a function (recursive or while loop). Note that in all cases mentioned above you first have to determine the "true" cluster numbers k_0, k_1, k_2.
- To get the cluster-size distribution you have to count the number of clusters of each size after reaching the end of the system.
- Plot the cluster-size distribution for different values of p.

Please consider the following points:

- The cluster-size distribution n_s is normalized per site, that is, the total number of clusters of each size divided by total number of sites (L^2). (For this reason, $n_s < 1$.)
- Depending on the value of p (below, at, or above p_c) a different behavior of the cluster-size distribution is observed.

2.3.5 Correlation Length

To further analyze percolation clusters, we now consider the correlation function $c(r)$ that describes connectivity correlations as a function of the radial distance r. For one cluster (e. g., the largest cluster), we can use density correlations since, by definition, all sites in a cluster are connected. Otherwise, if we want to study correlation effects between multiple clusters, we should focus on connectivity correlations. Large values of the correlation function $c(r)$ mean that two quantities strongly influence each other, whereas zero means that they are uncorrelated.

Numerically, we can obtain the correlation function $c(r)$ by placing concentric circles or spheres around a site in the center region of the cluster. The correlation function is defined as

$$c(r) = \frac{\Gamma\left(\frac{d}{2}\right)}{2\pi^{d/2} r^{d-1} \Delta r} \left[M(r + \Delta r) - M(r) \right] , \tag{2.8}$$

where $M(r)$ (mass of the cluster) denotes the number of occupied sites in a d-dimensional hypersphere of radius r. The prefactor in eq. (2.8) is the inverse of the volume of a d-dimensional annulus between r and $r + \Delta r$, and $\Gamma(\cdot)$ is the gamma function. We see that $c(r)$ is the number of occupied sites within an annulus of thickness Δr at a distance r from the center and normalized by the volume of the annulus. We apply this method to the largest percolation cluster for different occupation probabilities p. When we compute $c(r)$ for a given cluster (see Figure 2.15), we find that the correlation function usually decreases exponentially with distance r, eventually with an offset C. Mathematically, this means

Figure 2.15 Illustration of how to numerically determine the correlation function $c(r)$ of a percolation cluster. At a given radius r, we count the number of occupied sites within an annulus of thickness Δr. Note that the origin ($r = 0$) must belong to the cluster.

$$c(r) \propto C + \exp\left(-\frac{r}{\xi}\right) , \tag{2.9}$$

where the constant C is equal to the order parameter of percolation, $P(p)$, and thus vanishes for $p < p_c$. The newly introduced quantity ξ is the correlation length. It describes the typical length scale over which the correlation function decays. Below p_c, the correlation length ξ is proportional to the radius of a typical cluster.

When we analyze the dependence of the correlation length ξ on the occupation probability p (see Figure 2.16), we find that it diverges at the critical occupation probability p_c [39],

$$\xi(p) \propto |p - p_c|^{-\nu} \quad \text{where} \quad \nu = \begin{cases} \frac{4}{3} , & \text{in 2D} , \\ 0.8751(11) , & \text{in 3D} . \end{cases} \tag{2.10}$$

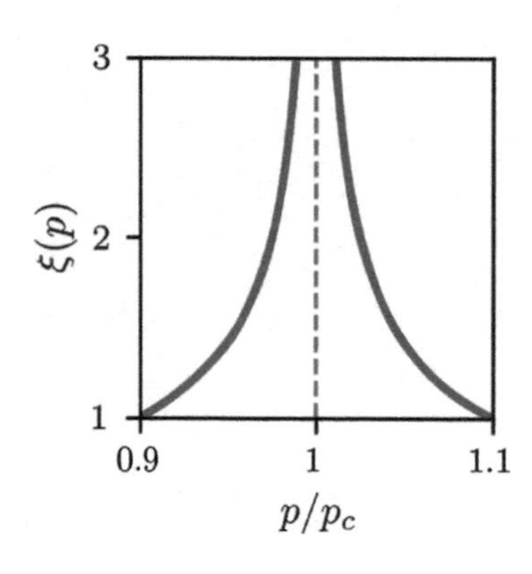

Figure 2.16 Singular behavior of ξ with respect to p in an infinite system.

Apparently, the assumption of an exponential behavior is no longer valid at p_c. In fact, at p_c the correlation function follows a power law [33]

$$c(r) \propto r^{-(d-2+\eta)} \quad \text{with} \quad \eta = \begin{cases} \frac{5}{24} , & \text{in 2D} , \\ -0.046(8) , & \text{in 3D} . \end{cases} \tag{2.11}$$

The exponents ν and η in Eqs. (2.10) and (2.11) are again universal critical exponents.

2.3.6 Finite Size Effects

We encounter problems when the system size L is smaller than the correlation length ξ. Instead of the singularity that was mentioned in eq. (2.10), the correlation length ξ only takes on finite values (see Figure 2.17). We consider the values p_1 and p_2 that are defined by correlation lengths $\xi(p_1)$ and $\xi(p_2)$ that are of the order of the linear system size L. The region between p_1 and p_2 is called the *critical region*. We cannot trust any quantity obtained numerically within this critical region. Based on

$$L = \xi(p_1) \propto (p_1 - p_c)^{-\nu} \tag{2.12}$$

and

$$p_1 - p_2 \approx 2(p_1 - p_c),$$

we find that the critical region shrinks with system size like

$$p_1 - p_2 \propto L^{-\frac{1}{\nu}}. \tag{2.13}$$

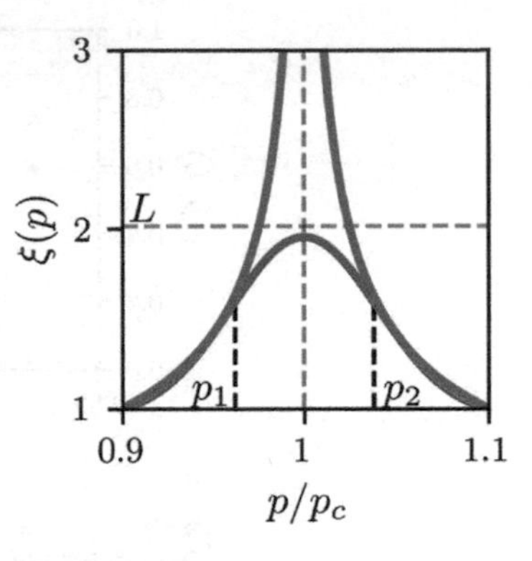

Figure 2.17 The finite size of the system leads to a cutoff in the correlation length (orange solid line). The critical region is defined as the interval that is confined by the two dashed black vertical lines.

In the limit $L \to \infty$, the critical region will vanish. We can obviously not realize such a limit on a computer. If we identify our numerically obtained effective critical occupation probability $p_{\text{eff}}(L)$ with p_1 ($\xi(p_1) \approx L$), we will always obtain $p_{\text{eff}}(L) < p_c$. We assume that we are close enough to p_c such that the deviation from the true p_c is proportional to $L^{-1/\nu}$. We thus find

$$p_{\text{eff}}(L) = p_c\left(1 - aL^{-\frac{1}{\nu}}\right), \tag{2.14}$$

where $a > 0$ is a constant. The best we can do at this point is using the data acquired for finite system sizes and extrapolate to the values of the infinite system. To do so, we can use eq. (2.14), plot p_{eff} as a function of $L^{-1/\nu}$, and extrapolate the data to the point at which the vertical axis is crossed (corresponding to the limit $L^{-1/\nu} \to 0$). This method enables us to find the critical occupation probability p_c.

2.3.7 Finite Size Scaling

In the neighborhood of the percolation threshold p_c, we find that the dependence of the second moment of the cluster-size distribution χ (see eq. (2.4)) on p and L can be reduced to a one variable function. This is a consequence of a general scaling law that exists around all critical points.

By plotting χ around p_c against the occupation probability p for several values of L, we obtain curves that differ strongly around the critical point. The most important difference is the height of the peaks which is found to grow as a power law with the linear system size L. Using this observation and the fact that the critical region shrinks like $L^{-1/\nu}$ as shown in eq. (2.13), we can rescale the horizontal and vertical axis of the left panel of Figure 2.18 to get a data collapse as shown in the corresponding right panel. Such a data collapse can be also observed in related models such as *kinetic gelation* (see Figure 2.19).[5]

Let us try to comprehend what this data collapse means. We have a function χ that was originally a function of two parameters (the occupation probability p and the system size L) and that now behaves as though it was a one-parameter function. Mathematically, the data collapse of χ as shown in Figures 2.18 and 2.19 can be written as

[5] In kinetic gelation, bonds are randomly occupied so as to form cross-linked chains in order to describe gelation through additive polymerization [40].

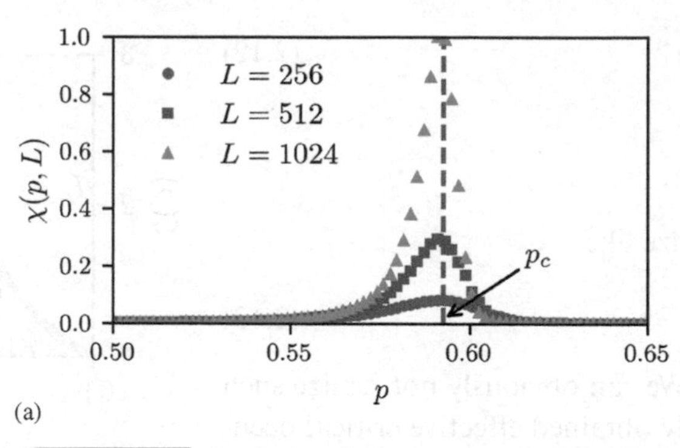

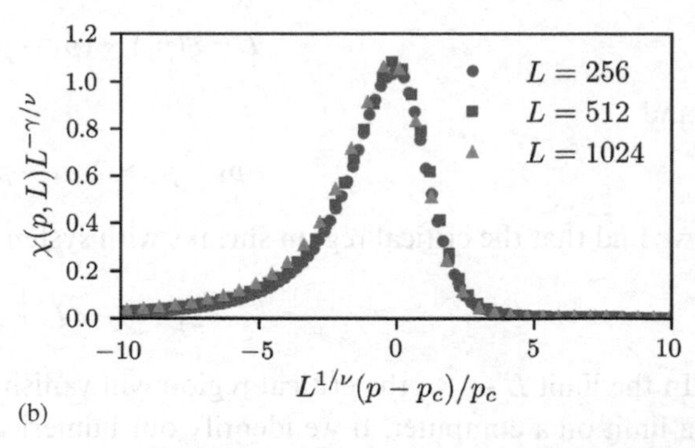

Figure 2.18 For site percolation on a square lattice with side length L, we show (left) the system-size dependence of the second moment of the cluster-size distribution (eq. (2.4)) and (right) the corresponding finite-size scaling (eq. (2.15)). The number of samples is 10^4. We normalized $\chi(p, L)$ by $\max_p[\chi(p, L = 1024)]$ in the left panel.

$$\chi(p, L) = L^{\frac{\gamma}{\nu}} \aleph_\chi \left[(p - p_c) L^{\frac{1}{\nu}} \right], \tag{2.15}$$

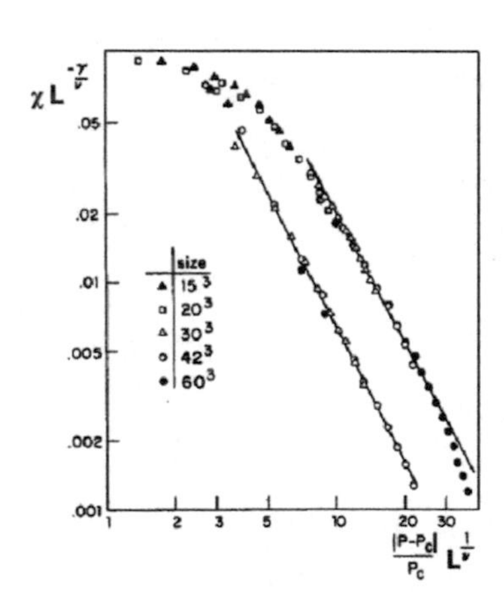

Figure 2.19 Finite size scaling of χ for the kinetic gelation model on a cubic lattice. We observe a data collapse when plotting $\chi L^{-\gamma/\nu}$ against $\frac{|p-p_c|}{p_c} L^{1/\nu}$. The right curve corresponds to $p < p_c$ and the left one to $p > p_c$. The straight lines have a slope of $-\gamma$ (see the critical exponents of χ for infinite systems, eq. (2.5)). The figure is taken from Ref. [40].

where $\aleph_\chi$ is called a scaling function of χ. This is an example of the finite size scaling first proposed in Ref. [41].

When we approach the critical occupation probability p_c, the scaling function $\aleph_\chi$ approaches a constant and we find that the peak $\chi_{\max}$ depends on the system size, so

$$\chi_{\max}(L) \propto L^{\frac{\gamma}{\nu}}. \tag{2.16}$$

This can be verified by plotting the maxima of the data of the left panel of Figure 2.18 as a function of L. These expressions are reminiscent of those previously introduced in the context of the cluster-size distribution in Section 2.3.4. We recall that a scaling function is a function that combines two variables (in this context p, L), which can be expressed in terms of only *one* variable (the so-called scaling field).

Based on eq. (2.5), we know that $\chi(p) \propto |p - p_c|^{-\gamma}$ in the vicinity of p_c. Thus, at p_c and in the limit $L \to \infty$, we find

$$\lim_{L\to\infty} \chi(p, L) = \lim_{L\to\infty} L^{\frac{\gamma}{\nu}} \aleph_\chi \left[(p - p_c) L^{\frac{1}{\nu}} \right] \propto |p - p_c|^{-\gamma}. \tag{2.17}$$

This implies that $\aleph_\chi(x) \propto |x|^{-\gamma}$ for $x \to \infty$.

2.3.8 Size Dependence of the Order Parameter

We are now going to consider the order parameter $P(p)$ at the critical occupation probability p_c. We denote the size of the largest cluster by s_∞ and the side length of the lattice by L. When we plot s_∞ against L, we notice that there is a power law at work,

$$s_\infty \propto L^{d_f}, \tag{2.18}$$

where the exponent d_f depends on the dimension of the system. We illustrate this behavior in Figure 2.20. For two-dimensional lattices, we find $d_f = 91/48$ and for three dimensions we find $d_f = 2.5226(1)$ [42]. The exponent d_f is called the *fractal dimension*, which will be explained in more detail in Section 2.4.

The size of the largest cluster is proportional to the product of system size L^d and $P(p)$ (i.e., the probability of a lattice site to belong to the largest cluster). That is,

$$s_\infty \propto L^d P(p)\,. \tag{2.19}$$

We now combine Eqs. (2.19) and (2.10) and obtain

$$s_\infty \propto L^d P(p) \propto L^d |p - p_c|^\beta \propto L^d L^{-\beta/\nu}\,, \tag{2.20}$$

where we used that the correlation length is of the order of L for a finite system in the vicinity of p_c. Based on Eqs. (2.18) and (2.20), we can establish the following connection:

$$d_f = d - \frac{\beta}{\nu}\,. \tag{2.21}$$

Analogous to eq. (2.15), the order parameter P also satisfies a finite-size scaling relation for $p > p_c$

$$P(p, L) = L^{-\frac{\beta}{\nu}} \aleph_P \left[(p - p_c) L^{\frac{1}{\nu}}\right]\,. \tag{2.22}$$

At p_c we obtain

$$P \propto L^{-\frac{\beta}{\nu}}\,. \tag{2.23}$$

When we combine all of this, we find

$$s_\infty \propto P L^d \propto L^{\left(-\frac{\beta}{\nu}+d\right)} \propto L^{d_f}\,,$$

in agreement with Eqs. (2.20) and (2.21).

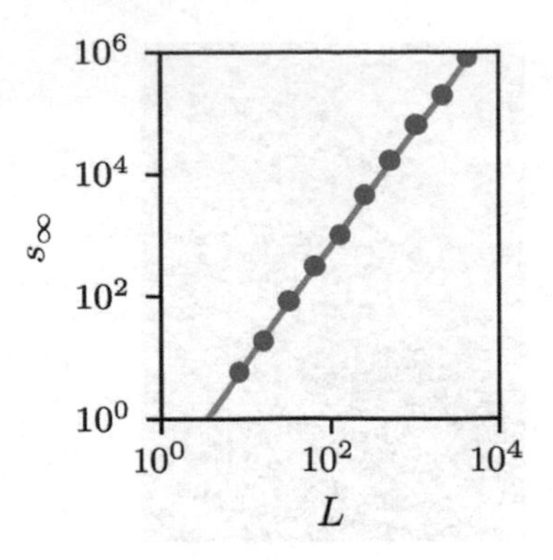

Figure 2.20 Log-log plot of the size dependence of the order parameter in a square lattice at the percolation threshold $p_c \approx 0.5972$. The slope of the gray solid line is the fractal dimension $d_f = 91/48 \approx 1.9$.

2.3.9 The Shortest Path

We have outlined how to decide whether there exists a spanning cluster by means of the burning algorithm, and we have determined and analyzed the cluster-size distribution with the Hoshen–Kopelman algorithm. One unanswered question, however, is how the shortest-path length of a spanning cluster behaves as a function of the system size. Simulations show that the shortest-path length of a spanning cluster satisfies

$$\ell \propto L^{d_{\min}}\,,$$

which is also a power law of the linear lattice size L. The exponent $d_{\min}$ depends on the dimension [44]:

$$d_{\min} = \begin{cases} 1.13077(2)\,, & \text{in 2D}\,, \\ 1.3756(6)\,, & \text{in 3D}\,, \\ 1.607(5)\,, & \text{in 4D}\,. \end{cases} \tag{2.24}$$

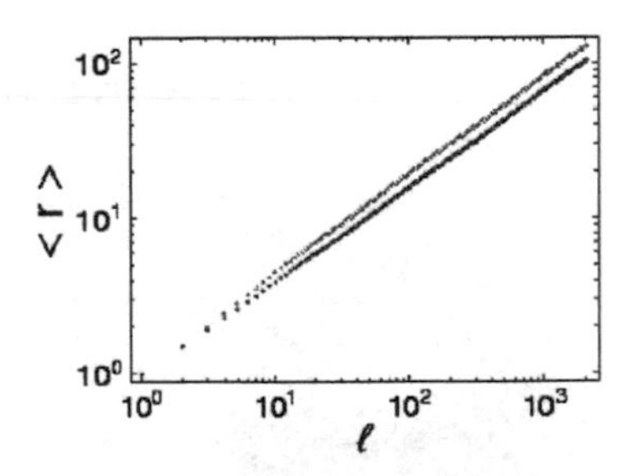

Figure 2.21 To illustrate the shortest-path dependence on the system size, we show the Euclidean distance $\langle r \rangle$ as a function of the shortest-path length ℓ for four-dimensional site (top curve) and bond (bottom curve) percolation. Both quantities are related through $\langle r \rangle \propto \ell^{1/d_{\min}}$. The figure is taken from Ref. [43].

Figure 2.22 Joan Adler is a professor emeritus at the Technion in Israel.

We can determine $d_{\min}$ for site percolation and bond percolation as well as for different lattice types. The quantity $d_{\min}$ is called the fractal dimension of the shortest path, a concept which we will discuss in Section 2.4. In Figure 2.21, we show the shortest path dependence on the system size for site and bond percolation for a four-dimensional system.

2.3.10 Bootstrap Percolation

Many variants of percolation have been invented to describe different physical phenomena. One of those is bootstrap percolation [45–47], which has been studied by Joan Adler (see Figure 2.22). In the canonical bootstrap model, sites are initially occupied randomly (with a given probability p) as in the percolation model. Next, all sites that do not have at least m occupied neighbors are removed. We then repeat this last step until the resulting lattice configuration does not change anymore. In the case of $m = 1$, all isolated sites are removed and for $m = 2$ all "dangling" sites are removed. This does not affect p_c. If we, however, choose $m \geq 3$, the percolation threshold p_c and the nature of the transition are strongly dependent on the lattice. On some lattices and for specific values of m, the order parameter even jumps at the transition (first order transition).

Bootstrap percolation has many applications in solid state physics and fluid flow in porous media. We show an example of bootstrap percolation in Figure 2.23.

2.4 Fractals

2.4.1 Self-Similarity

Figure 2.24 Benoit B. Mandelbrot (1924–2010) coined the term *fractal*. He worked at IBM and different academic institutions, including Harvard and Yale Universities.

Fractals are objects that exhibit *self-similarity*. That is, certain patterns occur repeatedly at different spatial scales. Benoit B. Mandelbrot (see Figure 2.24) coined the term *fractal* [48]. He also introduced the concept of *fractal dimension* to characterize fractal objects. Before taking a closer look at this concept, it may be useful to consider some examples of self-similarity. In a nutshell, we could define an object to be self-similar if it is built up of smaller copies of itself. Such objects occur both in mathematics and nature. Let us discuss a few examples.

The Sierpinski Triangle

The Sierpinski triangle is a mathematical object that is constructed by an iterative application of a simple operation. It was first described by the Polish mathematician Wacław Franciszek Sierpiński (1882–1969). To construct a Sierpinski triangle, we subdivide a triangle into four sub-triangles and discard the center triangle (see Figure 2.25). In the next step of the iteration, we subdivide each of the three

Figure 2.25 The Sierpinski triangle is a self-similar mathematical object that is created iteratively.

remaining triangles and remove each central triangle. This obviously produces an object that is exactly self-similar in the limit of infinitely many iterations. When zooming in, we see the exact same image at all scales. The object becomes mathematically a fractal in the limit of an infinite number of iterations.

Self-Similarity in Nature

Naturally occurring self-similar objects are usually only approximately self-similar. What we mean by that is that we cannot zoom into natural patterns indefinitely often and expect to find the same pattern at all scales. To better illustrate this, we consider a tree. A tree has different branches, and the whole tree looks similar to a branch connected to the tree trunk (see Figure 2.26). The branch itself resembles the smaller branches attached to it and so on. Evidently, this breaks down after a few iterations, when the leaves of the tree are reached. This example is based on a hierarchical tree structure. However, natural self-similar patterns are not necessarily hierarchical objects as illustrated by the following example.

Gold colloids were shown to arrange in fractals of fractal dimension 1.70 (the meaning of this will soon be explained) by David Weitz in 1984 [49]. Colloidal gold is a suspension of sub-micrometer-sized gold particles in a fluid (e. g. , water). These gold colloids aggregate to fractal structures as we show in Figure 2.27. As in the case of trees, gold colloids are also only approximately self-similar. If we zoom too deeply into the colloidal structure, we will end up seeing individual gold particles and not a colloidal structure (see top left panel of Figure 2.27). Moreover, the finite size of the aggregate colloids does not allow us to zoom out of the structure infinitely far. The diameter of gold particles and the finite size of aggregate colloids therefore define lower and upper cutoffs, respectively.

Figure 2.23 We show a freshly occupied square lattice with periodic boundary conditions having an occupation probability $p = 0.55$ (top) and the same lattice after removing all "dangling" sites with less than $m = 2$ occupied neighbors (bottom).

2.4.2 Fractal Dimension: Mathematical Definition

Keeping these examples in mind, we will now give the precise mathematical definition of *fractal dimension*. To determine the fractal dimension of an object, we can use the following (theoretical) procedure.

Consider all possible coverings of an object with d-dimensional spheres of radius $r_i \leq \epsilon$, where ϵ is the resolution. Let $N_\epsilon(C)$ be the number of spheres used in the covering C. Then, the volume of the covering is

Additional Information: Definition of Covering

A covering of an object $X \subseteq \mathbb{R}^d$ is a collection of sets whose union contains X as a subset. If we denote the collection of sets by $C = \{A_i | i \in I\}$, we call C a covering of X if

$$X \subseteq \bigcup_{i \in I} A_i .$$

Figure 2.26 An example of the hierarchical branching pattern of a baobab tree.

$$V_\epsilon(C) \propto \sum_{i=1}^{N_\epsilon(C)} r_i^d , \tag{2.25}$$

where d is the dimension of the space into which the object is embedded.

We define V_ϵ^* as the volume of the covering that, among all coverings with the smallest number of spheres, has minimal volume

$$V_\epsilon^* = \min_{V_\epsilon(C)} \left[\min_{N_\epsilon(C)} \left(V_\epsilon(C) \right) \right] . \tag{2.26}$$

The fractal dimension of the object is then defined as

$$d_\mathrm{f} := \lim_{\epsilon \to 0} \frac{\log\left(V_\epsilon^* / \epsilon^d \right)}{\log\left(L/\epsilon \right)} , \tag{2.27}$$

where L is the linear system dimension.

For most objects, we may simplify the definition of the fractal dimension in eq. (2.27) in the following way: When the length of the object is stretched by a factor of a, its volume (or mass) grows by a factor of a^{d_f}. We obtain this interpretation by rewriting eq. (2.27) (in the limit $\epsilon \to 0$) as

$$\frac{V_\epsilon^*}{\epsilon^d} = \left(\frac{L}{\epsilon} \right)^{d_\mathrm{f}} , \tag{2.28}$$

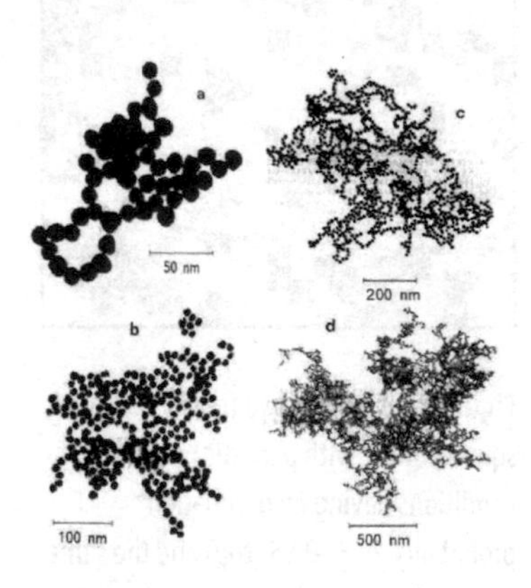

Figure 2.27 Gold colloids at different scales [50].

If we now consider a rescaling of L by a according to $L \to aL$, we find that $V_\epsilon^* \propto a^{d_\mathrm{f}}$. Thus, the volume V_ϵ^* scales as claimed.

As a simple example, we again consider the Sierpinski triangle (see Figure 2.28). Stretching its sides by a factor of two evidently increases its area by a factor of three. Using this observation and the aforementioned proportionality, we find that the Sierpinski triangle has fractal dimension $\log(3)/\log(2) \approx 1.585$.

2.4.3 The Box Counting Method

The box counting method is a method to numerically determine the fractal dimension of an object. It is conceptually easy because the underlying idea is close to the mathematical definition.

In the box counting method, we superimpose a grid with grid spacing ϵ on the fractal object. We define the number of boxes in the grid that are not empty (contain a

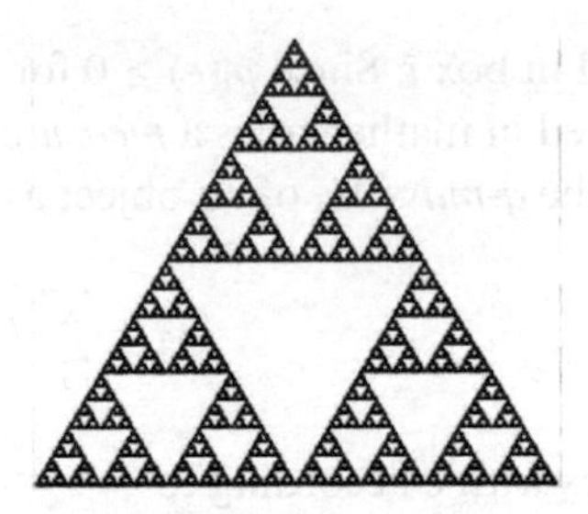

Rescaling the Sierpinski triangle. Stretching the length by a factor of two increases the black area by a factor of three. **Figure 2.28**

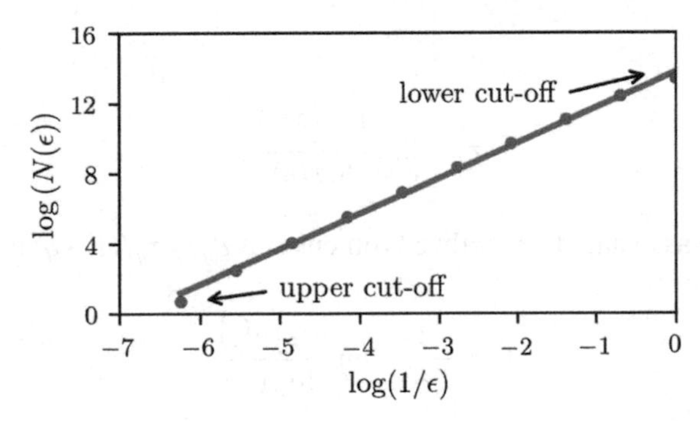

We apply the box counting algorithm to a percolation cluster on the square lattice at p_c and plot the number of nonempty boxes $N(\epsilon)$ against the inverse grid spacing ϵ^{-1} in a log-log scale. For $\epsilon = 1$, we reach the grid spacing of the underlying lattice (lower cutoff). For $\epsilon = L$, one box covers the whole lattice of linear size L (upper cutoff). The slope of the linear fit (blue solid line) corresponds to the measured fractal dimension. **Figure 2.29**

part of the fractal) as $N(\epsilon)$. We do this for a large range of ϵ and plot $N(\epsilon)$ as a function of ϵ in a log-log plot. We show a typical result in Figure 2.29.

We recognize a region where the slope is constant in the plot. In this region, the object is self-similar and the slope equals the fractal dimension. Outside this self-similar regime, the finite resolution and finite size of the object destroy the self-similarity. To convince ourselves that this indeed reproduces the fractal dimension that we defined before, recall that ϵ defines a characteristic length scale, while $N(\epsilon)$ is proportional to the volume of the fractal object. Numerically, the box counting method has rather pronounced finite-size effects, because at the border of an object very sparsely occupied cells can have an important weight.

Multifractality

By expanding the box counting method, we may observe multifractality, a generalization of the fractal dimension. Instead of only considering whether a box is completely empty or not, we now also count how much mass of the fractal, that is, how many occupied points or pixels there are in a certain box i; we will denote this number by $N_i(\epsilon)$. We introduce the occupation density $p_i(\epsilon) = N_i(\epsilon)/N$, where N is the total number of occupied points across all boxes. Thus, $p_i(\epsilon)$ is the fraction of the mass of the fractal contained in box i. Since $p_i(\epsilon) \geq 0$ for all i and $\sum_i p_i(\epsilon) = 1$, the functions $p_i(\epsilon)$ form what is called in mathematics a *measure*.

Let us define the *q-mass* M_q of an object as

$$M_q = \sum_i p_i(\epsilon)^q \,. \tag{2.29}$$

The q-mass varies with ϵ according to

$$M_q \propto \epsilon^{\tau_q} \,, \tag{2.30}$$

where

$$\tau_q = \lim_{\epsilon \to 0} \frac{\log\left(M_q\right)}{\log(\epsilon)} \,. \tag{2.31}$$

Furthermore, the associated generalized dimensions $d_q = \tau_q/(1-q)$ are [51–53]

$$d_q = \frac{1}{1-q} \lim_{\epsilon \to 0} \frac{\log\left(M_q\right)}{\log(\epsilon)} \,, \tag{2.32}$$

where d_0 is the fractal dimension (d_f) as defined before and d_1 is the information dimension. In information theory, the mathematical structure of d_q is similar to the *Rényi entropy*.

Additional Information: Information Dimension

To evaluate the term $M_q^{1/(1-q)}$ in the limit $q \to 1$, we set $q = 1 - q'$ and consider the limit $q' \to 0$:

$$\begin{aligned}
\lim_{q \to 1} M_q^{\frac{1}{1-q}} &= \lim_{q' \to 0} \left(\sum_i p_i^{1-q'} \right)^{\frac{1}{q'}} = \lim_{q' \to 0} \left(\sum_i p_i e^{-q' \log(p_i)} \right)^{\frac{1}{q'}} \\
&= \lim_{q' \to 0} \left[\sum_i p_i \left(1 - q' \log(p_i) - O(q'^2)\right) \right]^{\frac{1}{q'}} \\
&= \lim_{q' \to 0} \left[1 - q' \sum_i p_i \log(p_i) \right]^{\frac{1}{q'}} \\
&= \exp\left[- \sum_i p_i \log(p_i) \right] .
\end{aligned} \tag{2.33}$$

We find that

$$\lim_{q \to 1} d_q = - \lim_{\epsilon \to 0} \frac{\sum_i p_i \ln(p_i)}{\log(\epsilon)} \,. \tag{2.34}$$

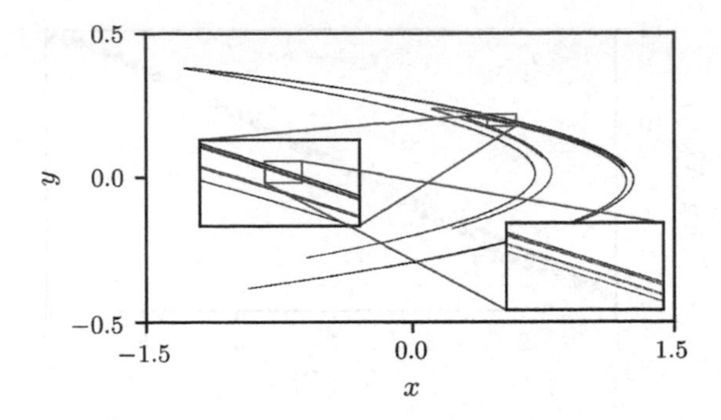

The strange attractor arising from a solution of eq. (2.35) for $a = 1.4$ and $b = 0.3$ ($(x_0, y_0) = (0.6, 0.2)$).

Figure 2.30

Additional Information: Information Dimension (*cont.*)

That is, for $q \to 1$, the generalized dimension d_q corresponds to the Shannon entropy normalized by $\log(\epsilon)$. Because of this connection to information theory, the generalized dimension d_1 is sometimes referred to as *information dimension*. For $q = 0$, the dimension d_0 is also called *capacity dimension*. A possible application of multifractal analysis is the characterization of the distribution of species in a certain habitat (i. e., measuring species diversity) [54].

For simple fractals, d_q does not depend on q and for *multifractals* d_q decreases with q. One example of a multifractal object is the attractor of the Hénon map, which corresponds to the set of values (x_n, y_n) that result from the iterations

$$\begin{pmatrix} x_{n+1} \\ y_{n+1} \end{pmatrix} = \begin{pmatrix} 1 - ax_n^2 + y_n \\ bx_n \end{pmatrix}, \tag{2.35}$$

where $n \in \{1, 2, \dots\}$. This map was proposed by Michel Hénon. When iterating this map of eq. (2.35) with certain a and b starting from any initial numbers (x_0, y_0), one ends up in an attractor, which for this specific case is called "strange attractor" because it is multifractal. We show an example for $a = 1.4$ and $b = 0.3$ in Figure 2.30.

2.4.4 The Sandbox Method

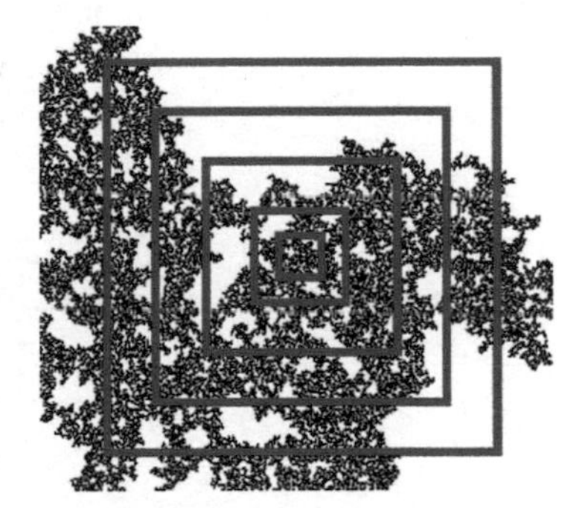

Figure 2.31 Illustration of the sandbox method for a percolation cluster on a square lattice: one measures the number of filled pixels in increasingly large concentrically placed boxes.

We will now return to simple fractals and leave aside further considerations of multifractality. The sandbox method is another technique that can be easily implemented to numerically determine the fractal dimension of an approximately self-similar object. In general, it converges faster than the box counting method. We show an illustration of this method in Figure 2.31. As in the case of the box counting method, we start with a fractal object. First, we choose a site belonging to the fractal which is located in its center region. Next, we place around this site a small box of size R on the fractal and count the number of occupied sites (or pixels) $N(R)$ in the box. We then successively

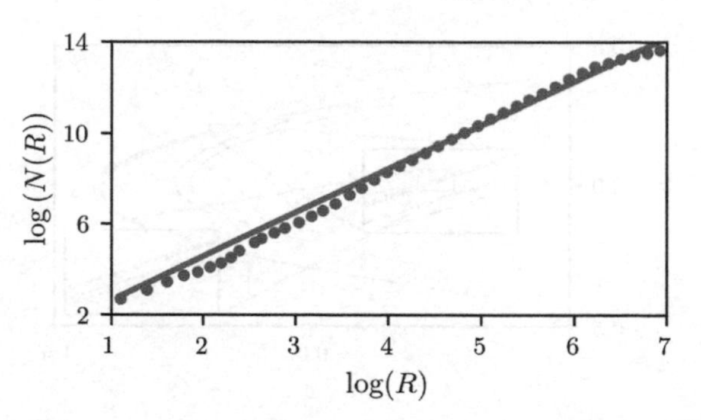

Figure 2.32 We apply the sandbox algorithm to one single percolation cluster at p_c on a square lattice and plot the number of nonempty sites $N(R)$ against the linear box size R in log-log scale. The slope of the linear fit (blue solid line) corresponds to the fractal dimension. To obtain a clearer linear dependence, one has to average each data point over several clusters.

increase the box size R in small steps until we cover the whole picture with our box, always storing $N(R)$. We finally plot $N(R)$ as a function of R in a log-log plot where the fractal dimension is the corresponding slope (see Figure 2.32).

2.4.5 The Ensemble Method

Another possibility to measure the fractal dimension having many objects of different sizes M is to determine their *radius of gyration*

$$R_g = \frac{1}{M(M-1)} \sqrt{\sum_{i \neq j} \left(\mathbf{r}_i - \mathbf{r}_j\right)^2}, \tag{2.36}$$

where $\mathbf{r}_i$ and $\mathbf{r}_j$ are the positions of lattice sites i and j, respectively. We can now extract the fractal dimension based on the proportionality

$$M \propto R_g^{d_f}. \tag{2.37}$$

This so-called *ensemble method* can, for instance, be used to determine the fractal dimension from many clusters of a percolation configuration at the critical occupation probability p_c.

Exercise: Fractal Dimension of the Percolating Cluster

Compute the fractal dimension of a percolating cluster at p_c and close to p_c.
To find the percolating cluster you may use the burning method:

- The procedure is exactly the same as in the burning algorithm, except that only one occupied site is set on fire.

Exercise: Fractal Dimension of the Percolating Cluster (*cont.*)

- If the fire reaches the other side of the cluster, the system percolates and all the burned sites belong to the percolating cluster. (Note: To find all sites of this cluster, you will have to burn further until the fire dies out.)
- If the fire dies out without reaching the other side, the chosen starting site did not belong to the percolating cluster. Choose another site from the first row. If the fire does not reach the other side starting from any (filled) site in the first row, throw the sample away, make a new one, and try again.

To compute the fractal dimension of the obtained spanning cluster, assign to all sites that belong to the spanning cluster a "1" and to all remaining sites a "0."

Task 1: Box Counting Method

Once the percolating (or largest) cluster is found, measure the fractal dimension of the percolating cluster at the critical point using the box counting algorithm:

1. Measure the number N of occupied cells for different grids. (A cell is considered to be occupied if it contains at least one site of the percolating cluster.)
2. The finest grid, here called the "fine grid" is the original lattice (lower cutoff). Coarser grids consist of larger cells which contain several cells of the finer one. A cell of a coarser grid is occupied if it contains at least one occupied cell of the finer grid.
3. Plot N as a function of the inverse size of the cell, the slope of which is the fractal dimension.

Task 2: Sandbox Method

Measure the fractal dimension of a percolating cluster at the critical point using the sandbox algorithm:

1. Compute the mass $N(R)$ (number of sites) of that part of the percolating cluster which lies within a square box of size R that is located at the center of the system, where $R \leq L$ (L: system size of a square lattice). (Note: For the sandbox method, the center of the boxes must be an occupied site.)
2. Vary R from small values $R \ll L$ (e.g., $R = 3$) to the maximal value $R = L$.
3. Plot $N(R)$ as a function of R and measure the fractal dimension (the slope in the log-log plot).

2.4.6 The Correlation Function Method

In the correlation function method, we consider the spatial correlation function of the density of an object. Such a function describes how the density is correlated over various distances. Large correlations mean that two quantities strongly influence each other, whereas no correlations (i.e., being equal to zero) would indicate that they are uncorrelated.

Additional Information: Scaling of $c(r)$ and $g(r)$

For percolation, we defined the correlation function $c(r)$ according to eq. (2.8), which led to the scaling $c(r) \propto r^{-(d-2+\eta)}$ at the percolation threshold p_c (see eq. (2.11)). In the definition of $c(r)$, the origin $(r = 0)$ is part of the percolation cluster, whereas the definition of $g(r)$ in eq. (2.38) is not based on this assumption. The density correlation function $g(r)$ can thus be seen as an average over $c(r)$ [55, 56]. Using $d_f = d - \beta/\nu$ (see eq. (2.21)) and the scaling relations of Section 3.2.4, we find that $c(r) \propto r^{-(d-2+\eta)} \propto r^{2(d_f-d)}$ at p_c, which is different from the scaling (2.39).

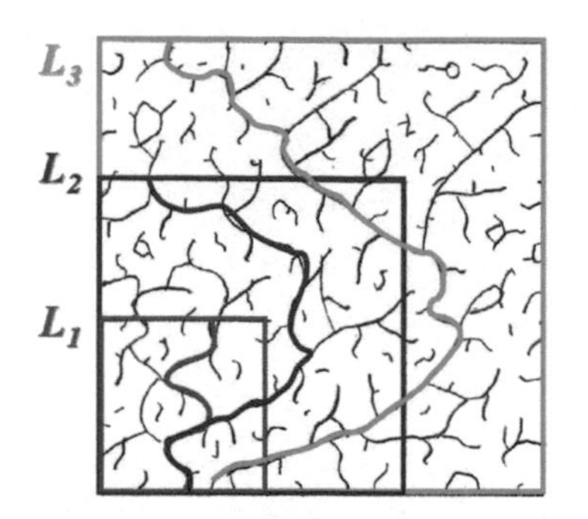

Figure 2.33 Example of a volatile fractal.

Let $\rho(x)$ be the density of the object at point x. The correlation function of the density at the origin and at a distance r is

$$g(r) = \langle \rho(0) \cdot \rho(r) \rangle \,. \tag{2.38}$$

The angle brackets $\langle \cdot \rangle$ denote a suitable averaging (for instance, over all pairs of points separated by a distance r). For a fractal object, the correlation function $g(r)$ decays like a power law as [57]

$$g(r) \propto r^{d_f - d} \,. \tag{2.39}$$

Therefore, in a double-logarithmic plot of $g(r)$ as a function of r, the slope of the (visually) linear part is the fractal dimension minus the dimension of the space the object is embedded into.

The correlation function method is mostly used, when determining the fractal dimension experimentally using scattering techniques, since the dependence of the scattered intensity is the Fourier transform of the density correlation function $g(r)$.

Figure 2.34 Robert Brown (1773–1858) was a Scottish botanist.

2.4.7 Volatile Fractals

As we have seen in Section 2.3.8, the spanning cluster of percolation at p_c is fractal and its fractal dimension is determined by the critical exponents of the order parameter and the correlation length as given in eq. (2.21). Percolation clusters are *volatile*. In a volatile fractal, the cluster is redefined at each scale. That is, the spanning cluster of a system of linear size $L_1 < L_2$ is not necessarily part of the spanning cluster in a system of size L_2 as we show in Figure 2.33.

2.5 Walks

2.5.1 Random Walks

Random walks are used to model many phenomena, including

- the motion of particles in liquids (Brownian motion, named after Robert Brown (see Figure 2.34)) [58, 59],
- stock price fluctuations [60, 61],

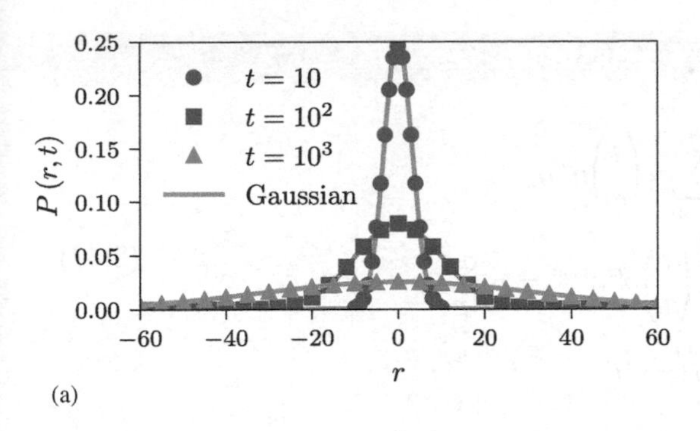

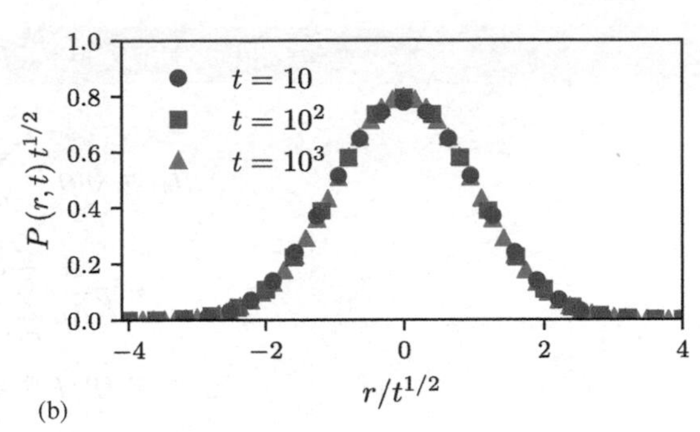

Figure 2.35 In the left panel, we show the binomial distribution (eq. (2.41)) for $t \in \{10, 10^2, 10^3\}$ and the corresponding Gaussian approximation (eq. (2.47)). We observe a data collapse when plotting $P(r, t)t^{1/2}$ against $r/t^{1/2}$ in the right panel.

- foraging behavior of animals [62], and
- polymer chains at the theta point [63, 64].

Moreover, random walks are also important in search algorithms such as the famous PageRank algorithm [65]. We could continue this list with many more examples from various disciplines [66]. For the interested reader, we refer to Refs. [67, 68].

Mathematically, a random walk is a stochastic process

$$X_t = X_0 + \sum_{j=1}^{t} Z_j , \tag{2.40}$$

where $Z_1, Z_2, \ldots, Z_t$ denote t independent random variables that can be scalars or vectors. We set $X_0 = 0$ and note that X_t can be defined either on a lattice ($\mathbb{Z}^d$) or on a continuous space $\mathbb{R}^d$. In one dimension, one possibility is to consider binary random variables Z_j. That is, $Z_j = 1$ with probability p and $Z_j = -1$ with probability $q = 1 - p$. Out of the t total steps, the walker moves l steps to the left ($Z_j = -1$) and m steps to the right ($Z_j = 1$). Thus, after t steps, the walker is at position $X_t = m - l = 2m - t$. The probability that the walker is at position $X_t = 2m - t$ is distributed according to a binomial distribution [68]

$$P\left(X_t = 2m - t\right) = \binom{t}{m} p^m q^{t-m} . \tag{2.41}$$

In Figure 2.35, we illustrate the binomial distribution of eq. (2.41) and the corresponding Gaussian approximation (see eq. (2.47)). We observe a data collapse when plotting $P(r, t)t^{1/2}$ against $r/t^{1/2}$.

Additional Information: Properties of 1D Random Walk

Based on eq. (2.41), we can now build some intuition about the behavior of the considered 1D random walk by computing the expectation value μ_r and variance σ_r^2 of the walker position $r = 2m - t$. Note that we use r instead of X_t for the sake of brevity. However, before doing so, we first have to determine the expectation value μ_m and variance σ_m^2 of m:

Additional Information: Properties of 1D random walk (*cont.*)

$$
\begin{aligned}
\mu_m &:= \langle m \rangle = \sum_{m=0}^{t} m \binom{t}{m} p^m q^{t-m} \\
&= p \frac{\mathrm{d}}{\mathrm{d}p} \sum_{m=0}^{t} \binom{t}{m} p^m q^{t-m} = p \frac{\mathrm{d}}{\mathrm{d}p} (p+q)^t \\
&= tp\,(p+q)^t = tp\,,
\end{aligned} \tag{2.42}
$$

and

$$
\begin{aligned}
\sigma_m^2 &:= \langle m^2 \rangle - \langle m \rangle^2 = \langle m^2 \rangle - (tp)^2 \\
&= \sum_{m=0}^{t} (m(m-1)+m) \binom{t}{m} p^m q^{t-m} - (tp)^2 \\
&= \left(p^2 \frac{\mathrm{d}^2}{\mathrm{d}p^2} + p \frac{\mathrm{d}}{\mathrm{d}p} \right) \sum_{m=0}^{t} \binom{t}{m} p^m q^{t-m} - (tp)^2 \\
&= \left(p^2 \frac{\mathrm{d}^2}{\mathrm{d}p^2} + p \frac{\mathrm{d}}{\mathrm{d}p} \right) (p+q)^t - (tp)^2 = tp - tp^2 = tpq\,.
\end{aligned} \tag{2.43}
$$

The relative width σ_m/μ_m decreases with t according to $\sigma_m/\mu_m \propto t^{-1/2}$. Now we are also able to determine the expectation value μ_r and the variance σ_r of the walker position $r = 2m - t$. Based on Eqs. (2.42) and (2.43), we obtain

$$
\mu_r := \langle r \rangle = 2\mu_m - t = t(p-q)\,, \tag{2.44}
$$

and

$$
\sigma_r^2 := \langle r^2 \rangle - \langle r \rangle^2 = 4\sigma_m^2 = 4tpq\,. \tag{2.45}
$$

If $p \neq q$, there exists a bias in the average walking direction $\mu_r \neq 0$. However, we are more interested in the undirected case where $p = q = 1/2$ such that $\mu_r = \langle r \rangle = 0$ and $\sigma_r^2 = \langle r^2 \rangle = t$. In this case, the mean-square displacement of the considered random walker is proportional to time t. In the limit of large t, we use Stirling's formula

$$
\ln t! = \left(t + \frac{1}{2} \right) \ln t - t + \frac{1}{2} \ln 2\pi + O\left(t^{-1}\right)\,, \tag{2.46}
$$

to approximate the binomial distribution of eq. (2.41) by a Gaussian [69]

$$
P(r,t) = \frac{2}{\sqrt{4\pi D t}} e^{-\frac{r^2}{4Dt}}\,, \tag{2.47}
$$

where $D = 1/2$ denotes the *diffusion coefficient*. Here we implicitly assumed that $r \in \{-t, \ldots, t\}$ and $t \in \{1, 2, \ldots\}$ are integers such that the difference between two consecutive values of r or t is $\Delta r = \Delta t = 1$.

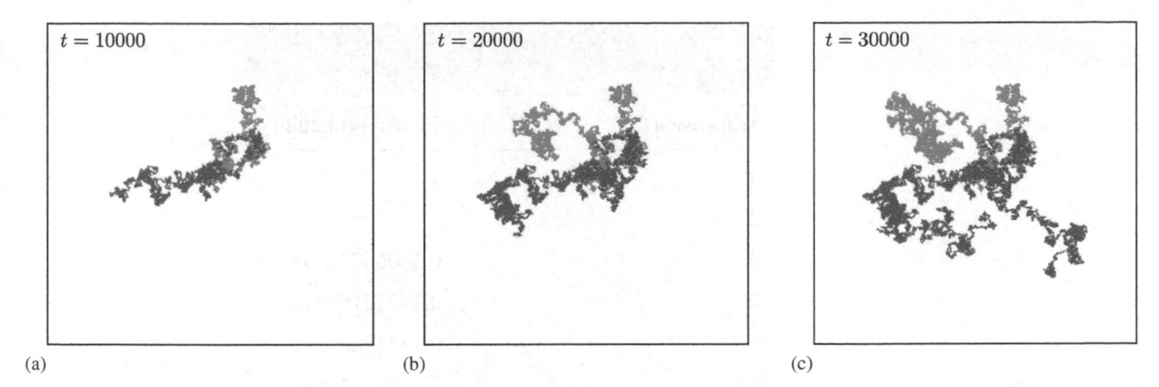

Snapshots of three realizations of random walks (blue, orange, and green) that start at the center of the lattice. **Figure 2.36**

On a computer, we can implement a random walk on a lattice as follows:

Random Walk

- Initialize the walker at a certain lattice site.
- Select one of the possible neighbors uniformly at random and move the walker to the new lattice site.
- Repeat the previous step until a certain target length is reached.

It is also possible to simulate an *off-lattice walk* by randomly selecting an angle $\phi \sim \mathcal{U}(0, 2\pi)$ instead of a certain neighboring lattice site. We show different realizations of two-dimensional random walks at different times in Figure 2.36. The probability to find the random walker at a distance $(r, r + \mathrm{d}r)$ from its starting point at $t = 0$ is

$$P(r,t)\,\mathrm{d}r = \frac{1}{\sqrt{4\pi Dt}} e^{-\frac{r^2}{4Dt}}\,\mathrm{d}r\,, \tag{2.48}$$

where D is the diffusion constant. In d dimensions, the normalizing factor is $(4\pi Dt)^{-d/2}$ and $P(\mathbf{r}, t)$ is a solution of the diffusion equation

$$\frac{\partial}{\partial t} P(\mathbf{r}, t) = D\nabla^2 P(\mathbf{r}, t)\,, \tag{2.49}$$

with initial condition $P(\mathbf{r}, 0) = \delta(\mathbf{r})$. This means that a random walk corresponds to a *regular diffusion* in the continuum limit. In dimensions $d \geq 2$, the fractal dimension of a random walk is $d_f = 2$ [68] and the mean-square displacement is proportional to t (i. e., $\langle r^2 \rangle \propto t^{2/d_f}$). Moreover, Mandelbrot conjectured that the fractal dimension of the boundary of a planar random walk is 4/3. This was proven by Lawler, Schramm, and Werner in 2000 [70].

Figure 2.37 George Pólya (1887–1985) was a professor for mathematics at ETH Zürich and at Stanford University. Photograph courtesy ETH Library.

In 1921, George Pólya (see Figure 2.37) demonstrated that the probability of a random walker to return to the origin equals 1 only in one and two dimensions. Table 2.2 summarizes some return probabilities for various dimensions [71]. In 1951, Dvoretzky

Table 2.2 The probability that a random walker will return to the origin for different dimensions

Dimension	Return probability
1	1
2	1
3	0.34053732955
4	0.19320167322
5	0.13517860982
6	0.10471549562
7	0.08584493411
8	0.07291264996
9	0.06344774965
10	0.05619753597

Note: From Ref. [71].

Figure 2.38 Paul Erdös (1913–1996) was one of the most influential mathematicians of the twentieth century.

and Erdös (see Figure 2.38) identified the following asymptotic behavior of the number of visited sites $N_{\text{cov}}(t)$ at time t [72, 73]:

$$N_{\text{cov}}(t) \propto \begin{cases} \sqrt{t}, & \text{d=1}, \\ t/\ln t, & \text{d=2}, \\ t, & \text{d>2}. \end{cases} \tag{2.50}$$

There also exists *anomalous diffusion* that is characterized by $\langle r^2 \rangle \propto t^{2/d_f}$ with $d_f \neq 2$. For $d_f < 2$, the process is referred to as enhanced (or super) diffusion and for $d_f > 2$ as subdiffusion. One example of an enhanced diffusion phenomenon is the Lévy flight [74] in which the random walker jumps from the current location to a location $r \geq r_{\text{min}} > 0$ that is distributed according to

$$p(r) \propto 1/r^{1+\kappa}, \tag{2.51}$$

where $\kappa > 0$. In d dimensions, the integral over the function $p(r)$ from r_{min} to ∞ can be normalized if $\kappa > d - 1$. The resulting fractal dimension is $d_f = \kappa$ for $\kappa < 2$ and $d_f = 2$ for $\kappa \geq 2$ [68].

Exercise: Random Walks

Task 1: Simple Random Walk

- Generate a two-dimensional random walk consisting of N positions (i. e., $N - 1$ segments). Within each step, the step size is fixed (length of the segments) but the angle has to be chosen at random.
- Measure the square of the end to end distance $\mathbf{R}^2$ (distance between position 1 and N).

Exercise: Random walks (*cont.*)

- Generate several configurations $k \in \{1, \ldots, M\}$ (random walks of same N, but different sequence of random numbers) and compute the average of $\mathbf{R}^2$ as well as the estimated error Δ:

$$\left\langle \mathbf{R}^2 \right\rangle = \frac{1}{M}\sum_{k=1}^{M} \mathbf{R}_k^2, \quad \Delta = \sqrt{\frac{1}{M}\left[\left\langle \left(\mathbf{R}^2\right)^2 \right\rangle - \left\langle \mathbf{R}^2 \right\rangle^2\right]}, \quad \left\langle \left(\mathbf{R}^2\right)^2 \right\rangle = \frac{1}{M}\sum_{k=1}^{M} \left(\mathbf{R}_k^2\right)^2 . \tag{2.52}$$

  Choose the number M of configurations such that the estimated error is below a desired value (e.g., 1 percent of the average value).
- Vary the number N of positions and check your result by comparing it to the analytical result.

Task 2: Chains of Spherical Particles
Now, consider chains consisting of N hard spherical particles of radius r in two dimensions (disks) or in three dimensions (spheres). The particles are not allowed to overlap and neighboring particles in a chain touch each other (self-avoiding random walk).

- Measure $\left\langle \mathbf{R}^2 \right\rangle$ and Δ (same as in task 1).
- Again, vary the number of particles N and compare the behavior to the result of task 1.

Please consider the following points:

- $\mathbf{R}^2$, $\left\langle \mathbf{R}^2 \right\rangle$, etc. depend on the number of particles N.
- The chain of nonoverlapping particles is created in the way similar to the simple random walk, but it has to be checked that there is no overlap with other particles. If an attempt is made which would create an overlap with any particle in the chain, the entire chain must be rejected, and one has to start all over again.

2.5.2 Self-Avoiding Walks

An important extension of a random walk is the concept of walks that cannot intersect their own trajectory (i.e., fulfill the condition of "excluded volume"). The most prominent example is the so-called *self-avoiding walk* (SAW). In one dimension, the walker would just move in one direction, and for dimensions $d \geq 4$, the intersection probability is vanishingly small, so we observe a regular random-walk behavior with fractal dimension $d_f = 2$ [68]. In two and three dimensions, the fractal dimensions are $d_f = 4/3$ and $d_f \approx 5/3$, respectively [75–77]. The SAW was first introduced by Paul Flory (see Figure 2.39) to describe polymers in a good solvent [75, 78].

Figure 2.39 Paul J. Flory (1910–1985) made important contributions to the field of polymer physics and chemistry. He was awarded the Nobel Prize in 1974.

For the SAW, all configurations of same chain length N have the same statistical weight [79]:

$$\Omega_N = \mu^N N^\theta , \tag{2.53}$$

where μ denotes a chemical potential and $\theta = 11/32$ for all two-dimensional lattices [76, 80]. The generating function is equivalent to a grand canonical partition function (see Section 3.1.4)

$$Z(x) = \sum_N \Omega_N x^N = \sum_N (\mu x)^N N^\theta , \tag{2.54}$$

where x is the fugacity which corresponds to the statistical weight of adding one element to the chain. The radius of convergence of $Z(x)$ defines the critical fugacity $x_c = 1/\mu$. The average chain length is

$$\langle N \rangle = \left. \frac{\partial \ln(Z)}{\partial x} \right|_{x=1} = \frac{\sum_N N\Omega_N}{\sum_N \Omega_N} = \begin{cases} \text{finite}, & \text{if } x_c > 1 , \\ \text{critical}, & \text{if } x_c = 1 , \\ \text{infinite}, & \text{if } x_c < 1 . \end{cases} \tag{2.55}$$

There also exist other models for non self-intersecting walks, like the *kinetic growth walk* [81] that have different statistical properties than the SAW.

Creating SAWs numerically, which satisfy the statistical properties of Eqs. (2.53)–(2.55) correctly is not completely trivial and we will present next a few such techniques. The simplest method is the rejection method. In this method, we grow a chain of length N by starting from one site and randomly move to one of the empty neighbors. We repeat this procedure until the required length is reached. If we try to move to an already occupied neighbor (self-intersection), we have to start all over again. This method is straightforward to implement, but the rejection rate becomes extremely high for longer walks. If we want to generate many samples of SAWs of a certain length N, this simple method is computationally too expensive.

To overcome that sampling issue, other methods have been developed that make the sampling of SAWs of length N more efficient. One method is the reptation[6] algorithm [82]:

Reptation Algorithm

Start with any nonintersecting walk of length N.

1. Choose one end of the walk.
2. Place last monomer from chosen end in a random direction at the other end.
3. If there is no overlap, retain this configuration otherwise, reject the move.
4. Repeat above until the desired number of configurations is reached.

All retained configurations correspond to samples of an SAW of length N. Another method that enables us to efficiently obtain samples of SAWs is the kink-jump algorithm [83]:

[6] The name derives from the term *reptile* because during the simulation, the chain moves back and forth like a snake.

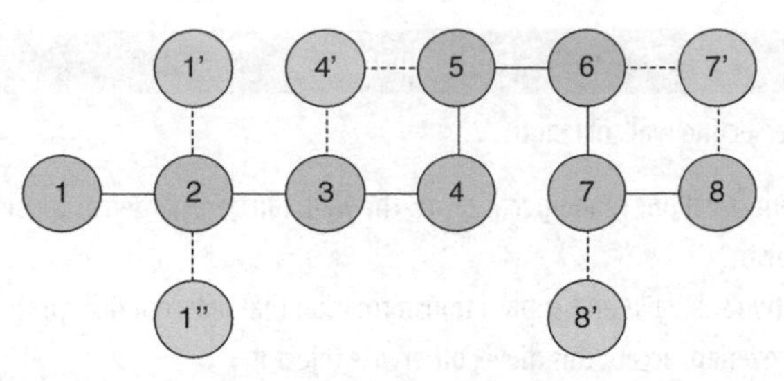

Figure 2.40 An illustration of possible kink-jump algorithm updates in a chain that consists of eight particles. Blue monomers correspond to the initial configuration, whereas gray monomers indicate possible positions for kink-jump movements.

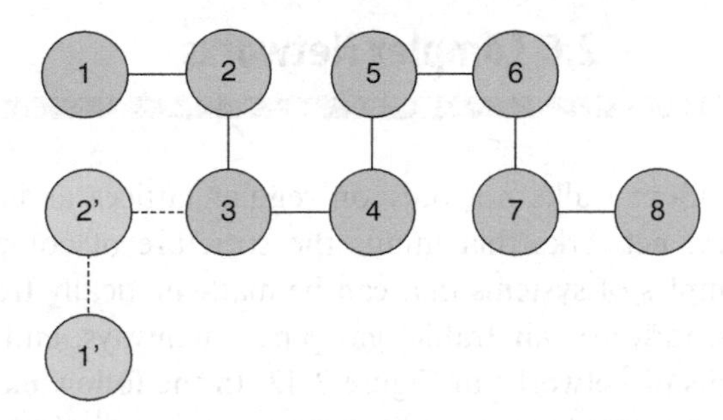

Figure 2.41 An illustration of a possible pivot algorithm update in a chain that consists of eight particles. Blue monomers correspond to the initial configuration, whereas the gray monomers correspond to a subwalk that was rotated counter-clockwise at an angle of 90°.

Kink-Jump Algorithm

Start with any nonintersecting walk of length N.

1. Uniformly at random select one monomer.
2. If the monomer lies at the end of the chain or at a corner, determine the possible new local positions that do not disrupt the chain and do not violate excluded volume condition.
3. If no new positions are available, keep the current chain.
4. If new positions are available, return to step 1 until the desired number of configurations is reached.

We illustrate possible kink-jump updates of monomer positions in Figure 2.40. Blue monomers correspond to the initial configuration and gray monomers indicate possible positions for updates according to the kink-jump algorithm. Instead of changing the location of only one monomer, we can also consider sub-chains. This brings us to the pivot algorithm [84] as described in the beginning of the next page.

Examples of sub-chain transformations include reflections and rotations about an axis. We show one example of a rotation transformation in Figure 2.41. In this example, we initially select the monomer with label "3" and then rotate the subwalk consisting of monomers "1" and "2" counterclockwise at an angle of 90°.

Pivot Algorithm

Start with any nonintersecting walk of length N.

1. Uniformly at random select one monomer as pivot. The walk can be regarded as consisting of two subwalks connected by the pivot.
2. Choose one of the two subwalks and apply a transformation that does not disrupt the chain.
3. If this produces no overlap, accept this move, otherwise reject it.
4. Return to step 2 and repeat until the desired number of configurations is reached.

2.6 Complex Networks

We can simulate random walks not only on regular lattices as in Section 2.5, but also on more general networks that mimic the structure of complex physical and social systems. Examples of systems that can be mathematically treated as networks include power grids, railways, air traffic, gas pipes, highways, and the Internet. We show two illustrations of networks in Figure 2.42. In the following sections, we provide an overview of concepts from network science, an interdisciplinary research field that combines methods from graph theory, statistical physics, and computer science. For further reading on network science, there exist many excellent textbooks [85–87].

2.6.1 Adjacency Matrix

Mathematically, a network $G(V, E)$ is an ordered pair, where V and E are the corresponding sets of nodes and edges. Connections between nodes are described by the adjacency matrix A. We set the matrix element $A_{uv} = 1$ if there exists an edge $e \in E$ between the node pair $\langle u, v\rangle \in V$ and otherwise we set $A_{uv} = 0$. In a *directed* network A_{uv} can be different from A_{vu}, whereas $A_{uv} = A_{vu}$ for *undirected* networks. The adjacency matrix of the (undirected) graph that we show in Figure 2.43 is

$$A = \begin{pmatrix} 0 & 1 & 1 & 1 & 0 & 1 & 1 \\ 1 & 0 & 1 & 0 & 1 & 1 & 0 \\ 1 & 1 & 0 & 1 & 0 & 0 & 1 \\ 1 & 0 & 1 & 0 & 1 & 0 & 0 \\ 0 & 1 & 0 & 1 & 0 & 1 & 0 \\ 1 & 1 & 0 & 0 & 1 & 0 & 1 \\ 1 & 0 & 1 & 0 & 0 & 1 & 0 \end{pmatrix}. \tag{2.56}$$

In this example, the adjacency matrix is sparse and it is computationally more efficient to use edge lists instead of storing the complete matrix. For complete graphs where all nodes are connected with each other, we have to store the whole adjacency matrix.

The elements (u, v) of the matrix powers A^k represent the number of paths of length k from node u to node v. For an *undirected network*, the degree of node u is

$$k_u = \sum_{v \in V} A_{uv} = \sum_{v \in V} A_{vu} . \tag{2.57}$$

For *regular* networks, all nodes have the same degree. An example of a regular network is the square lattice with periodic boundary conditions. For a *directed* network, we distinguish between in-degree (number of incoming edges) and out-degree (number of outgoing edges). Mathematically, we denote the in-degree of node u by

$$k_u^{\text{in}} = \sum_{v \in V} A_{vu} , \tag{2.58}$$

and the corresponding out-degree by

$$k_u^{\text{out}} = \sum_{v \in V} A_{uv} . \tag{2.59}$$

For undirected networks, summing over k_u yields

$$\sum_{u \in V} k_u = \sum_{u \in V} \sum_{v \in V} A_{uv} = 2|E| , \tag{2.60}$$

where $|E|$ denotes the number of edges of the network $G(V, E)$. This equation implies that every undirected network has an even number of nodes with an odd degree. If there would be an odd number of nodes with odd degrees, the right-hand side of eq. (2.60) would not be even (i.e., equal to $2|E|$). This is also known as the *handshaking lemma*. In a group ("network") of handshaking people, an even number of people ("nodes") must shake an odd number of other people's ("neighboring nodes") hands. For directed networks, we obtain

$$\sum_{u \in V} k_u^{\text{in}} = \sum_{u \in V} k_u^{\text{out}} = |E| . \tag{2.61}$$

Figure 2.42 An Apollonian network [88] with 1096 nodes and mean degree $\langle k \rangle \approx 5.99$ (a). A school friendship network (based on an Add Health survey) with 2539 nodes and mean degree $\langle k \rangle = 8.24$ (b). Both illustrations were created with Gephi [89]; different colors represent different communities.

2.6.2 Degree Distribution

Networks can be characterized by their degree distribution $P(k)$. For a square lattice, the degree distribution is $P(k) = \delta_{k4}$ (i.e., each node has degree $k = 4$). If the distribution follows a power law

$$P(k) \propto k^{-\gamma} , \tag{2.62}$$

the network is called scale-free [91]. Examples of networks that are approximately scale-free are the Internet and some social networks [92, 93]. One mechanism that produces scale-free networks is preferential attachment where each new node is more likely to attach to existing nodes with high degree. Mathematically, each new node will be attached to $m \leq m_0$ existing nodes and the attachment probability is proportional to the number of edges of the existing nodes. Here, m_0 is the initial number of nodes. Networks that result from this type of preferential attachment are called Barabási–Albert networks and their degree distribution is $P(k) \propto k^{-3}$. We show an example of a Barabási–Albert network in Figure 2.44.

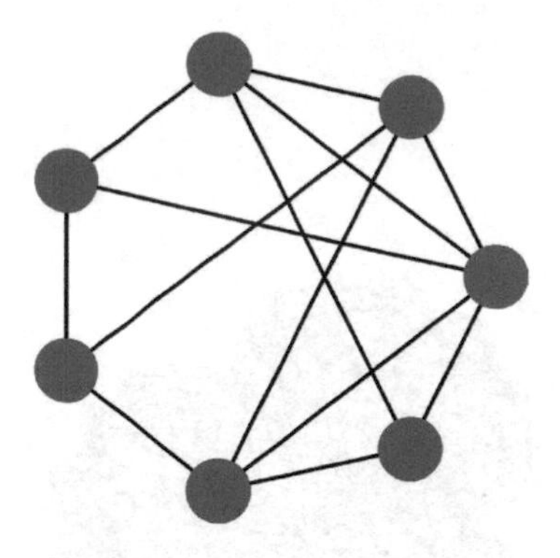

Figure 2.43 An example of a graph with seven nodes.

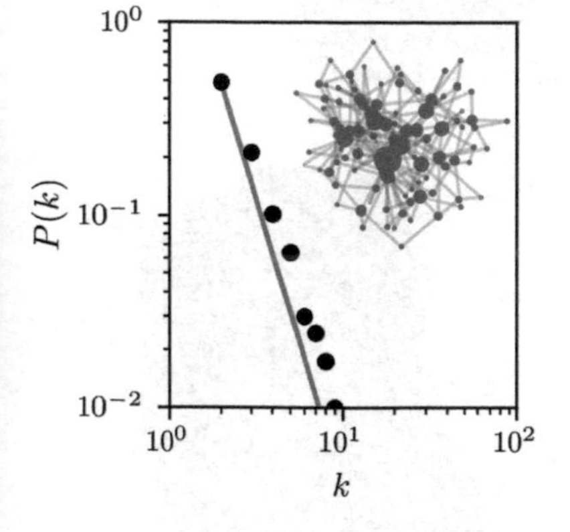

Figure 2.44 The degree distributions of a Barabási–Albert network. Black disks represent the degree distribution of a single network realization (N=1600 nodes), and the gray solid line is the corresponding analytic degree distribution (power law with exponent −3 [90]). The inset is an example of a Barabási–Albert network with 100 nodes. The sizes of nodes scale with their betweenness centralities (see eq. (2.68)).

Networks that are generated by a random process are also called random-graph models [94, 95]. Another example of a random-graph model are Erdős–Rényi networks [96]. To generate them, we start with N isolated nodes and add new edges between two uniformly at random selected nodes with probability p. The resulting degree distribution is binomial:

$$P(k) = \binom{N-1}{k} p^k (1-p)^{N-1-k}, \tag{2.63}$$

where N is the number of nodes.

Watts–Strogatz networks [97] also belong to the class of random-graph models and are generated by arranging nodes in a ring and connecting each node to its K nearest neighbors. The $K/2$ rightmost edges of each node are then rewired (i. e., reconnected) with probability p (see Figure 2.45). The degree distribution of a Watts–Strogatz network is [98]

$$P(k) = \sum_{l=0}^{f(k,K)} \binom{K/2}{l} (1-p)^l p^{K/2-l} \frac{(pK/2)^{k-K/2-l}}{(k-K/2-l)!} e^{-pK/2}, \tag{2.64}$$

where $f(k, K) = \min(k - K/2, K/2)$.

2.6.3 Small-World Networks

As a further network characteristic, we use $d_{\min}(s, t)$ to denote the shortest-path length between a source node s and a target node t for each node pair $\langle s, t\rangle \in V$. In a complete graph, all nodes are connected and the shortest-path length between two nodes s and t is $d_{\min}(s, t) = 1$. Certain networks such as Watts–Strogatz networks (see Figure 2.45) are so-called *small-world* networks. That is, the shortest-path length d between two randomly chosen nodes satisfies

$$d \propto \log(N), \tag{2.65}$$

while the clustering coefficient C is significantly larger than in an equivalent random network. Mathematically, the local clustering coefficient $c(u)$ of a node u with degree $k_u \geq 2$ is the number of triangles $T(G; u)$ that include that node divided by all possible triangles that could include u (i.e., the connected triples) [99]:

$$c(u) = \frac{2T(G; u)}{k_u(k_u - 1)}. \tag{2.66}$$

If the degree of a node is 0 or 1, the corresponding local clustering coefficient of that node is 0. The mean local clustering coefficient of a network is

$$C(G) = \frac{1}{|V|} \sum_{u \in V} c(u). \tag{2.67}$$

The interesting small world situation in which eq. (2.65) is valid (and the clustering coefficient is high) occurs for intermediate values of p (see Figure 2.45). In scale-free networks, one finds that $d \propto \log\log(N)$ [100]. These networks are said to have ultra small-world properties.

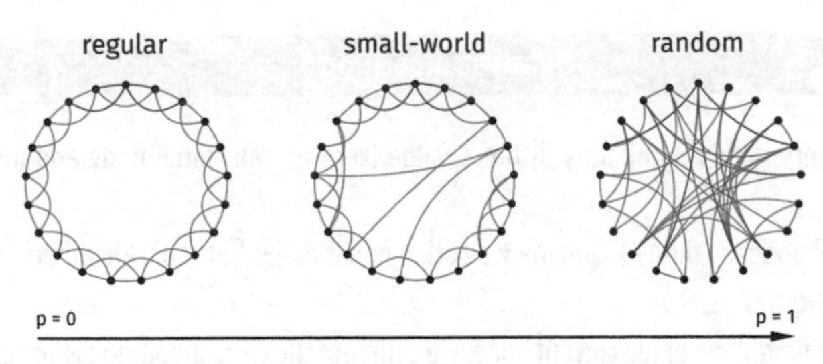

Different regimes in the Watts–Strogatz model.

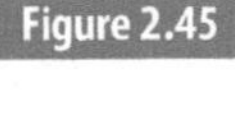
Figure 2.45

To analyze networks with respect to their shortest-path characteristics, we use σ_{st} to denote the total number of shortest paths from s to t. The number of shortest paths from s to t that traverse $v \in V$ is $\sigma_{st}(v)$. The geodesic node betweenness centrality c_B for a node v is the sum (over all node pairs $\langle s, t \rangle$ in a network) of the number of shortest paths from s to t that traverse v divided by the total number of shortest paths from s to t [87]. Including a normalization factor, the geodesic node betweenness centrality is defined as

$$c_B(v) = \sum_{\langle s,t \rangle \in V} \frac{\sigma_{st}(v)}{\sigma_{st}} \underbrace{\frac{2}{(|V|-1)(|V|-2)}}_{\text{normalization factor}} . \tag{2.68}$$

Note that $\sigma_{st} = 1$ if $s = t$ and that $\sigma_{st}(v) = 0$ if $v \in \{s, t\}$. We define the geodesic node betweenness of a network G as

$$C_B(G) = \frac{1}{|V|} \sum_{v \in V} c_B(v) . \tag{2.69}$$

That is, $C_B(G)$ is the mean geodesic node betweenness centrality over all nodes in a network.

A large geodesic betweenness centrality $c_B(v)$ indicates that many shortest paths traverse a particular node. This may help identifying important nodes in social, physical, and communication networks [87, 101–103].

Figure 2.46 Edsger W. Dijkstra (1930–2002) was a professor of computer science and mathematics at the University of Texas at Austin and the Eindhoven University of Technology.

2.6.4 Dijkstra's Algorithm

In the previous section, we defined the shortest path between two points on an arbitrary network $G(V, E)$, where V is a set of nodes and E is a set of edges. With the burning method (see Section 2.3.1), we have already seen an example of an algorithm that identifies the shortest-path length ("burning time") on a percolation cluster. For a general graph $G(V, E)$, Edsger Dijkstra (see Figure 2.46) proposed a greedy algorithm[7] in 1959.

[7] A greedy algorithm tries to find a global optimum through many local optimizations. At each step, the algorithm makes the locally optimal choice.

Dijkstra's Algorithm

1. We assign to every node a temporary distance value (zero for our initial node and infinity for all other nodes).
2. We set the initial node as "burned" and mark all other nodes as "unburned." All unburned nodes are stored in a corresponding set.
3. For all unburned neighbors of the current node, we compute the corresponding distances. If the computed distance is smaller than the current one, we assign the smaller value. Otherwise, we keep the current value.
4. After having considered all of the neighbors of the current node, we mark the current node as "burned" and remove it from the set of unburned nodes. We will not consider a burned node again.
5. We stop the algorithm if we reach the destination node or if the smallest tentative distance among the nodes in the unvisited set is infinity.
6. Otherwise, we select the "unburned" node that is marked with the smallest distance, consider it as current node, and return to step 3.

When combined with additional optimization techniques, the run time of Dijkstra's algorithm is proportional to $|E| + |V| \log |V|$. In addition to Dijkstra's algorithm, there also exists the Bellman–Ford–Moore algorithm that can also handle negative weights [104].

Exercise: Analyzing Networks

In this exercise, we generate (i) Erdős–Rényi and (ii) Barabási–Albert networks and analyze some of their properties.

(i) To build an Erdős–Rényi network, begin with a set of N isolated nodes and iterate over all possible node pairs. Two nodes are connected with probability p.

(ii) Barabási–Albert networks are constructed via preferential attachment after starting from a connected network with m_0 nodes. Preferential attachment means that a new node is connected to $m \leq m_0$ existing nodes and the probability p_i of being connected to an existing node i is proportional to the degree k_i of that node (i. e., $p_i = k_i / \sum_j k_j$).

Task 1: You can start with networks that have $N = 100$ nodes and later extend your analyses to networks with 1000 and 10,000 nodes. Vary the connection probability p for Erdős-Rényi networks and the parameter m for Barabási–Albert networks. What is the influence of these variations on the degree distribution?

Task 2: Implement Dijkstra's algorithm and determine the average path length between two nodes for (i) and (ii) as a function of the number of nodes N. What do you observe?

Equilibrium Systems 3

3.1 Classical Statistical Mechanics

In the first part of this chapter, our goal is to provide an overview of methods to computationally study systems that can be described within the framework of equilibrium statistical mechanics. We therefore first introduce the most important concepts from classical statistical mechanics that enable us to mathematically capture the macroscopic properties of interacting microscopic units. Historically, important contributions to the microscopic formulation of thermodynamics were made by Boltzmann and Gibbs [105, 106]. In particular, Gibbs's notion of a statistical ensemble enables us to interpret macroscopic physical quantities as averages over a large number of different configurations of a system's interacting units.

Figure 3.1 Ludwig Boltzmann (1844–1906) is one of the fathers of statistical mechanics and formulated one version of the ergodicity hypothesis.

3.1.1 Phase Space

Let us consider a three-dimensional classical physical system that consists of N particles. We need a three-component vector to describe the position of a certain particle and another three-component vector to describe its momentum. Therefore, we denote the canonical coordinates and corresponding conjugate momenta of all N particles by $q_1, \ldots, q_{3N}$ and $p_1, \ldots, p_{3N}$, respectively. The $6N$-dimensional space Γ that results from the set of canonical coordinates defines the *phase space*. This concept has been introduced by Ludwig Boltzmann (see Figure 3.1). The considered N particles could simply be uncoupled harmonic oscillators. In this case, the phase space of each single particle would look like the one we show in Figure 3.2. By keeping certain external parameters such as temperature and pressure constant, we can measure a macroscopic physical quantity by computing the *time average*

$$\langle Q \rangle_T = \lim_{T \to \infty} \frac{1}{T} \int_0^T Q(p(t), q(t))\, \mathrm{d}t \tag{3.1}$$

over different realizations of the underlying microscopic states. However, computing the time average of a certain macroscopic quantity makes it necessary to determine the time evolution of all microscopic states. If the number of involved particles is large, this approach would be computationally infeasible. Instead of computing time averages, another possibility is to consider an average over an ensemble of systems in different microstates (i.e., specific microscopic configurations of a certain system) under the same macroscopic conditions.

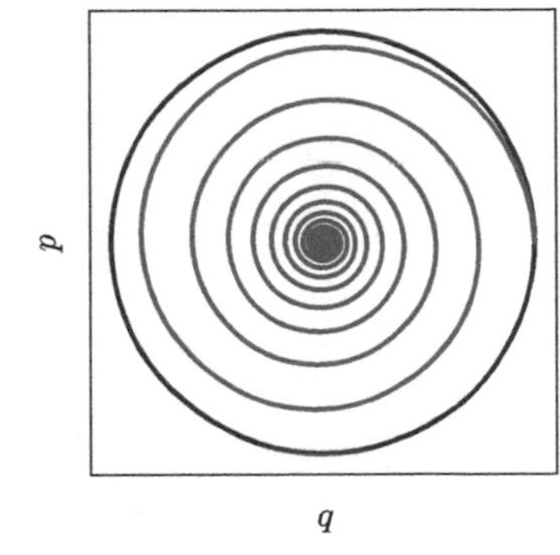

Figure 3.2 Orbits in phase space of an undamped (blue) and a damped (orange) harmonic oscillator.

The assumption that all states in an ensemble are reached by the time evolution of the corresponding system is referred to as *ergodicity hypothesis*. Next, we define the *ensemble average* of a quantity $Q(p,q)$ as

$$\langle Q\rangle = \frac{\int Q(p,q)\rho(p,q)\,\mathrm{d}p\mathrm{d}q}{\int \rho(p,q)\,\mathrm{d}p\mathrm{d}q}\,, \tag{3.2}$$

where $\rho\,(p,q)$ denotes the *phase space density* and $\mathrm{d}p\mathrm{d}q$ is a shorthand notation for $\mathrm{d}p^{3N}\mathrm{d}q^{3N}$.

3.1.2 Liouville Theorem

The dynamics of the considered N particles is described by their *Hamiltonian* $\mathcal{H}(p,q)$. The equations of motion are

$$\dot{p}_i = -\frac{\partial\mathcal{H}}{\partial q_i} \quad \text{and} \quad \dot{q}_i = \frac{\partial\mathcal{H}}{\partial p_i} \quad (i = 1,\ldots,3N)\,. \tag{3.3}$$

Moreover, the temporal evolution of a phase space element of volume V and boundary ∂V is given by

$$\frac{\partial}{\partial t}\int_V \rho\,\mathrm{d}V + \int_{\partial V} \rho v\,\mathrm{d}A = 0\,, \tag{3.4}$$

where $v = (\dot{p}_1,\ldots,\dot{p}_{3N},\dot{q}_1,\ldots,\dot{q}_{3N})$ is a generalized velocity vector. By applying the divergence theorem to eq. (3.4), we find that ρ satisfies the continuity equation

$$\frac{\partial\rho}{\partial t} + \nabla\cdot(\rho v) = 0\,, \tag{3.5}$$

where $\nabla = (\partial/\partial p_1,\ldots,\partial/\partial p_{3N},\partial/\partial q_1,\ldots,\partial q_{3N})$. We can further simplify eq. (3.5) with the help of the identity

$$\nabla\cdot v = \sum_{i=1}^{3N}\left(\frac{\partial\dot{q}_i}{\partial q_i} + \frac{\partial\dot{p}_i}{\partial p_i}\right) = \sum_{i=1}^{3N}\underbrace{\left(\frac{\partial}{\partial q_i}\frac{\partial\mathcal{H}}{\partial p_i} - \frac{\partial}{\partial p_i}\frac{\partial\mathcal{H}}{\partial q_i}\right)}_{=0} = 0\,. \tag{3.6}$$

We use Poisson brackets[1] to rewrite eq. (3.5) and obtain *Liouville's Theorem*,

$$\frac{\partial\rho}{\partial t} = \{\mathcal{H},\rho\}\,, \tag{3.8}$$

which describes the time evolution of the phase space density ρ.

[1] The Poisson bracket is defined as

$$\{u,v\} = \sum_i\left(\frac{\partial u}{\partial q_i}\frac{\partial v}{\partial p_i} - \frac{\partial u}{\partial p_i}\frac{\partial v}{\partial q_i}\right) = -\{v,u\}. \tag{3.7}$$

3.1.3 Thermal Equilibrium

In *thermal equilibrium*, the system reaches a steady state in which the distribution of configurations is time independent. That is, the phase space density satisfies $\partial\rho/\partial t = 0$. Liouville's theorem (eq. (3.8)) then leads to

$$v \cdot \nabla\rho = \{\mathcal{H}, \rho\} = 0\,. \tag{3.9}$$

One possibility to satisfy this equation is to take a phase space density ρ that depends on quantities such as energy or the number of particles, which are conserved during the time evolution of the system. We may then use such a phase space density to replace the *time average* (eq. (3.1)) by a corresponding ensemble average (eq. (3.2)).

In the subsequent sections, we also consider discrete configurations X for which we define the ensemble average as

$$\langle Q\rangle = \frac{1}{\Omega}\sum_X Q(X)\rho(X)\,, \tag{3.10}$$

where Ω is the normalizing volume such that $\Omega^{-1}\sum_X \rho(X) = 1$. With the help of ensemble averages, systems can be described by macroscopic quantities such as temperature, energy, and pressure.

3.1.4 Ensembles

We perform measurements to determine some observable quantities that characterize a given physical system while keeping other parameters constant. Mathematically, such measurements can be seen as an "ensemble average" (Section 3.1.3), which is based on a phase space density ρ that only depends on conserved quantities. However, we should bear in mind that it is not possible to independently adjust the values of all parameters. As an example, we may consider a classical gas. The dynamics of the gas is described by perfect elastic collisions and therefore it is impossible to compress the volume by keeping pressure and temperature unchanged at thermal equilibrium. The system behaves differently depending on which quantities are kept constant. Some quantities such as volume V and pressure p are conjugate to another. Either V can be held constant or p, not both.

Other examples are energy E and temperature T, particle number N and chemical potential μ, magnetization M and magnetic field H. Depending on the parameters held constant, the system is described by a

- microcanonical ensemble: constant E, V, N,
- canonical ensemble: constant T, V, N,
- canonical pressure ensemble: constant T, p, N,
- grandcanonical ensemble: constant T, V, μ.

The notion of a physical ensemble was introduced by Josiah W. Gibbs (see Figure 3.3). In the subsequent sections, we consider microcanonical and canonical systems.

Figure 3.3 Josiah W. Gibbs (1839–1903) introduced the concept of statistical ensembles and derived the laws of thermodynamics from statistical mechanics.

3.1.5 Microcanonical Ensemble

In a *microcanonical ensemble*, number of particles, volume, and energy are constant. Any configuration X of the system has a constant energy $E(X)$. The phase space density is

$$\rho(X) = \frac{1}{Z_{\mathrm{mc}}}\delta(\mathcal{H}(X) - E)\,, \tag{3.11}$$

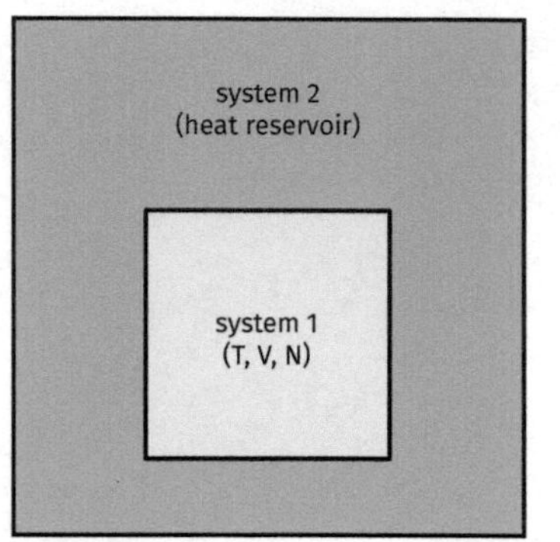

Figure 3.4 In a canonical ensemble, we consider a system (system 1) that is coupled to a heat reservoir (system 2). The heat reservoir guarantees a constant temperature.

where Z_{mc} is the *partition function* of the microcanonical ensemble:

$$Z_{\mathrm{mc}} = \sum_X \delta(\mathcal{H}(X) - E)\,. \tag{3.12}$$

3.1.6 Canonical Ensemble

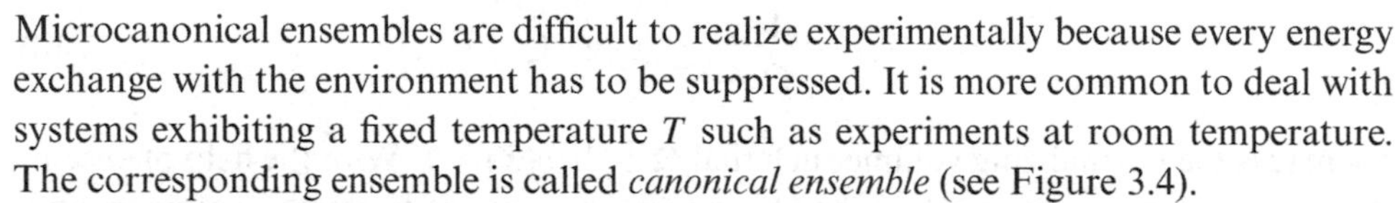

Microcanonical ensembles are difficult to realize experimentally because every energy exchange with the environment has to be suppressed. It is more common to deal with systems exhibiting a fixed temperature T such as experiments at room temperature. The corresponding ensemble is called *canonical ensemble* (see Figure 3.4).

At a given temperature T, the probability for a system to be in a certain configuration X with energy $E(X)$ is

$$\rho(X) = \frac{1}{Z_T}\exp\left[-\frac{E(X)}{k_B T}\right], \tag{3.13}$$

with the corresponding canonical partition function

$$Z_T = \sum_X \exp\left[-\frac{E(X)}{k_B T}\right]. \tag{3.14}$$

According to eq. (3.10), the ensemble average of a quantity Q in the canonical ensemble is

$$\langle Q \rangle = \frac{1}{Z_T}\sum_X Q(X)\, e^{-\frac{E(X)}{k_B T}}\,. \tag{3.15}$$

3.1.7 Classifying Phase Transitions

Phase transitions are omnipresent in many physical, biological, and ecological systems [107]. Important examples of phase transitions include the solid–liquid and liquid–gas transitions of water and other liquids. In Section 3.2, we will see that phase transitions also occur in models of magnetism. In general, phase transitions occur in systems in which underlying processes lead from a disordered to an ordered phase. In an ordered phase, the constituents of the system (molecules, spins, etc.) are in a state that exhibits some symmetry. In a thermodynamic system whose interactions are defined by a Hamiltonian, phase transitions result from the competition between internal energy U (or enthalpy H) and entropy S. To mathematically characterize this competition, we use the Helmholtz free energy $F = U - TS$ and the Gibbs free energy $G = H - TS$ as relevant thermodynamic potentials. The temperature T acts as a control parameter to go from a disordered phase ("high temperature") to an ordered phase ("low temperature"). The transition occurs at a temperature T_c. Based on a more historical classification of Paul Ehrenfest, we say that a phase transition is of nth order if [108]

$$\left(\frac{\partial^m F}{\partial T^m}\right)_V\bigg|_{T=T_{c^+}} = \left(\frac{\partial^m F}{\partial T^m}\right)_V\bigg|_{T=T_{c^-}} \tag{3.16}$$

and

$$\left(\frac{\partial^m F}{\partial V^m}\right)\bigg|_{T=T_{c^+}} = \left(\frac{\partial^m F}{\partial V^m}\right)\bigg|_{T=T_{c^-}}, \tag{3.17}$$

for $m \leq n-1$, and otherwise

$$\left(\frac{\partial^n F}{\partial T^n}\right)_V\bigg|_{T=T_{c^+}} \neq \left(\frac{\partial^n F}{\partial T^n}\right)_V\bigg|_{T=T_{c^-}} \tag{3.18}$$

and

$$\left(\frac{\partial^n F}{\partial V^n}\right)\bigg|_{T=T_{c^+}} \neq \left(\frac{\partial^n F}{\partial V^n}\right)\bigg|_{T=T_{c^-}}. \tag{3.19}$$

We defined a phase transition of nth order based on the Helmholtz free energy F (constant volume and temperature ensemble) because we will also use this potential in our discussion of magnetic systems in Section 3.2. For systems at constant pressure and temperature, it may be useful to use the Gibbs free energy G for which Eqs. (3.16)–(3.19) can be used as well to define a phase transition of nth order. We again note that the Ehrenfest classification is a more historical one. The majority of phase transitions are of first and second order and modern classifications mainly distinguish between these two cases. Second-order phase transitions with $n > 1$ are also called critical points.

For other systems whose interactions are not defined by a Hamiltonian, it is possible to classify phase transitions according to the behavior of the order parameter [109]. If the order parameter goes continuously to zero at T_c, the phase transition is of second (or higher) order. For discontinuous changes of the order parameter, the phase transition is said to be of first order.

3.2 Ising Model

Because of its historical relevance for the study of phase transitions in statistical physics and its broad applicability in many other fields, we now apply the outlined terminology from Section 3.1 to the Ising model. Wilhelm Lenz proposed this model to his doctoral student Ernst Ising (see Figure 3.5) to describe systems that are composed of magnetic dipole moments. Each dipole moment can be in one of two states (+1 or −1). The original goal was to model phase transitions in magnetic materials. As part of his doctoral thesis in 1924, Ernst Ising showed that the one-dimensional model exhibits no phase transition. For that reason, it was expected that this model was of no particular use. Then in 1944, Lars Onsager (see Figure 15.4) derived the partition function of the two-dimensional Ising model without magnetic field, finding a critical point at a finite temperature T_c [111] and later in 1949 published an equation to describe the temperature dependence of the magnetization. Onsager provided no proof for his formula, but it is known that he derived it using Toeplitz determinants [112]. A few years passed until a proof was established by Yang in 1952 [113]. Since then, the Ising model has been successfully applied to a large number of physical and nonphysical problems such as magnetic systems, binary mixtures, and models of opinion formation. Also when compared with experimental data, the Ising model has been found to generally agree very well with observations made for certain magnetic materials [114]. To date, no general analytical solution for the Ising model in three dimensions is known. This is the reason why this mathematical model has been studied so intensively from a numerical perspective [115–119], using tools from statistical physics – some of which we describe in this section.

Figure 3.5 Ernst Ising (1900–1998) was a professor at Bradley University. Ising was born in Cologne and migrated to the United States. For more information about his biography, see Ref. [110].

We start our discussion of the properties of the Ising model by considering a two-dimensional lattice with sites being in binary states $\sigma_i \in \{1, -1\}$ that only interact with their nearest neighbors (see Figure 3.6). This restriction can be relaxed by letting the sites also interact with the next-nearest neighbors and other extended neighborhoods. The interaction of spins is described by the Hamiltonian

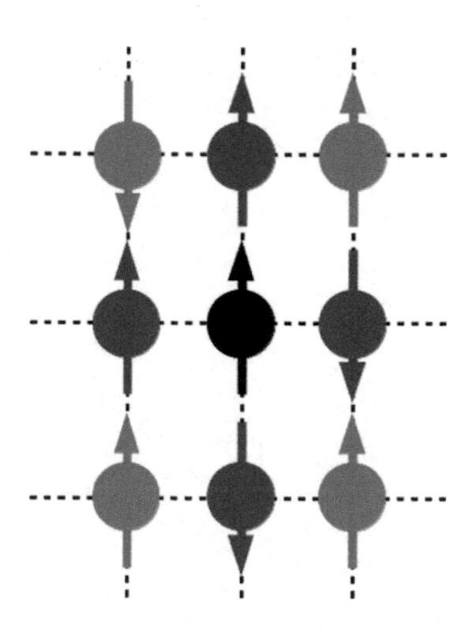

Figure 3.6 An illustration of the interaction of a magnetic dipole (black) with its nearest neighbors (dark gray) on a square lattice.

$$\mathcal{H}(\{\sigma\}) = -J \sum_{\langle i,j \rangle} \sigma_i \sigma_j - H \sum_{i=1}^{N} \sigma_i, \tag{3.20}$$

where the first term denotes the interaction between all nearest neighbors (represented by a sum over $\langle i, j \rangle$), and the second term accounts for the interaction of each site with an external magnetic field H, favoring the $\sigma_i = +1$ state if $H > 0$. In the ferromagnetic case ($J > 0$), the energy is lowered by an amount of $-J$ if two spins are parallel, whereas the energy is increased by an amount of $+J$ if two spins are antiparallel. For $H = 0$, there exist two ground states at $T = 0$, namely, all spins up or all spins down. If $J < 0$, the system is called antiferromagnetic, because antiparallel spins lead to an energy minimization. No interaction occurs for $J = 0$. In the case of a ferromagnetic Ising system, the first term in eq. (3.20) tries to create order by minimizing the overall energy as a consequence of aligning spins in the same direction. The second term tends to align the spins in the direction of the external field H. While the energy is lower in an ordered state, thermal fluctuations tend to destroy the order by flipping single spins.

For temperatures above a critical value T_c, such fluctuations dominate the system and there is no long-range alignment of spins observable anymore. However, the domains of aligned spins grow as the temperature decreases below the critical value. This transition, like any other phase transition (occurring in superconductors, superfluids, and sol–gel systems) between ordered and disordered states, can be characterized by an *order parameter* (see also Section 2.3.3). For the Ising model, the order parameter is the spontaneous magnetization, which we define in the next section. The value of the critical temperature depends on the underlying lattice. For a two-dimensional square lattice, the critical temperature ($k_B T_c / J = 2/\ln\left(1 + \sqrt{2}\right) \approx 2.269$)

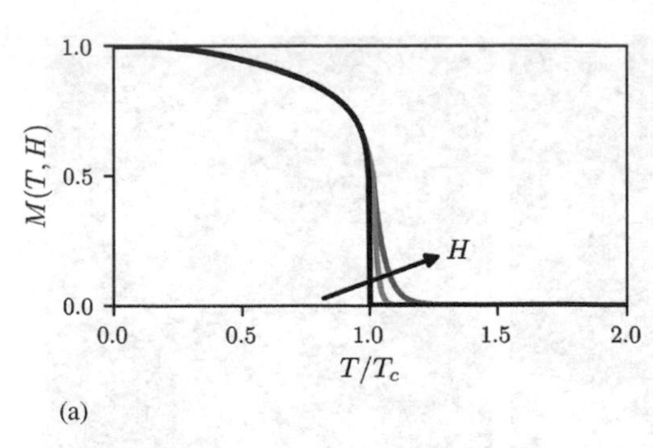

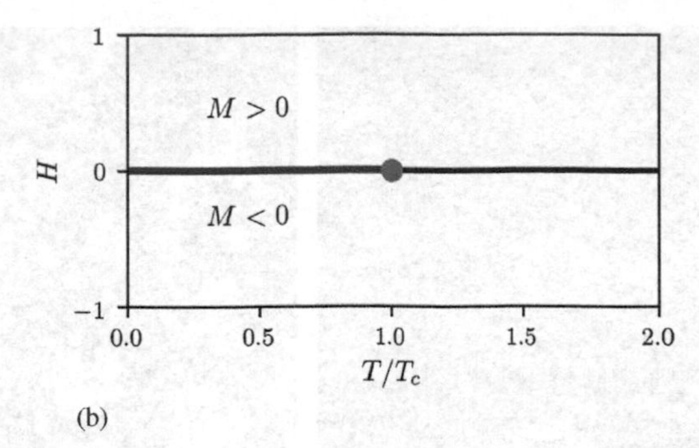

Figure 3.8 (a) The magnetization $M(T,H)$ for different fields $H \geq 0$ (indicated by the arrow) as a function of T. The black solid line represents the spontaneous magnetization $M_S(T)$ for $H = 0$ and should be interpreted in the sense that $\lim_H \rightarrow 0^+ M(T,H)$. (b) Phase diagram with the first-order transition line (blue) that results from a sign change of the external field.

has been computed analytically by Kramers and Wannier in 1941 [120] and Onsager (see Figure 15.4) in 1944 [111]. In three dimensions, extensive Monte Carlo simulations have been used by David Landau (see Figure 3.7) and collaborators to determine the critical temperature to $T_c = 4.5115232(1)$ $(J = k_B = 1)$ [117]. In addition to the second-order phase transition, there also exists a first-order transition in the Ising model. It occurs below the critical temperature if the direction of the external magnetic field changes. We illustrate this behavior in Figure 3.8. We observe a jump in the magnetization at $H = 0$ when we cross the first-order transition line. This jump is twice the order parameter.

Figure 3.7 David Landau (University of Georgia) has made many high-precision Monte Carlo calculations.

3.2.1 Order Parameter

The magnetization is

$$M(T,H) = \left\langle \frac{1}{N} \sum_{i=1}^{N} \sigma_i \right\rangle, \tag{3.21}$$

and corresponds to the ensemble average of the mean value of all spins. If the external field vanishes, the Hamiltonian is invariant under a simultaneous reversal of all spins. In other words, a certain equilibrated configuration of spins would also have the same energy if we would change the sign of every single spin.

Thus, the ensemble average defined by eq. (3.21) would not be a good measure of the magnetization because it corresponds to an ensemble average over all possible configurations. For every temperature, $M(T)$ vanishes because for every configuration there exists a configuration of opposite sign which neutralizes it. Therefore, we use the so-called spontaneous magnetization

$$M_S(T) = \lim_{H \to 0^+} \left\langle \frac{1}{N} \sum_{i=1}^{N} \sigma_i \right\rangle, \tag{3.22}$$

as the *order parameter* of the Ising model. In the definition of $M_S(T)$, the symmetry of the Ising model is broken by applying a vanishing positive field H that aligns the

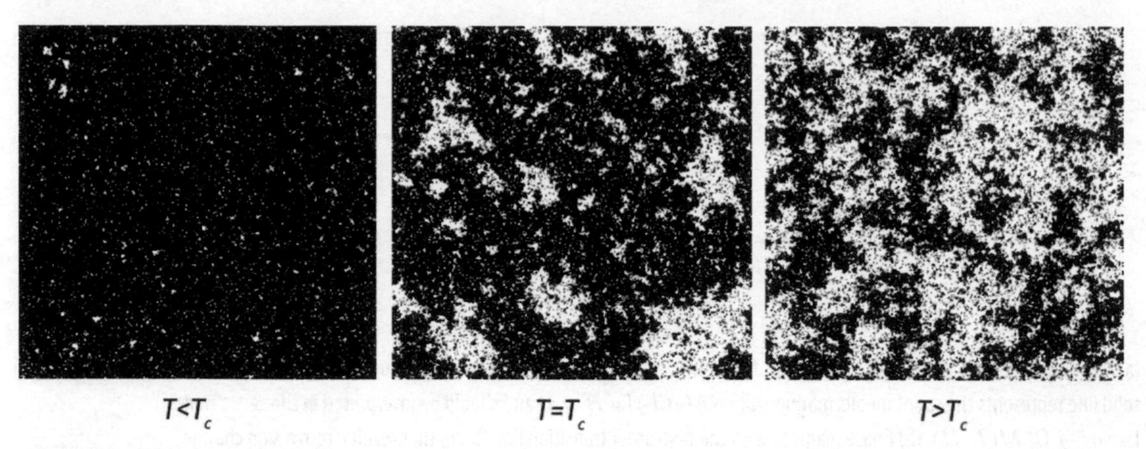

Figure 3.9 We show the formation of magnetic domains in the Ising model for different temperatures. For larger temperatures, the configurations are disordered due to thermal fluctuations. The simulations were performed on a square lattice with 512 × 512 sites. Black means that spins are pointing upward, and white indicates spins that point downward.

spins in one direction. Another possibility of breaking the symmetry can be realized by keeping the boundaries of the lattice in a certain state. This is, however, impracticable if periodic boundaries are being used. In Figure 3.9, we illustrate typical domain configurations for different temperatures. If $T > T_c$, one observes that thermal fluctuations lead to disordered configurations. On the other hand, magnetic domains form for small enough temperatures. Similar to eq. (2.1) in the case of percolation, in the vicinity of the critical temperature for $T < T_c$, the spontaneous magnetization scales as [109]

$$M_S(T) \propto (T_c - T)^\beta \,, \tag{3.23}$$

where $\beta = 1/8$ in two and $\beta = 0.326419(3)$ in three dimensions [118]. For $T = T_c$ and $H \to 0$, we find [109]

$$M(T = T_c, H) \propto H^{1/\delta} \,, \tag{3.24}$$

where $\delta = 15$ in two and $\delta = 4.78984(1)$ in three dimensions [118]. The exponents β and δ are *critical exponents* and characterize, together with other exponents, the underlying phase transition. Different techniques such as series expansions, field-theoretic methods, and very sophisticated Monte Carlo algorithms exist to determine critical exponents and critical temperatures with high precision.

3.2.2 Response Functions

Response functions are the second derivatives of the free energy and describe the sensitivity of the thermodynamic functions to fluctuations. The magnetic susceptibility

$$\chi(T) = \frac{\partial M(T,H)}{\partial H}, \tag{3.25}$$

quantifies the change of the magnetization M in response to an applied magnetic field H. Fluctuations of the spontaneous magnetization can be used to determine the susceptibility using the *fluctuation-dissipation theorem*:

$$\chi(T) = \frac{N}{k_B T}\left[\left\langle M_S(T)^2\right\rangle - \langle M_S(T)\rangle^2\right]. \tag{3.26}$$

Analogously, the specific heat is connected to energy fluctuations via

$$C(T) = \frac{\partial \langle E\rangle}{\partial T} = \frac{1}{k_B T^2}\left[\left\langle E(T)^2\right\rangle - \langle E(T)\rangle^2\right]. \tag{3.27}$$

Additional Information: Proof of the Fluctuation-Dissipation Theorem for the Susceptibility of the Ising Model

To derive eq. (3.26) for the Ising model, we use the definition of the spontaneous magnetization (see eq. (3.22)), and plug it into eq. (3.25) to obtain

$$\begin{aligned}\chi(T) &= \lim_{H\to 0^+}\frac{\partial\langle M(T,H)\rangle}{\partial H}\\ &= \lim_{H\to 0^+}\frac{\partial}{\partial H}\frac{1}{N}\frac{\sum_{\{\sigma\}}\sum_{i=1}^N \sigma_i \exp\left(\frac{E_0+H\sum_{i=1}^N\sigma_i}{k_BT}\right)}{\underbrace{\sum_{\{\sigma\}}\exp\left(\frac{E_0+H\sum_{i=1}^N\sigma_i}{k_BT}\right)}_{=Z_T(\mathcal{H})}}.\end{aligned}$$

Here we used the definition of the ensemble average (see eq. (3.15)) with $E_0 = J\sum_{\langle i,j\rangle}\sigma_i\sigma_j$ and the canonical partition function of the Ising Hamiltonian $Z_T(\mathcal{H})$. Using the product rule yields

$$\begin{aligned}\chi(T) &= \lim_{H\to 0^+}\frac{1}{Nk_BT}\frac{\sum_{\{\sigma\}}\left(\sum_{i=1}^N\sigma_i\right)^2\exp\left(\frac{E_0+H\sum_{i=1}^N\sigma_i}{k_BT}\right)}{Z_T(\mathcal{H})}\\ &\quad-\frac{1}{Nk_BT}\frac{\left[\sum_{\{\sigma\}}\sum_{i=1}^N\sigma_i\exp\left(\frac{E_0+H\sum_{i=1}^N\sigma_i}{k_BT}\right)\right]^2}{\left[Z_T(\mathcal{H})\right]^2}\\ &= \frac{N}{k_BT}\left[\left\langle M_S(T)^2\right\rangle - \langle M_S(T)\rangle^2\right] \geq 0.\end{aligned}$$

Equations (3.26) and (3.27) describe the variance of magnetization and energy, respectively. Similar to the power law scaling of the spontaneous magnetization defined in eq. (3.23), we find for the magnetic susceptibility in the vicinity of T_c

$$\chi(T) \propto |T_c - T|^{-\gamma}, \tag{3.28}$$

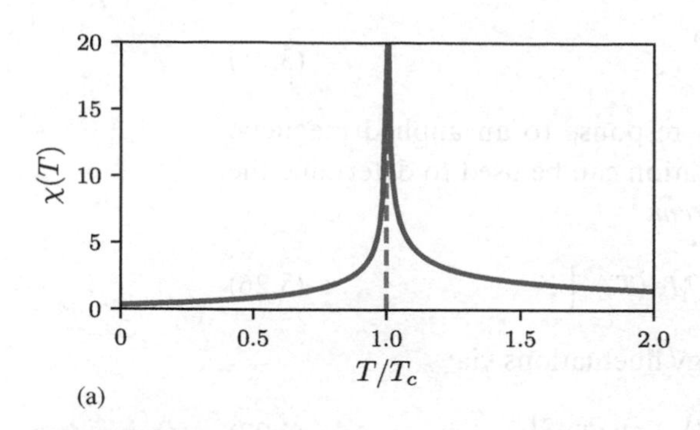

(a)

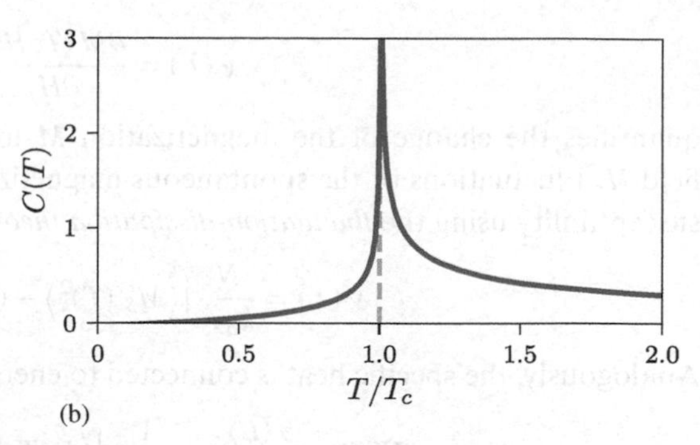

(b)

Figure 3.10 Schematic plots of the susceptibility $\chi(T)$ and specific heat $C(T)$ as functions of temperature. In the thermodynamic limit, both quantities diverge at the critical temperature T_c.

where $\gamma = 7/4$ in two and $\gamma = 1.23701(28)$ in three dimensions [117], and note the analogy with eq. (2.5) in the case of percolation. Moreover, the specific heat exhibits the following scaling:

$$C(T) \propto |T_c - T|^{-\alpha}, \tag{3.29}$$

where $\alpha = 0$ in two[2] and $\alpha = 0.11008(1)$ in three dimensions [118]. The temperature dependence of susceptibility and specific heat is illustrated in Figure 3.10.

3.2.3 Correlation Length

The spin–spin correlation function is

$$G(r_1, r_2; T, H) = \langle\sigma_1\sigma_2\rangle - \langle\sigma_1\rangle\langle\sigma_2\rangle, \tag{3.30}$$

where the vectors r_1 and r_2 point to lattice sites 1 and 2. If the system is translationally and rotationally invariant, the correlation function only depends on $r = |r_1 - r_2|$. At the critical point, the correlation function decays as

$$G(r; T_c, 0) \propto r^{-(d-2+\eta)}, \tag{3.31}$$

where η is another critical exponent and d is the dimension of the system. This scaling behavior is similar to the one of percolation (see eq. (2.11)). In two dimensions, $\eta = 1/4$ and $\eta = 0.036298(2)$ in three dimensions [118]. For temperatures different from the critical temperature, the correlation function exhibits an exponential decay

$$G(r; T, 0) \propto r^{-\vartheta} e^{-r/\xi}, \tag{3.32}$$

[2] The exponent of $\alpha = 0$ corresponds here to a logarithmic decay since

$$\lim_{s\to 0} \frac{|x|^{-s} - 1}{s} = -\ln|x|.$$

However, numerically, it is difficult to decide if the exponent is zero or just has a small value [121].

Table 3.1 The critical exponents of the Ising model in different dimensions

Exponent	$d = 2$	$d = 3$	$d \geq 4$
α	0	0.11008(1) [118]	0
β	1/8	0.32630(22) [117]	1
γ	7/4	1.237075(10) [118]	1
δ	15	4.78984(1) [118]	3
η	1/4	0.036298(2) [118]	0
ν	1	0.629971(4) [118]	1/2

where ξ defines the *correlation length*. In the two-dimensional Ising model, the exponent ϑ equals 2 above and 1/2 below the transition point [121, 122]. In the vicinity of T_c, the correlation length ξ diverges according to

$$\xi(T) \propto |T_c - T|^{-\nu}, \tag{3.33}$$

with $\nu = 1$ in two dimensions and $\nu = 0.629912(86)$ in three dimensions [117]. Note the similarities with Eqs. (2.10) and (2.11) for percolation.

3.2.4 Critical Exponents and Universality

The aforementioned six critical exponents are connected by four scaling relations [109]

$$\alpha + 2\beta + \gamma = 2 \quad \text{(Rushbrooke)}, \tag{3.34}$$

$$\gamma = \beta(\delta - 1) \quad \text{(Widom)}, \tag{3.35}$$

$$\gamma = (2 - \eta)\nu \quad \text{(Fisher)}, \tag{3.36}$$

$$2 - \alpha = d\nu \quad \text{(Josephson)}, \tag{3.37}$$

that have been derived in the context of the phenomenological scaling theory for ferromagnetic systems [109, 123]. Due to these relations, the number of independent exponents reduces to two. The Josephson law includes the spatial dimension d of the system and thus defines a so-called *hyperscaling* relation. Such relations are not valid above the upper critical dimension which is equal to $d_c = 4$ for the Ising model. Above the upper critical dimension, the critical exponents are equal to those obtained with the mean-field approximation and are thus independent of dimension (see Table 3.1).

The importance of critical exponents becomes more clear when we study different systems exhibiting phase transitions. Critical control parameters such as T_c in the Ising model sensitively depend on the interaction details. However, critical exponents only

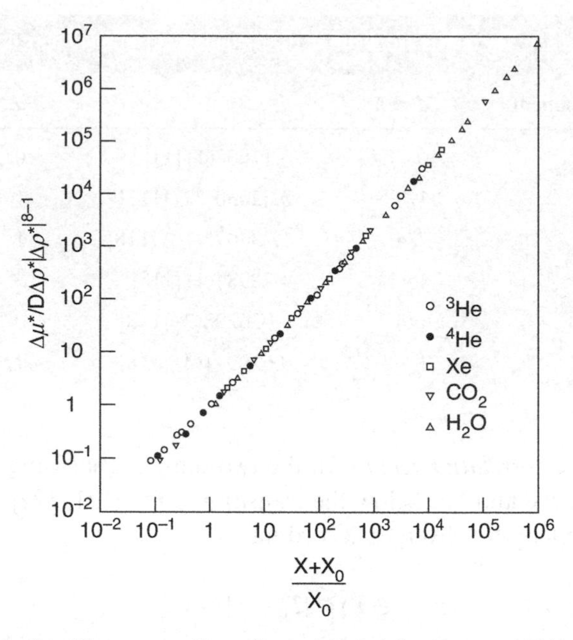

Figure 3.11 Widom scaling function for five different gases. The scaling variable is defined as $x = \Delta T\,|\Delta\rho|^{-1/\beta}$, and x_0 depends on the amplitude B of the power law for the coexistence curve $\Delta\rho = B\Delta T^{\beta}$ [121]. The figure is taken from Ref. [125].

depend on fundamental system properties such as dimension or symmetries and are therefore said to be *universal*. Based on these observations, the universality hypothesis states that for equilibrium systems all critical phenomena can be reduced to a small number of *universality classes* [124]. All systems belonging to a certain universality class share the same critical exponents and the same scaling behavior near the critical point. Universality is not a mere theoretical concept, but can be also observed experimentally.

In Figure 3.11, we show an example of five different fluids undergoing a liquid-gas transition. All substances exhibit different interatomic interactions, and still we observe a clear data collapse for the rescaled chemical potential.

4 Monte Carlo Methods

Monte Carlo methods are broadly applicable to different problems that are (i) impossible to solve analytically and (ii) difficult to handle for other numerical techniques due to the large computational complexity [126]. For example, it may be difficult to numerically explore the complete phase space of a certain model. Instead, we could apply appropriate random sampling techniques (i.e., Monte Carlo methods) and only explore relevant parts of phase space. In this chapter, we focus on Monte Carlo algorithms to numerically study the properties of the Ising model and other Hamiltonian systems. The basic idea behind Monte Carlo is that random sampling of the phase space, instead of averaging over all states, may be sufficient to compute ensemble averages of certain thermodynamic quantities. If the number of samples is large enough, the computed estimate converges toward the correct value. The main steps of Monte Carlo sampling are as follows.

Monte Carlo Sampling

1. Randomly choose a new configuration in phase space (using a Markov chain).
2. Accept or reject the new configuration, depending on the strategy used (e.g., Metropolis or Glauber dynamics).
3. Compute the physical quantity and add it to the averaging procedure.
4. Repeat the previous steps.

4.1 Computation of Integrals

To compute the ensemble average

$$\langle Q \rangle = \frac{\int Q(p,q)\rho(p,q)\,\mathrm{d}p\mathrm{d}q}{\int \rho(p,q)\,\mathrm{d}p\mathrm{d}q} \tag{4.1}$$

of a thermodynamic quantity Q (e.g., the energy of a gas), we have to approximate a high-dimensional integral. We therefore first focus on Monte Carlo methods that are applicable to integrals, particularly higher-dimensional integrals (we will see later on that Monte Carlo is in fact the most efficient way to compute higher-dimensional integrals).

Let us consider the integral of a function $g(x)$ in an interval $[a,b]$. We may approximate the integral by choosing N points x_i (regularly or randomly) on the x-axis with

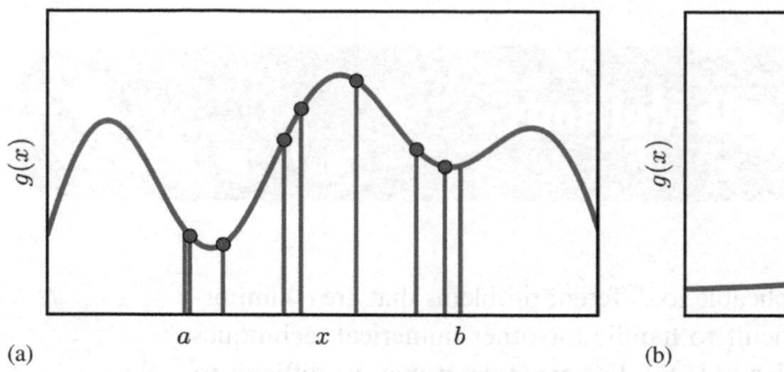

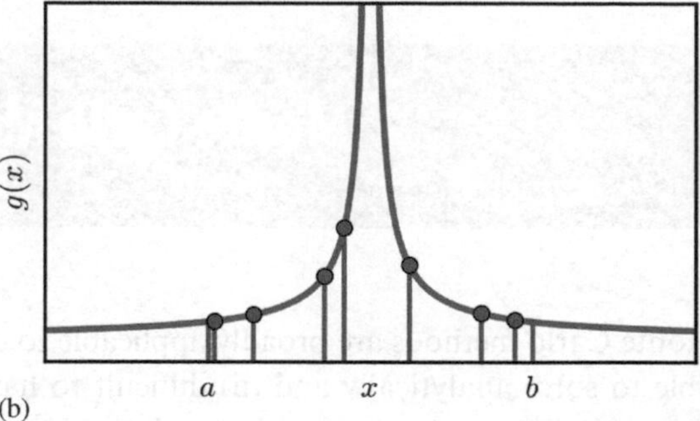

Figure 4.1 Examples of Monte Carlo sampling (orange dots) for a smooth function (a) and for a function with a singularity (b).

their corresponding values $g(x_i)$, summing over these sampled results, and multiplying the resulting expression with the length of the interval:

$$\int_a^b g(x)dx \approx (b-a)\frac{1}{N}\sum_{i=1}^{N} g(x_i). \tag{4.2}$$

If we choose the points x_i uniformly at random, the process is called *simple sampling*, which works very well if $g(x)$ is smooth.

But what if we cannot make the assumption that $g(x)$ is smooth? Let us consider, for instance, a function with a singularity at a certain value (see Figure 4.1). Due to the limited number of points that we would usually choose around the singularity (in simple sampling), we would not be able to really appreciate and reproduce its behavior and eq. (4.2) would be a bad approximation. We thus need more precision which goes beyond the possibilities of simple sampling. An improvement is provided by a second function $p(x)$ for which $\frac{g(x)}{p(x)}$ is smooth. The sampling points are now distributed according to $p(x)$, and we have

$$\int_a^b g(x)\,\mathrm{d}x = \int_a^b \frac{g(x)}{p(x)} p(x)\,\mathrm{d}x \approx (b-a)\left[\frac{1}{N}\sum_{i=1}^{N}\frac{g(x_i)}{p(x_i)}\right], \tag{4.3}$$

where the numbers x_i have been sampled according to $p(x)$. Thus, we have changed our way of sampling by using the distribution function $p(x)$. This manifests itself in the summand above. One could state that $p(x)$ helps us to pick our sampling points according to their importance (e. g., select more points close to a singularity); in fact, this kind of sampling is called *importance sampling*.

4.2 Integration Errors

We have outlined that the aforementioned Monte Carlo integration methods provide approximations of certain integrals and not exact solutions. Naturally, the question arises what error one has to expect in the course of such an approximation.

Error in Conventional Methods

Let us first consider the error in conventional methods in which one uses a regular spacing. As an example, let us focus on the trapezoidal rule: Consider the Taylor series expansion of an integral over a function $f(x)$ from x_0 to $x_0 + \Delta x$:

$$\begin{aligned}\int_{x_0}^{x_0+\Delta x} f(x)\,\mathrm{d}x &= f(x_0)\Delta x + \frac{1}{2}f'(x_0)\Delta x^2 + \frac{1}{6}f''(x_0)\Delta x^3 + \ldots \\ &= \left[\frac{1}{2}f(x_0) + \frac{1}{2}(f(x_0) + f'(x_0)\Delta x + \frac{1}{6}f''(x_0)\Delta x^2 + \ldots) + \ldots\right]\Delta x \\ &= \frac{1}{2}\left[f(x_0) + f(x_0+\Delta x)\right]\Delta x + \mathcal{O}\left(\Delta x^3\right).\end{aligned}$$

This approximation, represented by $\frac{1}{2}[f(x_0) + f(x_0 + \Delta x)]\Delta x$ is called the *trapezoidal rule* (see Figure 4.2 for a geometric interpretation). We see that the error is of order $\mathcal{O}\left(\Delta x^3\right)$. Suppose we take Δx to be one third its previous value, then the error will decrease by a factor of 27! At the same time though, the size of the domain would also be divided by three. The net factor is thus only 9 and not 27.

We now subdivide our interval $[x_0, x_1]$ into N subintervals of size $\Delta x = \frac{x_1 - x_0}{N}$. The trapezoidal rule approximation of the whole integral is therefore

$$\begin{aligned}\int_{x_0}^{x_1} f(x)\,\mathrm{d}x &\approx \frac{\Delta x}{2}\sum_{j=0}^{N-1} f(x_0 + j\Delta x) + f(x_0 + (j+1)\Delta x) \\ &= \frac{\Delta x}{2}[f(x_0) + 2f(x_0+\Delta x) + 2f(x_0 + 2\Delta x) + \ldots \\ &\quad + 2f(x_0 + (N-1)\Delta x) + f(x_1)].\end{aligned}$$

We thus see that while the error for each step is of order $\mathcal{O}\left(\Delta x^3\right)$, the cumulative error is N times this error and thus of order $\mathcal{O}\left(\Delta x^2\right) \propto \mathcal{O}\left(N^{-2}\right)$.

The generalization to two dimensions is rather straightforward. Instead of each interval contributing an error of $\mathcal{O}\left(\Delta x^3\right)$, we now have a two-dimensional grid and find for each cell of the grid an error that is proportional to $\mathcal{O}\left(\Delta x^4\right)$. The cumulative error is then proportional to $N\mathcal{O}\left(\Delta x^4\right) \propto \mathcal{O}\left(\Delta x^2\right)$ since $N \propto \Delta x^{-2}$.

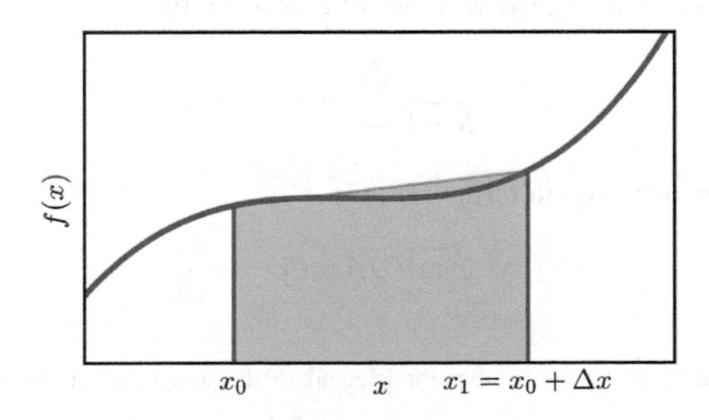

Graphical interpretation of the trapezoidal rule.

Figure 4.2

We can do the same for a d-dimensional domain of the integral. The grid is now d-dimensional, and the error for each segment will be $O(\Delta x^{d+2})$ with the sum over all "subintervals" giving us $NO\left(\Delta x^{d+2}\right)$ with $N \propto \Delta x^{-d}$. Thus, the error is again proportional to $NO\left(\Delta x^{d+2}\right) \propto O\left(\Delta x^2\right)$ and independent of the dimension. We conclude that the error in conventional methods is of order $O\left(\Delta x^2\right)$ and given that $N \propto \Delta x^{-d}$, we have $\Delta x \propto N^{-\frac{1}{d}}$ and the error is of order $O\left(\Delta x^2\right) \propto N^{-\frac{2}{d}}$.

Monte Carlo Error

Let us first consider the simplified case of a one-dimensional function of one variable, $g : [a, b] \to \mathbb{R}$.

If we consider a set of random points $\{x_i\}$ with $i \in \{1, \dots, N\}$ that are uniformly chosen from the interval $[a, b]$, the estimate for the integral over $g(x)$ is

$$Q = \int_a^b g(x)\,\mathrm{d}x \approx \frac{b-a}{N} \sum_{i=1}^{N} g(x_i) = (b-a)\bar{g}\,, \tag{4.4}$$

where $\bar{g}$ is the mean of the considered samples. For M samples, we denote the ensemble mean by

$$\langle g \rangle = \frac{1}{M} \sum_{j=1}^{M} \bar{g}_j\,. \tag{4.5}$$

We estimate the variance of $\bar{g}$ with

$$\sigma_{\bar{g}}^2 = \frac{1}{M-1} \sum_{j=1}^{M} \left(\bar{g}_j - \langle g \rangle\right)^2\,, \tag{4.6}$$

where the denominator is $M - 1$ as we want to obtain an unbiased estimator of the variance.[1] According to the central limit theorem, the variance of the ensemble average $\langle Q \rangle$ of the integral Q is

$$\mathrm{var}(\langle Q \rangle) = (b-a)^2 \sigma_{\bar{g}}^2 = \frac{(b-a)^2}{N^2} \sum_{i=1}^{N} \sigma_g^2 = (b-a)^2 \frac{\sigma_g^2}{N}\,. \tag{4.7}$$

An unbiased estimator of σ_g, the variance of g, is given by

$$\sigma_g^2 = \frac{1}{N-1} \sum_{i=1}^{N} \left(g(x_i) - \bar{g}\right)^2\,. \tag{4.8}$$

We estimate the error of $\langle Q \rangle$ in terms of

$$\delta \langle Q \rangle \approx \sqrt{\mathrm{var}(\langle Q \rangle)} = (b-a) \frac{\sigma_g}{\sqrt{N}}\,. \tag{4.9}$$

[1] The expected value of an unbiased estimator is equal to the true value of the quantity to be estimated [127].

We can now generalize this argument to multidimensional integrals:

$$Q = \int_{a_1}^{b_1} \mathrm{d}x_1 \int_{a_2}^{b_2} \mathrm{d}x_2 \ldots \int_{a_n}^{b_n} \mathrm{d}x_n g(x_1, \ldots, x_n). \tag{4.10}$$

The previous integration interval $[a, b]$ becomes a hypercube with volume $V \subseteq \mathbb{R}^n$. Instead of the interval $[a, b]$ appearing in the variance, we now use the hypercube volume V to obtain

$$\mathrm{var}(\langle Q \rangle) = V^2 \sigma_{\bar{g}}^2 = V^2 \frac{\sigma_g^2}{N}, \tag{4.11}$$

and the error estimate becomes

$$\delta \langle Q \rangle \approx \sqrt{\mathrm{var}(\langle Q \rangle)} = V \frac{\sigma_g}{\sqrt{N}}. \tag{4.12}$$

As we can see, the proportionality to $\frac{1}{\sqrt{N}}$ still remains.

Comparison of Errors

We have seen that for the trapezoidal rule, the error of computing integrals converges as $N^{-\frac{2}{d}}$ and thus depends on the dimension, while the error in Monte Carlo methods is independent of the dimension. We can determine the dimension at which Monte Carlo methods become more efficient,

$$N^{-\frac{2}{d_0}} \stackrel{!}{=} \frac{1}{\sqrt{N}}, \tag{4.13}$$

and we thus find $d_0 = 4$. We conclude that for $d > 4$, Monte Carlo methods are more efficient than the trapezoidal rule and should therefore be used whenever such higher-dimensional integrals are calculated. Note that for higher-order numerical integration methods (e.g., Simpson's method), the corresponding value of d_0 will be larger.

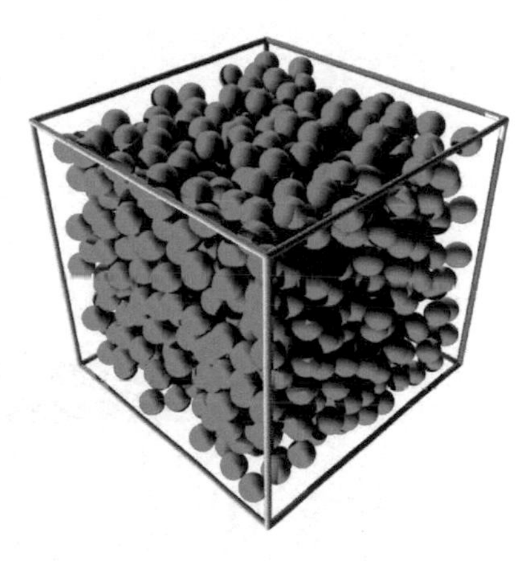

Figure 4.3 A cube filled with spheres to illustrate the sphere-packing problem.

4.3 Hard Spheres in a Box

Let us now look at an example of higher-dimensional integration. We consider a sphere-packing problem, in which N hard spheres of radius R are randomly arranged in a 3D box of volume V (see Figure 4.3). Our goal is two compute the average distance between these spheres. For more information on sphere packings, we refer the interested reader to the textbook *Sphere Packings, Lattices and Groups* (3rd ed.), by John Conway and Neil Sloane [128]

We describe the centers of the spheres by their position vector $\mathbf{x}_i = (x_i, y_i, z_i)$, $1 \leq i \leq N$. The distance between two spheres is

$$r_{ij} := \sqrt{\left(x_i - x_j\right)^2 + \left(y_i - y_j\right)^2 + \left(z_i - z_j\right)^2}. \tag{4.14}$$

Note that the spheres are hard (i.e., they cannot overlap). Thus, the minimal distance between two neighboring spheres is the distance when the spheres are in contact, which is equal to $2R$. This translates into the following condition for r_{ij}:

$$r_{ij} \overset{!}{\geq} 2R\,. \tag{4.15}$$

So far, this does not seem to have too much to do with integration because we are simply placing some spheres in a volume V and measuring some distance r_{ij}. Let us say that we are interested in the average distance between the centers of spheres in the box. Then we need to evaluate the following integral to obtain an analytical result:

$$\langle r_{ij} \rangle = \frac{1}{Z} \int_V \frac{2}{N(N-1)} \sum_{i<j} r_{ij}\, \mathrm{d}^3 r_1 \ldots \mathrm{d}^3 r_N \quad \text{with} \quad Z = \int_V \mathrm{d}^3 r_1 \ldots \mathrm{d}^3 r_N\,. \tag{4.16}$$

We implicitly assume that the integrals in eq. (4.16) use distances r_{ij} that satisfy the "no overlap" condition (eq. (4.15)). Let us stay with this formula for a moment and have a look at the different factors. The first factor, Z^{-1}, is a normalization factor. The following factor is of combinatorial origin: We start with N choices (the whole population) and pick one sphere uniformly at random. There are only $(N-1)$ choices left for the second one (as it cannot be the first one again). This gives a total number of $N(N-1)$ possible combinations for picking two spheres. As they are indistinguishable, there is an additional factor of $\frac{1}{2}$. Note that we would need to correct with an additional factor of $\frac{1}{2}$ if we would sum over $i \neq j$, since in that case we would have counted each distance twice which was avoided by simply summing over $i < j$. The Monte Carlo approach to approximate eq. (4.16) is relatively simple:

Hard Spheres in a Box

- Choose a particle position (i. e., the center of the new sphere).
- Make sure that the new sphere does not overlap with any other sphere (see condition on r_{ij}). If it does overlap, reject the position and try again.
- Once all the spheres have been placed, calculate the distances r_{ij}.
- Calculate the average over all r_{ij}.

This procedure will converge and give a value that approximates the integral under the right conditions. It is important to note that our result will, of course, depend on the number of spheres positioned: Imagine we had taken a ridiculously small number of spheres (let us say only two) in a rather large volume and carried out the distance measurement. There would not be much to average over and on top of that, due to the relative liberty of positioning the two spheres in the volume, the distance (and with it the average distance) would fluctuate wildly depending on the setup. If we repeat this n times and average, the result will thus not converge as nicely as it would with a relatively large number N of spheres where the space left between spheres would be small. Placing many spheres in the volume would improve convergence but will slow down the algorithm as with more spheres comes a higher rate of rejected positions. For large N, it is even possible that the last few spheres cannot be placed at all!

Exercise: Hard Spheres in a 3D Box

Consider n hard spheres in a three-dimensional cube with edge length L. The radius of the spheres is R and they are not allowed to overlap. Thus, the positions $\mathbf{r}_i$ are not independent of each other and the integral $\int \dots d^3r_1 \dots d^3r_n$ used to calculate average quantities (e.g., the average distance) over all possible configurations is a conditioned integration. It is not easy to solve this integral numerically with conventional methods. We are now employing Monte Carlo methods for approximating integrals in N dimensions:

$$\int f(x_1 \dots x_N)\,\mathrm{d}x_1 \dots \mathrm{d}x_N = Z\frac{1}{M}\sum_{k=1}^{M} f^{(k)}(x_1 \dots x_N) \quad \text{with} \quad Z = \int \mathrm{d}x_1 \dots \mathrm{d}x_N\,. \tag{4.17}$$

The mean distance between the particles averaged over all possible configurations is defined by the conditioned integration

$$\langle r_{\text{mean}} \rangle = \frac{1}{Z}\int r_{\text{mean}}\,\mathrm{d}^3r_1 \dots \mathrm{d}^3r_n \quad \text{with} \quad Z = \int \mathrm{d}^3r_1 \dots \mathrm{d}^3r_n\,, \tag{4.18}$$

where the mean particle distance for one configuration is given by

$$r_{\text{mean}} = \frac{2}{n(n-1)}\sum_{i<j} r_{ij} \quad \text{with} \quad r_{ij} = \sqrt{(x_i - x_j)^2 + (y_i - y_j)^2 + (z_i - z_j)^2}\,. \tag{4.19}$$

Calculate $\langle r_{\text{mean}} \rangle$ by using a Monte Carlo method which is relatively easy to implement:

- Choose n particles with random position, such that they are not overlapping:
  - choose particle position
  - reject it if the excluded volume condition is not satisfied
  - repeat until n particles are placed
- For the given configuration k, calculate $r^{(k)}_{\text{mean}}$.
- Repeat for M configurations, thus calculating

$$\langle r_{\text{mean}} \rangle \simeq \frac{1}{M}\sum_{k=1}^{M} r^{(k)}_{\text{mean}}\,. \tag{4.20}$$

Note that this equation can be easily derived from eq. (4.18) when applying eq. (4.17).

- Vary the number M of configurations and plot $\langle r_{\text{mean}} \rangle$ versus the number M to check the convergence of the integral.
- Vary the number n of particles to check the influence of the dimensionality of the integral on the convergence. For that you should also try very small numbers of particles (e.g., $n = 2$, leading to a six-dimensional integration). For this purpose, you may also consider a "one-dimensional box" (instead of the three-dimensional one) to further reduce the dimension of the integral.

Please consider the following points:

- The total sphere volume $n \cdot 4/3\pi R^3$ should be small enough compared to the total volume of the box so that the particles can easily fit into the box.
- Choose different volume fractions $\nu = (n \cdot 4/3\pi R^3)/L^3$ (total sphere volume divided by total volume). What are the highest volume fractions you can achieve?

4.4 Markov Chains

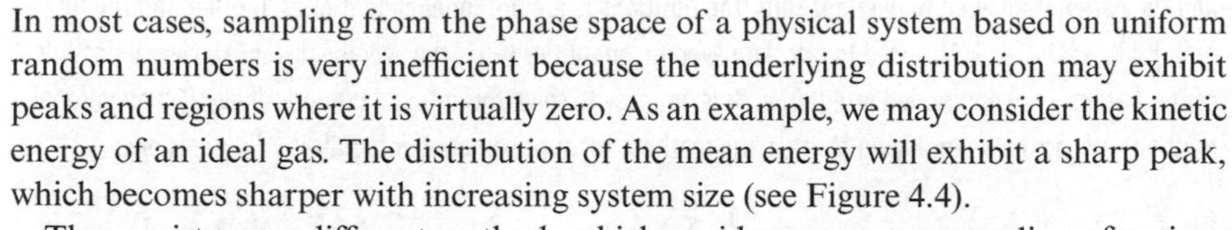
In most cases, sampling from the phase space of a physical system based on uniform random numbers is very inefficient because the underlying distribution may exhibit peaks and regions where it is virtually zero. As an example, we may consider the kinetic energy of an ideal gas. The distribution of the mean energy will exhibit a sharp peak, which becomes sharper with increasing system size (see Figure 4.4).

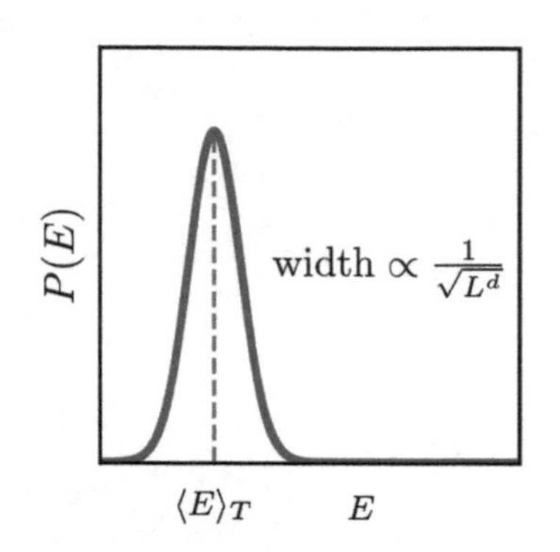

Figure 4.4 Example of an energy distribution whose width is proportional to $1/\sqrt{L^d}$, where d is the system dimension and L is the linear system size.

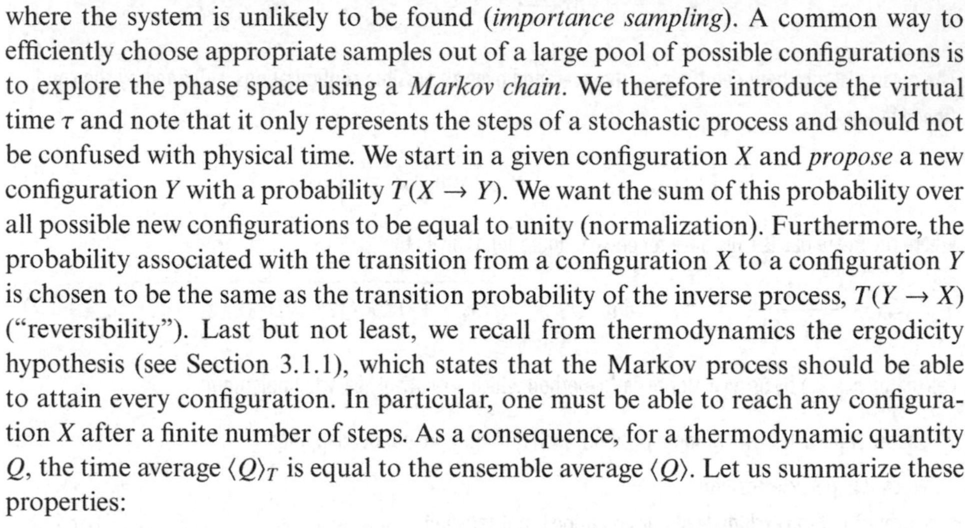
There exist many different methods which avoid unnecessary sampling of regions where the system is unlikely to be found (*importance sampling*). A common way to efficiently choose appropriate samples out of a large pool of possible configurations is to explore the phase space using a *Markov chain*. We therefore introduce the virtual time τ and note that it only represents the steps of a stochastic process and should not be confused with physical time. We start in a given configuration X and *propose* a new configuration Y with a probability $T(X \to Y)$. We want the sum of this probability over all possible new configurations to be equal to unity (normalization). Furthermore, the probability associated with the transition from a configuration X to a configuration Y is chosen to be the same as the transition probability of the inverse process, $T(Y \to X)$ ("reversibility"). Last but not least, we recall from thermodynamics the ergodicity hypothesis (see Section 3.1.1), which states that the Markov process should be able to attain every configuration. In particular, one must be able to reach any configuration X after a finite number of steps. As a consequence, for a thermodynamic quantity Q, the time average $\langle Q\rangle_T$ is equal to the ensemble average $\langle Q\rangle$. Let us summarize these properties:

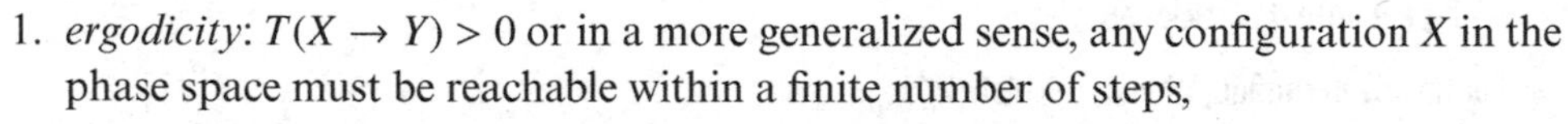
1. *ergodicity*: $T(X \to Y) > 0$ or in a more generalized sense, any configuration X in the phase space must be reachable within a finite number of steps,
2. *normalization*: $\sum_Y T(X \to Y) = 1$,
3. *reversibility*: $T(X \to Y) = T(Y \to X)$.

In order to implement importance sampling, not every new configuration Y will be *accepted*. To illustrate this, let us consider the Ising model: We may propose a new configuration that will increase the energy of the system, but depending on the temperature it may not be likely that the new configuration will also be accepted. At low temperatures, it is unlikely that a spin flip will occur, so we will handle this with an acceptance probability A. That is, we propose a new configuration and use an acceptance probability A such that we only sample the important regions in phase space. We use $A(X \to Y)$ to denote the probability of acceptance of a new configuration Y starting from a configuration X. In practice, we are interested in the overall probability of a configuration actually making it through these two steps; this probability is the product of the transition probability $T(X \to Y)$ and the acceptance probability $A(X \to Y)$ and is called the *probability of a Markov chain*:

$$W(X \to Y) = T(X \to Y) \cdot A(X \to Y)\,. \tag{4.21}$$

If we are interested in the evolution of the probability $p(X, \tau)$ that the system is in state X at time τ, we can derive the evolution equation in the following way:

- a configuration X is reached by coming from Y (this will contribute positively to $p(X, \tau)$);
- a configuration X is left by going to some other configuration Y (this will decrease the probability $p(X, \tau)$).

The first of these two processes is proportional to the probability for a system to be in Y, $p(Y)$, while the second one needs to be proportional to the probability for a system to be in X, $p(X)$. When we combine all of this, we obtain the so-called *master equation*:

$$\frac{\mathrm{d}p(X,\tau)}{\mathrm{d}\tau} = \sum_{Y \neq X} p(Y)W(Y \to X) - \sum_{Y \neq X} p(X)W(X \to Y) . \tag{4.22}$$

We now summarize the properties of $W(X \to Y)$: Similarly to $T(X \to Y)$, the first property is ergodicity. Any configuration X must be reachable after a finite number of steps, which can be achieved through the sufficient condition $W(X \to Y) > 0 \; \forall X, Y$.

If we sum over all possible configurations, we should obtain unity (normalization).

The third property, homogeneity, is new. It tells us that if we sum over all "initial configurations," the product of the probability for a system to be in Y multiplied by its Markov probability for the transition to X is given by the probability for a system to be in X. More comprehensibly, the probability for a system to be in X is simply a result of systems coming from other configurations to X.

Thus, the properties of $W(X \to Y)$ are as follows:

1. *ergodicity:* $\forall X, Y \quad W(X \to Y) > 0$,
2. *normalization:* $\sum_Y W(X \to Y) = 1$,
3. *homogeneity:* $\sum_Y p(Y)W(Y \to X) = p(X)$.

A Markov chain that satisfies conditions 1–3 converges toward a unique stationary distribution p_{st} which does not change in time. To efficiently sample the relevant regions of the phase space, the probability of the Markov chain $W(\cdot)$ has to depend on the Hamiltonian. To achieve that, we set the stationary distribution p_{st}, defined by

$$\frac{\mathrm{d}p_{\mathrm{st}}(X,\tau)}{\mathrm{d}\tau} = 0 , \tag{4.23}$$

equal to the desired equilibrium distribution of the considered physical system p_{eq}:

$$p_{\mathrm{st}} \stackrel{!}{=} p_{\mathrm{eq}} . \tag{4.24}$$

It then follows from the stationary state condition of the Markov chain that

$$\sum_{Y \neq X} p_{\mathrm{eq}}(Y)W(Y \to X) = \sum_{Y \neq X} p_{\mathrm{eq}}(X)W(X \to Y) . \tag{4.25}$$

A sufficient condition for this equation to hold is

$$p_{\mathrm{eq}}(Y)W(Y \to X) = p_{\mathrm{eq}}(X)W(X \to Y) . \tag{4.26}$$

We will refer to this condition as *condition of detailed balance*. As an example, in the canonical ensemble at fixed temperature T, the equilibrium distribution is given by the Boltzmann distribution (eq. (3.13))

$$p_{\mathrm{eq}}(X) = \frac{1}{Z_T}\exp\left[-\frac{E(X)}{k_B T}\right], \tag{4.27}$$

with the partition function $Z_T = \sum_X \exp\left[-\frac{E(X)}{k_B T}\right]$.

Additional Information: Markov Chain Probability

The attentive reader might object that as $W(X \to Y) = A(X \to Y)T(X \to Y)$, with both $A(X \to Y) \leq 1$ and $T(X \to Y) \leq 1$, W cannot satisfy $\sum_Y W(X \to Y) = 1$. We now explain that the conditions $A(X \to Y) \leq 1$ and $T(X \to Y) \leq 1$ are only seemingly in contradiction to $\sum_Y W(X \to Y) = 1$. We therefore note that we are looking at independent events here; the event of choosing a new configuration is independent of the event of accepting it. The corresponding probabilities $A(X \to Y)$ and $T(X \to Y)$ satisfy

$$\sum_Y A(X \to Y) = 1 \quad \text{and} \quad \sum_Y T(X \to Y) = 1\,. \tag{4.28}$$

Furthermore, for

$$\sum_i a_i = 1 \quad \text{and} \quad \sum_i b_i = 1, \tag{4.29}$$

we find

$$\begin{aligned} \sum_{ij} a_i b_i &= (a_1 + a_2 + \ldots)(b_1 + b_2 + \ldots) \\ &= a_1 b_1 + a_1(1 - b_1) + a_2 b_2 + a_2(1 - b_2) + \ldots \\ &= a_1 + a_2 + \cdots = 1\,, \end{aligned} \tag{4.30}$$

where we used $\sum_i b_i = 1$ in the second step. For two independent events, we therefore obtain $\sum_{ij} a_i b_i = 1$ again. When we apply this procedure to our example involving $A(X \to Y)$ and $T(X \to Y)$, we obtain

$$\sum_Y T(X \to Y)A(X \to Y) = \sum_Y W(X \to Y) = 1\,. \tag{4.31}$$

4.5 M(RT)^2 Algorithm

One possible choice of the acceptance probability satisfying the detailed balance condition is

$$A\,(X \to Y) = \min\left[1, \frac{p_{\mathrm{eq}}\,(Y)}{p_{\mathrm{eq}}\,(X)}\right], \tag{4.32}$$

which can be verified by rewriting eq. (4.26) and distinguishing between two cases. This algorithm was developed in 1953 at Los Alamos National Laboratory and first implemented by Arianna W. Rosenbluth (see Figure 4.5) on a MANIAC I computer. It is often referred to as Metropolis or M(RT)^2 algorithm. The original algorithm has been extended by Wilfred K. Hastings in 1970 to account for nonsymmetric proposal distributions [129]. For the canonical ensemble with $p_{\text{eq}}(X) = \frac{1}{Z_T}\exp\left[-\frac{E(X)}{k_B T}\right]$, the acceptance probability becomes

$$A(X \to Y) = \min\left[1, \exp\left(-\frac{\Delta E}{k_B T}\right)\right], \tag{4.33}$$

Figure 4.5 Arianna W. Rosenbluth (1927–2020) first implemented the M(RT)^2 algorithm. M(RT)^2 is an abbreviation of the last names of the authors of the original paper [130] that introduced the algorithm. *RT* is squared because except for Nicholas C. Metropolis, the other four authors of the paper formed two married couples and therefore carried the same family names (Rosenbluth and Teller). The real contribution of some of the authors (in particular of Metropolis and of A. H. Teller) is subject to controversy [131, 132]. It has been even stated by Roy Glauber and Emilio Segré that the original algorithm was invented by Enrico Fermi [133].

where $\Delta E = E(Y) - E(X)$ is the associated change of energy. Equation (4.33) implies that the Monte Carlo (MC) update is always accepted if the energy decreases, and if the energy increases, it is accepted with probability $\exp\left(-\frac{\Delta E}{k_B T}\right)$. It is straightforward to show that the M(RT)^2 update rule satisfies the detailed balance condition (see eq. (4.26)). We observe that the detailed balance condition is clearly satisfied for $\Delta E = 0$ and thus distinguish between the remaining two cases: (i) $\Delta E < 0$ and (ii) $\Delta E > 0$. For $\Delta E < 0$, we find that $A(X \to Y) = 1$ and $A(Y \to X) = \exp\left(\frac{\Delta E}{k_B T}\right)$. The detailed balance condition (see eq. (4.26)) is indeed satisfied since

$$p_{\text{eq}}(Y)\underbrace{A(Y \to X)}_{=\exp\left(\frac{\Delta E}{k_B T}\right)} = p_{\text{eq}}(X)\underbrace{A(X \to Y)}_{=1} \tag{4.34}$$

and thus

$$\frac{p_{\text{eq}}(Y)}{p_{\text{eq}}(X)} = \exp\left(-\frac{\Delta E}{k_B T}\right). \tag{4.35}$$

Case (ii) with $\Delta E > 0$ is symmetric to case (i).

We use the M(RT)^2 algorithm to explore the phase space of the Ising model by flipping spins according to the acceptance probability of eq. (4.33). In summary, the steps of one MC update of the M(RT)^2 algorithm when applied to the Ising model are as follows.

One MC Update of the M(RT)^2 Algorithm

- Randomly choose a lattice site i.
- Compute $\Delta E = E(Y) - E(X) = 2J\sigma_i h_i$ with $h_i = \sum_{\langle i,j \rangle} \sigma_j$ and $E = -J\sum_{\langle i,j \rangle} \sigma_i \sigma_j$.
- Flip the spin if $\Delta E \leq 0$, otherwise accept it with probability $\exp\left(-\frac{\Delta E}{k_B T}\right)$.

This Monte Carlo update is performed for many (randomly or regularly) chosen sites and the time τ of the Markov process is measured in units of MC updates per site. We should bear in mind that there is only a limited number of possible energy differences in a lattice. To speed up the simulation, we can therefore create a lookup

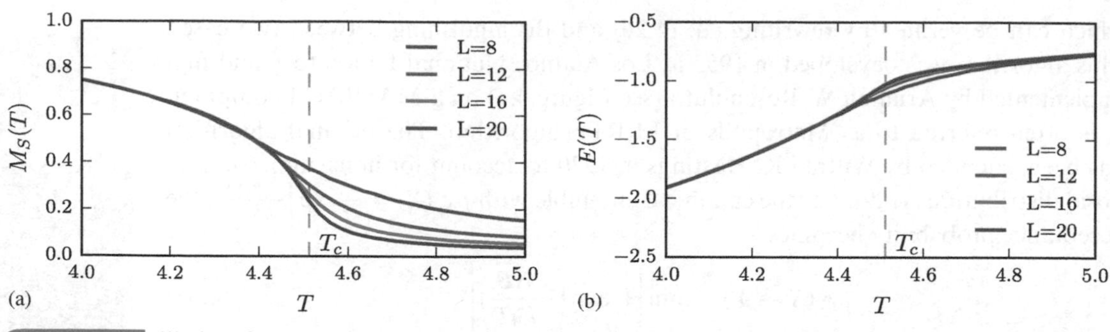

Figure 4.6 We show the spontaneous magnetization $M_S(T)$ and the energy per spin $E(T)$ of the three-dimensional Ising model for different temperatures T and linear system sizes L.

table that acts as a storage for all possible nearest-neighbor combinations. For example $h_i \in \{0, \pm 2, \pm 4\}$ for an Ising model simulation on a square lattice. In Figure 4.6, we show the spontaneous magnetization $M_S(T)$ and the energy per spin $E(T)$ of the three-dimensional Ising model. We performed these simulations with the M(RT)^2 algorithm.

Exercise: 3D Ising Model and M(RT)^2 Algorithm

Write a program for a Monte Carlo simulation to solve the 3D Ising model with periodic boundary conditions and implement the M(RT)^2 algorithm. Note that only a few energy values are possible. Therefore, it is helpful to create look-up tables to avoid recomputing the Boltzmann factor.

1. Measure and plot *energy* E, *magnetization* M, *susceptibility* χ and *heat capacity* C at different temperatures T.
2. Determine the critical temperature T_c.
3. Study how your results depend on the system size (start with small systems).

For getting the average energy/magnetization you should measure the current values (e.g., each system sweep) and average over them (see also remarks below).
Please consider the following points:

- To have an idea of how many single Monte Carlo steps are reasonable, considering system sweeps may be helpful.
- As a starting configuration you may consider a ground state, the system will then relax to the state corresponding to a given temperature.
- For equilibration at a given temperature you may have to apply a couple of system sweeps. Take about 100 system sweeps to be relatively safe. You may also plot the relaxation process for some temperatures (e.g., M vs. MC-steps/sweeps).
- Using the quantities after each system sweep for getting the average magnetization and energy will at most temperatures be enough to get more or less uncorrelated configurations, except close to the critical point ("critical slowing down").

Exercise: Fractal dimension of the percolating cluster (*cont.*)

- The critical temperature is $T_c \simeq 4.51$.
- To speed up your code, you may consider the following points:
  - You can store the values of the exponential function (only a few discrete values are needed).
  - Use integer variables inside your main loop (e. g., for the values of magnetization M and energy E).
  - The current values of magnetization M and energy E can be easily updated after each spin flip, using the magnetization change due to the spin flip and ΔE, respectively.
  - Reuse equilibrated systems from the previous temperature (T) when going to the next temperature ($T + \Delta T$). This helps avoiding throwing away a lot of configurations to reach equilibrium.

4.6 Glauber Dynamics (Heat Bath Dynamics)

The M(RT)^2 algorithm is not the only possible choice to satisfy the detailed balance condition. Roy J. Glauber (see Figure 4.7) proposed the acceptance probability

$$A_G\left(X \to Y\right) = \frac{\exp\left(-\frac{\Delta E}{k_B T}\right)}{1 + \exp\left(-\frac{\Delta E}{k_B T}\right)}. \tag{4.36}$$

Figure 4.7 Roy J. Glauber (1925–2018) received the Nobel Prize for his work on optical coherence.

In contrast to the M(RT)^2 acceptance probability, updates with $\Delta E = 0$ are not always accepted but only with probability $1/2$. We illustrate this behavior in Figure 4.8. To prove that eq. (4.36) satisfies the condition of detailed balance, we have to show that

$$p_{\rm eq}(Y) A_G(Y \to X) = p_{\rm eq}(X) A_G(X \to Y) \tag{4.37}$$

since $T(Y \to X) = T(X \to Y)$. Equation (4.37) is equivalent to

$$\frac{p_{\rm eq}(Y)}{p_{\rm eq}(X)} = \frac{A_G(X \to Y)}{A_G(Y \to X)}, \tag{4.38}$$

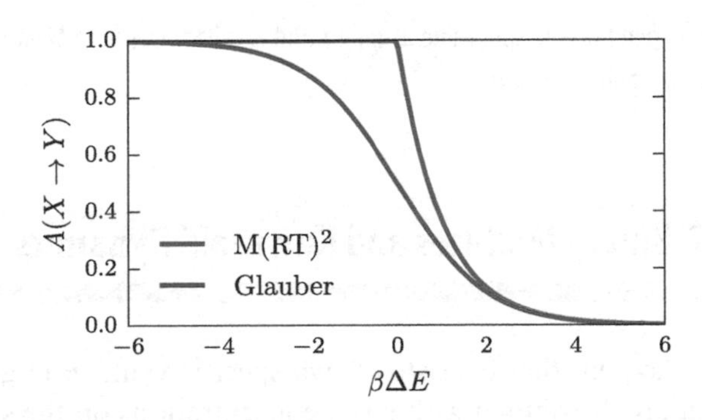

A comparison of the acceptance probabilities of M(RT)^2 and Glauber dynamics (see Eqs. (4.32) and (4.36)). **Figure 4.8**

which is satisfied since

$$\frac{p_{\text{eq}}(Y)}{p_{\text{eq}}(X)} = \exp\left(-\frac{\Delta E}{k_B T}\right) \tag{4.39}$$

and

$$\frac{A_G(X \to Y)}{A_G(Y \to X)} = \frac{\exp\left(-\frac{\Delta E}{k_B T}\right)}{1 + \exp\left(-\frac{\Delta E}{k_B T}\right)} \left[\frac{\exp\left(\frac{\Delta E}{k_B T}\right)}{1 + \exp\left(\frac{\Delta E}{k_B T}\right)}\right]^{-1} = \exp\left(-\frac{\Delta E}{k_B T}\right). \tag{4.40}$$

As in the M(RT)^2 algorithm, only the local configuration around one lattice site is relevant for the update procedure. Furthermore, $E = -J \sum_{\langle i,j \rangle} \sigma_i \sigma_j$ for the Ising model and thus the probability to flip a spin σ_i is

$$A_G(X \to Y) = \frac{\exp\left(\frac{-2J\sigma_i h_i}{k_B T}\right)}{1 + \exp\left(\frac{-2J\sigma_i h_i}{k_B T}\right)}, \tag{4.41}$$

where $h_i = \sum_{\langle i,j \rangle} \sigma_j$ is the local field. A possible implementation is

$$\sigma_i(\tau + 1) = -\sigma_i(\tau) \cdot \operatorname{sign}(A_G(\sigma_i) - z), \tag{4.42}$$

where $z \in (0, 1)$ denotes a uniformly distributed random number.

Additional Information: Equivalence of Glauber Dynamics and Heat Bath Monte Carlo

For Glauber dynamics, the spin flip and no flip probabilities can be expressed as

$$p_{\text{flip}} = \begin{cases} p_i & \text{for } \sigma_i = -1 \\ 1 - p_i & \text{for } \sigma_i = +1 \end{cases} \quad \text{and} \quad p_{\text{no flip}} = \begin{cases} 1 - p_i & \text{for } \sigma_i = -1, \\ p_i & \text{for } \sigma_i = +1, \end{cases} \tag{4.43}$$

where

$$p_i = \frac{\exp(2J\beta h_i)}{1 + \exp(2J\beta h_i)}. \tag{4.44}$$

This gives

$$\sigma_i(\tau + 1) = \begin{cases} +1 & \text{with probability } p_i, \\ -1 & \text{with probability } 1 - p_i. \end{cases} \tag{4.45}$$

This method does not depend on the spin value at time t and is called *heat bath Monte Carlo*. As we have shown, it is equivalent to Glauber dynamics.

4.7 Binary Mixtures and Kawasaki Dynamics

We now consider a system that consists of two species A and B (e.g., two different gas molecules) that are distributed with given concentrations on the sites of a lattice. These two species could, for instance, represent two sorts of metallic atoms whose numbers are conserved. We show an illustration of such a system in Figure 4.9. This binary mixture can be modeled by defining

- E_{AA} as the energy of an $A - A$ bond,
- E_{BB} as the energy of a $B - B$ bond,
- E_{AB} as the energy of an $A - B$ bond.

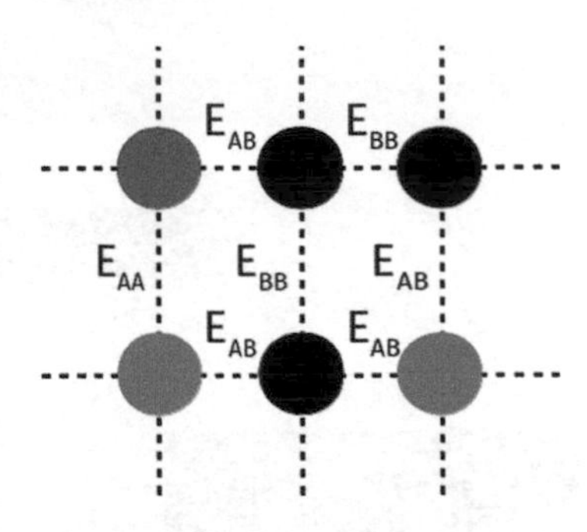

Figure 4.9 An example of a binary mixture consisting of two different atoms A and B.

If we set $E_{AA} = E_{BB} = 0$ and $E_{AB} = 1$, this model corresponds to an Ising model with constant magnetization. Systems with constant magnetization can be simulated with Kawasaki dynamics.

Kawasaki dynamics corresponds to an M(RT)^2 or Glauber algorithm with constant numbers of spins in each population. The corresponding algorithm was introduced by Kyozi Kawasaki (see Figure 4.10) and works as follows:

Kawasaki Dynamics

- Choose an $A - B$ bond.
- Compute ΔE for $A - B \rightarrow B - A$.
- Metropolis: If $\Delta E \leq 0$ flip, else flip with probability

$$p = \exp\left(-\frac{\Delta E}{k_B T}\right).$$

- Glauber: Flip with probability

$$p = \exp\left(-\frac{\Delta E}{k_B T}\right) / \left[1 + \exp\left(-\frac{\Delta E}{k_B T}\right)\right].$$

Figure 4.10 Kyozi Kawasaki received the Boltzmann Medal in 2001 for his contributions to the "understanding of dynamic phenomena in condensed matter systems." Photograph courtesy Helmut R. Brand.

This procedure is very similar to the previously discussed update schemes. The only difference is that the magnetization is kept constant.

4.8 Creutz Algorithm

Until now, we only considered canonical Monte Carlo algorithms for simulating systems at constant temperature. In 1983, Michael Creutz (see Figure 4.11) suggested a method to perform microcanonical Monte Carlo simulations of systems at constant energy [134]. The algorithm is therefore usually referred to as *Creutz* algorithm. The condition of energy conservation is slightly relaxed in this algorithm and energy is not fully conserved.

The movement in phase space is therefore not strictly constrained to a subspace of constant energy, but there is a certain additional volume in which the system can freely move. The condition of constant energy is softened by introducing a so-called *demon*.

The demon corresponds to a small reservoir of energy E_D that can store a certain maximum energy E_{max}. The Creutz algorithm is defined by the following steps:

Creutz Algorithm

- Choose a site.
- Compute ΔE for the spin flip.
- Accept the change if $E_{max} \geq E_D - \Delta E \geq 0$ and update the demon energy according to $E_D \rightarrow E_D + \Delta E$.

Figure 4.11 Michael Creutz works at the Brookhaven National Laboratory. Photograph courtesy Brookhaven National Laboratory.

Besides the fact that we can randomly choose a site, this method involves no random numbers and is thus *deterministic* and therefore *reversible*. However, the temperature of the system is not known.

It is possible to estimate the temperature with the Boltzmann distribution. By plotting a histogram of the energies E_D one observes a distribution $P(E_D) \propto \exp\left(-\frac{E_D}{k_B T}\right)$. We show an example of a distribution $P(E_D)$ in Figure 4.12. The fit is not optimal due to the small number of different values of E_D. The method is faster for larger values of E_{max}, because the condition of constant energy is relaxed and the exploration of phase space is less restricted to certain regions, but then the ensemble becomes "less" microcanonical.

Exercise: Microcanonical Monte Carlo and Q2R

Simulate a microcanonical Ising system using the Creutz algorithm:

1. Start with an initial spin configuration X of a given energy E and define a demon energy E_D such that $E_{max} \geq E_D \geq 0$.
2. Choose a spin at random and flip it to obtain the configuration Y.
3. Calculate the energy difference ΔE between configurations X and Y.
4. If $E_{max} \geq E_D - \Delta E \geq 0$ choose a new spin and repeat the process. If not, revert the spin flip and choose a new spin.

Determine the corresponding temperature T using

$$P(E_D) \propto e^{-\frac{E_D}{k_B T}} .$$

Additional tasks:

1. Compute T for different E. Plot energy and magnetization as a function of temperature and compare your results to the results obtained with the M(RT)^2 algorithm.
2. Repeat the above tasks for different system sizes and compare your results.
3. What happens in the case $E_{max} = 0$ (Q2R algorithm)? Discuss the issue of ergodicity.

4.9 Boundary Conditions

Computer simulations are always restricted to finite domains. Therefore, one of the finite-size effects that we have to take into account are boundary effects of the considered domain. We can set the values of the boundary spins to a certain value or just have no boundary spins at all. Alternatively, we may introduce periodic boundaries. In some situations, certain boundary conditions also correspond to a real physical situation. For finite lattices, the following boundary conditions may be used:

- open boundaries (i. e., no neighbors at the edges of the system),
- fixed boundary conditions,
- periodic boundaries.

Usually, periodic boundary conditions converge faster toward the thermodynamic limit compared to open ones, while fixed boundary conditions typically induce a surface field. It is also possible to identify the last element of a row (or column) with the first element of the next row (or column), which leads to a so-called *helical boundary condition*. This is done to store the system in a one-dimensional array, for instance, to increase the length of the inner loop in a computer program. There are more effects related to systems of finite size that we describe in more detail in Chapter 5.

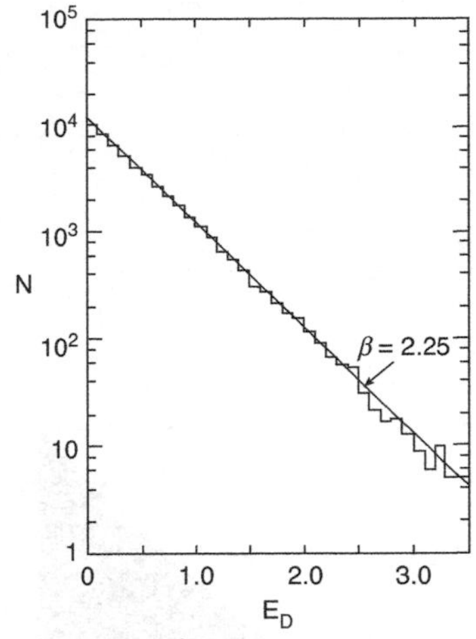

Figure 4.12 The demon energy E_D is exponentially distributed. From the Boltzmann factor, it is possible to extract the inverse temperature $\beta = (k_B T)^{-1} = 2.25$. The figure is taken from Ref. [134].

4.10 Application to Interfaces

The Ising model is used not only as a model of ferromagnetism, but has also found applications in various other areas. One application is the simulation of interfaces (i. e., boundaries between regions that are occupied by different sorts of matter). Let us consider two materials A and B with an arbitrary interface (see Figure 4.13). The surface tension γ is given by the difference in free energy between the mixed system $(A + B)$ and the pure system A:

$$\gamma = f_{A+B} - f_A \,.$$

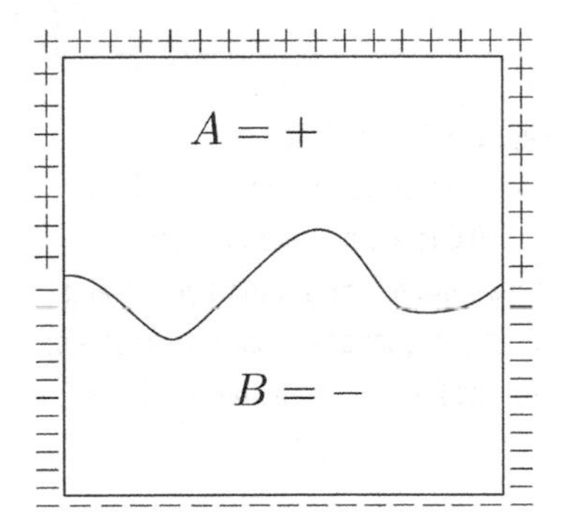

Figure 4.13 Interface between two materials A and B.

We start with binary variables $\sigma = \pm 1$ again, but we apply fixed boundaries, namely, "+" in the upper half for A and "-" in the lower half for B. We use Kawasaki dynamics (see Section 4.7) to simulate the evolution of the interface at temperature T since the number of particles A and B is constant. The Hamiltonian is an Ising Hamiltonian without field

$$\mathcal{H}(\{\sigma\}) = -J \sum_{\langle i,j \rangle}^{N} \sigma_i \sigma_j \,. \tag{4.46}$$

Each bond of the interface has a height h measured from the bottom and we can thus define the width W of the interface according to

$$W = \sqrt{\frac{1}{N}\sum_i (h_i - \bar{h})^2}\,, \tag{4.47}$$

where $\bar{h}$ is the average height and N is the number of bonds belonging to the interface.

A special case is the flat surface, where all points at the interface have the same height, $h_i = \bar{h}$ and thus $W = 0$. In general, the width W increases very quickly with temperature T, exhibiting a singularity at the critical temperature T_c. As we get closer to T_c, the interface becomes more and more diffuse and the system starts to "ignore" the boundary conditions, placing "+" entries next to a boundary where we had imposed "−." The transition is known as roughening transition.

4.10.1 Next-Nearest Neighbors

So far, we have only considered nearest-neighbor interactions. One obvious extension of this simplified model is to also consider next-nearest-neighbor interactions:

$$\mathcal{H}(\{\sigma\}) = -J\sum_{i,j\ \mathrm{nn}} \sigma_i\sigma_j - K\sum_{i,j\ \mathrm{nnn}}^{N} \sigma_i\sigma_j\,, \tag{4.48}$$

where nnn denotes the next-nearest neighbors.

The new term introduces stiffness to the interface and thus counteracts curvature. Curved surfaces will have an increased energy due to this new term, which makes the system relax to a flatter configuration.

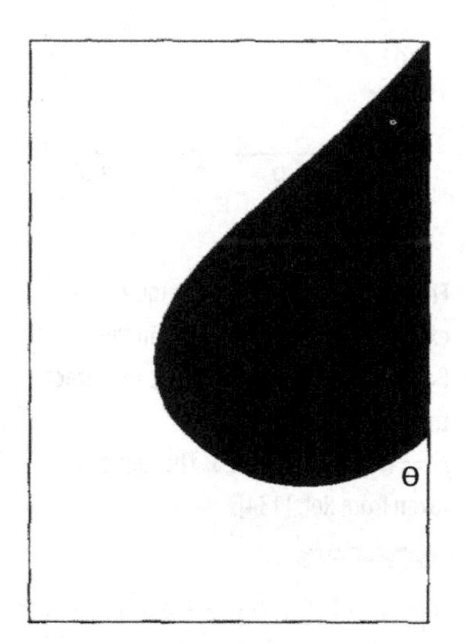

Figure 4.14 Simulation of a two-dimensional drop with $L = 257$, $g = 0.001$ forming a contact angle θ. Black indicates spins with value $+1$, and white corresponds to spins with a value of -1. The figure is taken from Ref. [135].

4.10.2 Shape of a Drop

We can go even further and also include gravity, which will, for instance, help us reproduce drops that are attached to a wall. We start with a block of +1 sites attached to a wall of an $L \times L$ system filled with −1s. The Hamiltonian is given by

$$\mathcal{H}(\{\sigma\}) = -J\sum_{i,j,\,nn}^{N} \sigma_i\sigma_j - K\sum_{i,j,\,\mathrm{nnn}}^{N} \sigma_i\sigma_j - \sum_j h_j \sum_{\mathrm{line}\ j} \sigma_i\,, \tag{4.49}$$

where gravity g is introduced via a height-dependent magnetic field

$$h_j = h_1 + \frac{(j-1)(h_L - h_1)}{L-1} \quad \text{and} \quad g = \frac{h_L - h_1}{L}\,.$$

To generate interface configurations, we use Kawasaki dynamics (mass conservation) and do not permit disconnected clusters of +1s (i. e., isolated water droplets in the air). In Figure 4.14 we show an example of a drop simulation ($L = 257$, $g = 0.001$ after 5×10^7 MC updates averaged more than 20 samples). Based on such simulations, we can measure the contact angle Θ, which is a function of the temperature and goes to zero when approaching the critical temperature T_c.

5 Phase Transitions

5.1 Temporal Correlations

Each time we accept a spin flip in our sampling chain, a new configuration is generated. Unfortunately, both new and previous configurations are strongly correlated. As a consequence, we cannot expect that the error of our Monte Carlo method scales with $1/\sqrt{N}$, if we average over all N sampled configurations. We thus have to find a measure that tells us whether we are already in equilibrium or not, and which enables us to generate uncorrelated configurations. According to the definition of a Markov chain, the dependence of a quantity A on virtual time τ is [136]

$$\langle A(\tau)\rangle = \sum_X p(X,\tau)\,A(X) = \sum_X p(X,\tau_0)\,A(X(\tau))\,. \tag{5.1}$$

In the second step of eq. (5.1), we used that the average is taken over an ensemble of initial configurations $X(\tau_0)$ that evolve according to eq. (4.22) [136].

If one begins the simulation at a time $\tau_0 \leq \tau$, the *nonlinear correlation function*

$$\Phi_A^{\mathrm{nl}}(\tau) = \frac{\langle A(\tau)\rangle - \langle A(\infty)\rangle}{\langle A(\tau_0)\rangle - \langle A(\infty)\rangle} \tag{5.2}$$

is a measure to quantify the deviation of $A(\tau)$ from the steady state value $A(\infty)$ relative to the deviation of $A(\tau_0)$ from $A(\infty)$. Due to its normalization, the nonlinear correlation function $\Phi_A^{\mathrm{nl}}(\tau)$ equals 1 at τ_0 and 0 for $\tau \to \infty$. Strictly speaking, $\Phi_A^{\mathrm{nl}}(\tau)$ is not a correlation function, but a measure to investigate correlations with respect to the initial configuration. The *nonlinear* correlation time τ_A^{nl} describes the relaxation toward equilibrium and is defined as[1]

$$\tau_A^{\mathrm{nl}} = \int_0^\infty \Phi_A^{\mathrm{nl}}(\tau)\,\mathrm{d}\tau\,. \tag{5.3}$$

[1] If we consider an exponential decay of $\Phi_A^{\mathrm{nl}}(\tau)$, we find that this definition yields its characteristic time since

$$\int_0^\infty \exp\left(-\tau/\tau_A^{\mathrm{nl}}\right)\mathrm{d}\tau = \tau_A^{\mathrm{nl}}\,.$$

Close to a critical point, we observe the so-called *critical slowing down* of the dynamics, which means that the nonlinear correlation time diverges as a power law

$$\tau_A^{\mathrm{nl}} \propto |T - T_c|^{-z_A^{\mathrm{nl}}}, \tag{5.4}$$

where z_A^{nl} is the nonlinear dynamical critical exponent. This critical slowing down causes difficulties because eq. (5.4) implies that the time needed to reach equilibrium becomes infinite at T_c.

The linear correlation function of two quantities A and B in equilibrium is

$$\Phi_{AB}(\tau) = \frac{\langle A(\tau_0)B(\tau)\rangle - \langle A\rangle\langle B\rangle}{\langle AB\rangle - \langle A\rangle\langle B\rangle}, \tag{5.5}$$

where

$$\langle A(\tau_0)B(\tau)\rangle = \sum_X p(X,\tau_0)A(X(\tau_0))B(X(\tau)).$$

As τ goes to infinity, $\Phi_{AB}(\tau)$ decreases from unity to zero. If $A = B$, we call eq. (5.5) the *autocorrelation function*. For the spin–spin correlations in the Ising model, we obtain

$$\Phi_\sigma(\tau) = \frac{\langle \sigma(\tau_0)\sigma(\tau)\rangle - \langle \sigma(\tau_0)\rangle^2}{\langle \sigma^2(\tau_0)\rangle - \langle \sigma(\tau_0)\rangle^2}. \tag{5.6}$$

The *linear* correlation time τ_{AB} describes the relaxation in equilibrium

$$\tau_{AB} = \int_0^\infty \Phi_{AB}(\tau)\,\mathrm{d}\tau. \tag{5.7}$$

As for the nonlinear correlation time, in the vicinity of T_c, we observe *critical slowing down* that is described by

$$\tau_{AB} \propto |T - T_c|^{-z_A}, \tag{5.8}$$

where z_A is the *linear* dynamic critical exponent. The dynamic exponents for spin correlations of the Ising model for Glauber and $\mathrm{M(RT)}^2$ updating turn out to be [137–140]

$$\begin{aligned} z_\sigma &= 2.166(7)\ (2\mathrm{D}), \\ z_\sigma &= 2.055(10)\ (3\mathrm{D}). \end{aligned} \tag{5.9}$$

There exist conjectured relations between these Ising critical exponents and the critical dynamic exponents for spin and energy correlations. The relations [141]

$$\begin{aligned} z_\sigma - z_\sigma^{\mathrm{nl}} &= \beta, \\ z_E - z_E^{\mathrm{nl}} &= 1 - \alpha \end{aligned} \tag{5.10}$$

are numerically well established, however, not yet proven.

5.2 Decorrelated Configurations

The behavior of the correlation length described in the previous Section 5.1 is valid for an infinite lattice and diverges at T_c (see eq. (3.33)). In a finite system, however, we cannot observe a quantity that diverges toward infinity. As we schematically illustrate in Figure 5.1, the correlation ξ must be less than the system size L at T_c (i.e., $\xi \approx L$ at T_c).

We connect the behavior of $\xi(T)$ close to the critical temperature with its scaling law $\xi(T) \propto |T - T_c|^{-\nu}$ at T_c (see eq. (3.33)) and the corresponding correlation-time scaling of eq. (5.7):

$$\tau_{AB} \propto |T - T_c|^{-z_{AB}} \propto L^{\frac{z_{AB}}{\nu}} . \tag{5.11}$$

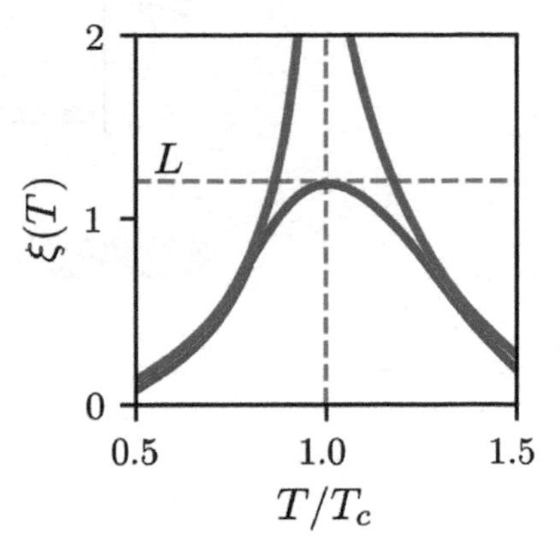

Figure 5.1 Schematic behavior of the correlation length $\xi(T)$ as a function of temperature close to T_c. The correlation length diverges in an infinite system (blue solid line) at T_c according to eq. (3.33). In a finite system (orange solid line), however, we observe a roundoff since the correlation length cannot become larger than the system size L.

This equation implies that the number of samples that have to be discarded to obtain uncorrelated configurations is finite, although it increases with the system size L. When we study large system sizes, the computation may take very long.

To ensure to sample uncorrelated configurations, we should

- first reach equilibrium (discard $n_0 = c\tau^{\mathrm{nl}}(T)$ configurations),
- use only every $n_e^{\mathrm{th}} = c\tau(T)$ configuration for averaging, and
- at T_c use $n_0 = cL^{\frac{z_{\mathrm{nl}}}{\nu}}$ and $n_e = cL^{\frac{z}{\nu}}$,

where $c \approx 3$ is a "safety factor" to make sure to discard enough samples. A trick for reducing critical slowing down is to use cluster algorithms, which we introduce in Chapter 6.

5.3 Finite-Size Scaling

In the Ising model, spins tend to form clusters and spatial correlations emerge in addition to the temporal correlations that we discussed in Section 5.1. The spatial correlation function decays exponentially with distance and the correlation length defines a characteristic length scale (see Section 3.2.3). According to eq. (3.33), the correlation length ξ diverges with an exponent of $-\nu$ at T_c. Moreover, susceptibility and heat capacity also exhibit a divergent behavior at T_c (see Eqs. (3.28) and (3.29)). The larger the system size, the more pronounced is the divergence.

In finite systems, we cannot observe a divergence toward infinity but a peak of some finite height. The height of the susceptibility peak scales with $L^{\frac{\gamma}{\nu}}$ while the width of the critical region shrinks as $L^{-\frac{1}{\nu}}$. If we rescale the values for different system sizes accordingly, we obtain a data collapse. That is, we observe that all the values fall onto a single curve. This can be used to compute critical exponents.

The finite size scaling relation of the susceptibility is [41]

$$\chi(T, L) = L^{\frac{\gamma}{\nu}} F_\chi\left[(T - T_c) L^{\frac{1}{\nu}}\right] , \tag{5.12}$$

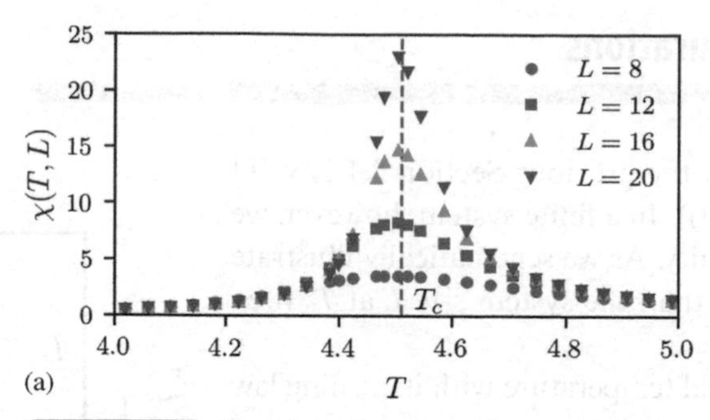

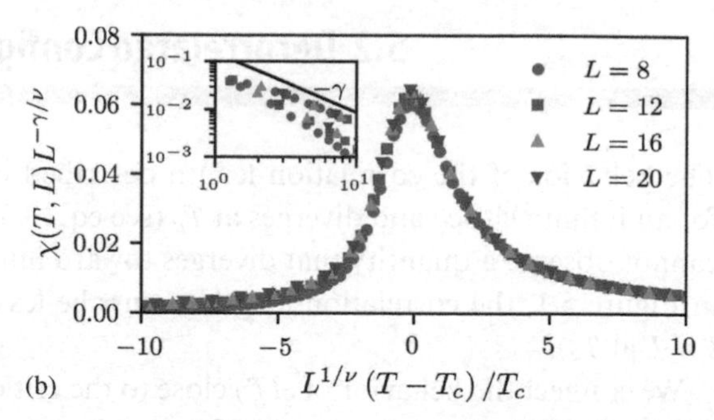

Figure 5.2 System-size dependence of the susceptibility (a) and the corresponding finite-size scaling of eq. (5.12) (b). The inset in panel (b) shows the two branches of $\chi(T, L) L^{-\gamma}$ for $T > T_c$ and $T < T_c$ in a log-log plot; the black solid line is a guide to the eye with slope $-\gamma \approx -1.237$. The data are based on a simulation of the Ising model on a cubic lattice for different linear system sizes L. The number of samples, which we generated with the Wolff algorithm (see Section 6.5), is 25×10^3.

where F_χ is the *scaling function* of the susceptibility.[2] We show an example of a finite-size scaling data collapse of the susceptibility in Figure 5.2. For the spontaneous magnetization, the corresponding finite-size scaling relation is

$$M_S(T, L) = L^{-\frac{\beta}{\nu}} F_{M_S}\left[(T - T_c) L^{\frac{1}{\nu}}\right]. \tag{5.13}$$

Exercise: Finite-Size Scaling

Perform simulations of the three-dimensional Ising model for different system sizes to determine the critical exponents γ and ν of the magnetic susceptibility χ.

1. Use "decorrelated configurations" to measure the temperature dependence of the susceptibility for different system sizes.
2. The critical temperature is $T_c \simeq 4.51$.
3. Vary γ/ν and $1/\nu$ until you get the best possible data collapse. The best way to judge the quality of the data collapse is by eye. (Within the linear regions of the scaling function, the slope in a log-log plot should be $-\gamma$. Check if the slope is correct as a verification of your data collapse.)

You may find the following points useful as well:

- You can get an estimate for the ratio γ/ν by plotting the maximal value of χ as a function of the system size in a log-log plot.
- You may try a more systematic way of determining the exponents:
  1. You can get an estimated value γ/ν as a starting value. Try to get the best possible data collapse by tuning $1/\nu$ only.
  2. Within the linear regions of the scaling function, you can measure the slope $-\gamma$.
  3. Take the new values of γ and ν as new starting value for γ/ν.

5.4 Binder Cumulant

In the previous section, we have outlined that the finite size scaling relation of the susceptibility (see eq. (5.12)) constitutes a method to determine critical exponents. However, we still need a way to determine the critical temperature more precisely. To overcome this hurdle, we make use of the so-called *Binder cumulant*

$$U_L(T) = 1 - \frac{\langle M^4 \rangle_L}{3 \langle M^2 \rangle_L^2}, \tag{5.14}$$

which is independent of the system size L at T_c since

$$\frac{\langle M^4 \rangle_L}{\langle M^2 \rangle_L^2} = \frac{L^{-\frac{4\beta}{\nu}} F_{M4}\left[(T-T_c)L^{\frac{1}{\nu}}\right]}{\left\{L^{-\frac{2\beta}{\nu}} F_{M2}\left[(T-T_c)L^{\frac{1}{\nu}}\right]\right\}^2} = F_C\left[(T-T_c)L^{\frac{1}{\nu}}\right], \tag{5.15}$$

Figure 5.3 Kurt Binder (Johannes Gutenberg-Universität Mainz) made numerous important contributions to the field of computational statistical physics.

where we used eq. (5.13) in the first step. The Binder cumulant was introduced by Kurt Binder [142] (see Figure 5.3). At the critical temperature ($T = T_c$), the scaling function F_C, which is nothing but the ratio of two other scaling functions, is a system size-independent constant. As shown in the left panel of Figure 5.4, for $T > T_c$, the magnetization exhibits a Gaussian distribution

$$P_L(M) = \sqrt{\frac{L^d}{2\pi\sigma_L^2}} \exp\left[-\frac{M^2 L^d}{2\sigma_L^2}\right], \tag{5.16}$$

with $\sigma_L^2 = k_B T \chi_L$. Since, for Gaussian distributions, the fourth moment equals three times the second moment squared,

$$\langle M^4 \rangle_L = 3 \langle M^2 \rangle_L^2, \tag{5.17}$$

it follows that $U_L(T)$ must be zero for $T > T_c$. Below the critical temperature ($T < T_c$), there exist two ground states, one with positive and one with negative magnetization; the corresponding distribution is [143]

$$P_L(M) = \frac{1}{2}\sqrt{\frac{L^d}{2\pi\sigma_L^2}} \left\{\exp\left[-\frac{(M-M_S)^2 L^d}{\sigma_L}\right] + \exp\left[-\frac{(M+M_S)^2 L^d}{\sigma_L}\right]\right\}, \tag{5.18}$$

as illustrated in the right panel of Figure 5.4.For this distribution, $\langle M^4 \rangle_L = \langle M^2 \rangle_L^2$ and therefore $U_L(T) = \frac{2}{3}$ for $T < T_c$. In summary, we obtained

$$U_L(T) = \begin{cases} \frac{2}{3}, & \text{for } T < T_c, \\ \text{const.}, & \text{for } T = T_c, \\ 0, & \text{for } T > T_c, \end{cases} \tag{5.19}$$

[2] Based on eq. (3.28), we can infer that $F_\chi[x] \propto x^{-\gamma}$ as $L \to \infty$.

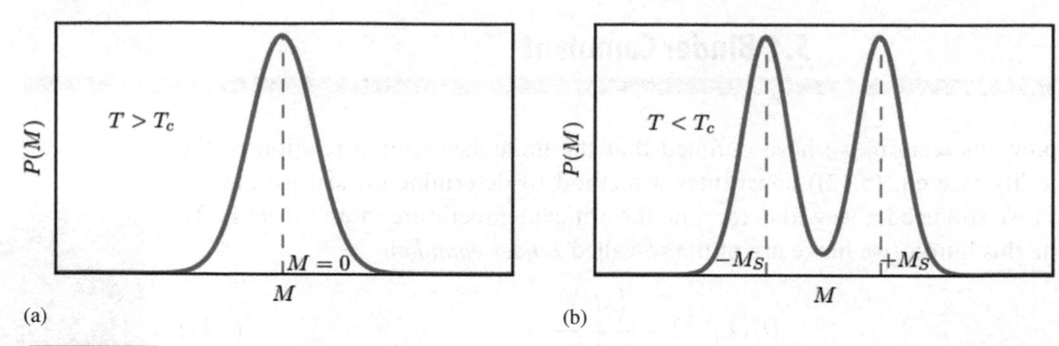

Figure 5.4 Schematical distribution $P(M)$ of the magnetization M above and below the critical temperature T_c.

for $L \to \infty$. In Figure 5.5, we show the behavior of the Binder cumulant for the three-dimensional Ising model and different system sizes. This constitutes a very efficient way to compute the critical temperature based on the very temperature sensitive Binder cumulant U_L. For infinite systems, the cumulant exhibits a jump at T_c as described by eq. (5.19).

All our discussions related to the behavior of different thermodynamic quantities at the critical temperature assumed a pure power law behavior as, for example, those described by Eqs. (3.23), (3.28), and (3.29). However, we can observe such a power law dependence only for temperature values very close to the critical temperature. Farther away from T_c, we cannot observe a clear power law behavior anymore, and corrections to scaling have to be included:

$$M(T) = A_0 (T_c - T)^{\beta} + A_1 (T_c - T)^{\beta_1} + \ldots, \tag{5.20}$$

$$\xi(T) = C_0 (T_c - T)^{-\nu} + C_1 (T_c - T)^{-\nu_1} + \ldots, \tag{5.21}$$

with $\beta_1 > \beta$ and $\nu_1 < \nu$ being universal correction to scaling exponents. These corrections can become very important for high-quality data, where the errors are small and the deviations become visible. The scaling functions must also be generalized as

$$M(T, L) = L^{-\frac{\beta}{\nu}} F_M \left[(T - T_c) L^{\frac{1}{\nu}} \right] + L^{-x} F_M^1 \left[(T - T_c) L^{\frac{1}{\nu}} \right] + \ldots, \tag{5.22}$$

where $x = \max\left[\frac{\beta_1}{\nu}, \frac{\beta}{\nu_1}, \frac{\beta}{\nu} - 1\right]$.

These corrections to scaling are responsible for the fact that the curves for Binder cumulants do not cross at exactly the same point for small systems as shown in the right panel of Figure 5.5.

5.5 First-Order Transitions

Until now, we only focused on the temperature dependence of thermodynamic quantities in the context of the second-order phase transition of the Ising model. For $T < T_c$,

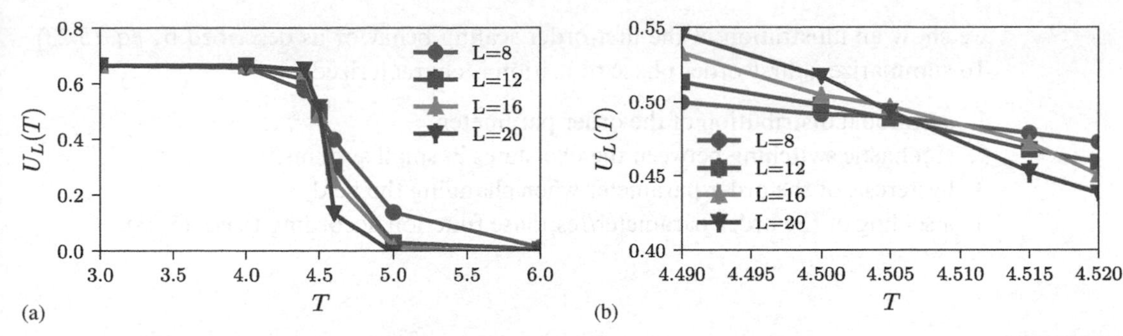

Figure 5.5 The temperature dependence of the Binder cumulant for the three-dimensional Ising model for different system sizes. Panel (a) shows the crossing point of panel (b) with higher resolution.

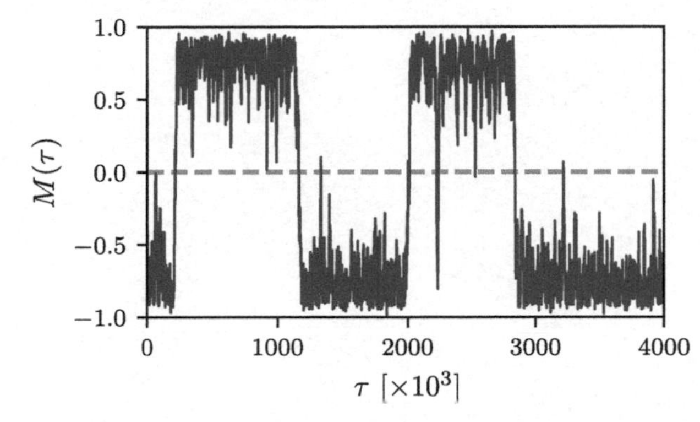

Figure 5.6 For small system sizes and below T_c, the magnetization $M(\tau)$ switches sign after a certain number of Monte Carlo time steps τ if the magnetic field is zero. We simulated a 2D Ising system with 16×16 spins.

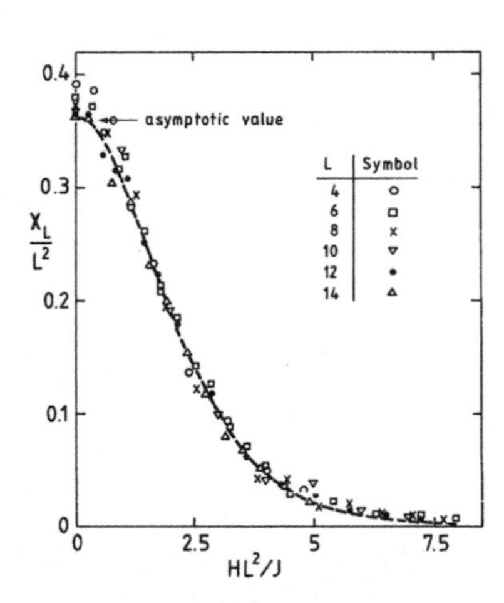

Figure 5.7 The first-order scaling behavior as described by eq. (5.23) for an Ising model on the square lattice at $k_BT/J = 2.1$. The figure is taken from Ref. [143].

the Ising model exhibits a jump in the magnetization, which is a first derivative of the free energy, at $H = 0$. As outlined in Section 3.1.7, this is a transition of first order. Because of the bimodal distribution of the magnetization (see Figure 5.4), we can observe stochastic switching between the two magnetic phases for small system sizes (see Figure 5.6). The typical switching time is also called "ergodic time" and increases with system size and decreases with temperature.

Kurt Binder showed that the magnetization as a function of the field H is described by $\tanh\left(\alpha L^d\right)$ if the distribution of the magnetization is given by eq. (5.18) [143]. In particular, he showed that the magnetization and susceptibility are given by [143]

$$\begin{aligned} M(H) &= \chi_L^D H + M_L \tanh\left(\beta H M_L L^d\right), \\ \chi_L(H) &= \frac{\partial M}{\partial H} = \chi_L^D + \frac{\beta M_L L^d}{\cosh^2\left(\beta H M_L L^d\right)}. \end{aligned} \tag{5.23}$$

Similar to the scaling of a second-order transition, we can scale the maximum of the susceptibility ($\chi_L(H = 0) \propto L^d$) and the width of the peak ($\Delta\chi_L \propto L^{-d}$). In Figure 5.7, we show an illustration of the first-order scaling behavior as described by eq. (5.23). To summarize, a first-order phase transition is characterized by

1. a bimodal distribution of the order parameter,
2. stochastic switching between the two states in small systems,
3. hysteresis of the order parameter when changing the field,
4. a scaling of the order parameter/response function according to eq. (5.23).

Cluster Algorithms 6

Based on our discussion of correlation effects in Sections 5.1 and 5.2, we have learned that larger system sizes require longer simulation times close to critical points due to critical slowing down (see Section 5.2). All algorithms we considered so far are based on single spin flips only. As a consequence of this update scheme, we have to wait sufficiently long before adding a new uncorrelated configuration to our averaging procedure. Otherwise, the considered samples may not be statistically independent. The aim of cluster algorithms is to reduce the computation time by flipping multiple spins at the same time. It is essential that the group of spins to flip is chosen with a sufficiently large acceptance probability. To do this, we will generalize the Ising model and adapt this generalization to our needs.

6.1 Potts Model

The Potts model [144, 145] was introduced by Renfrey B. Potts as a generalization of the Ising model with $q \geq 2$ states. It is a very versatile model due to its applications in many fields including sociology, biology and material science. The Hamiltonian of the Potts model is

$$\mathcal{H}(\{\sigma\}) = -J \sum_{\langle i,j \rangle} \delta_{\sigma_i \sigma_j} - H \sum_i \sigma_i \,, \tag{6.1}$$

where $\sigma_i \in \{1, \ldots, q\}$ and $\delta_{\sigma_i \sigma_j}$ is unity when nodes i and j are in the same state (Kronecker delta). The Potts model exhibits a *first order transition* at the critical temperature in two dimensions for $q > 4$, and for $q > 2$ for dimensions larger than three.[1] For $q = 2$, the Potts model is equivalent to the Ising model. Moreover, there exists a connection between the Potts model and bond percolation. Kasteleyn (see Figure 6.1) and Fortuin demonstrated that the two models have related partition functions [147]. A thermodynamic system is characterized by its partition function from which all thermodynamic quantities can be derived. Therefore, the partition function completely describes the thermodynamic properties of a system. If two systems have the same partition function (up to a multiplicative constant), we consider these two systems as equivalent. We will derive the relation between the Potts model and bond percolation in the following section.

[1] In three dimensions, the transition from second to first order is at $q_c = 2.620(5)$ [146].

6.2 The Kasteleyn and Fortuin Theorem

Figure 6.1 Pieter W. Kasteleyn (1924–1996) worked at the University of Leiden. Photograph courtesy KNAW.

We consider the Potts model not just on a square lattice but on an arbitrary graph that consists of nodes and bonds ν. Each node can be in q possible states and each connection leads to an energy cost of unity if two connected nodes are in a different state and of zero if they are in the same state:

$$E = J \sum_{\nu} \epsilon_{\nu} \quad \text{with} \quad \epsilon_{\nu} = \begin{cases} 0\,, & \text{if endpoints are in the same state,} \\ 1\,, & \text{otherwise}\,. \end{cases} \tag{6.2}$$

As we show in Figure 6.2, we introduce *contraction* and *deletion* operations of a bond ν on the graph. As shown in detail in the gray box on the next page, the partition function of the Potts model can be split up into two partition functions, one for the deleted and one for the contracted graph. After applying these operations to every bond, the graph is reduced to a set of separated points corresponding to clusters of nodes that are connected and in the same state out of q states. The partition function of the Potts model $Z = \sum_X e^{-\beta E(X)}$ reduces to

$$Z = \sum_{\substack{\text{configurations of} \\ \text{bond percolation}}} q^{\#\text{ of clusters}} p^c (1-p)^d = \left\langle q^{\#\text{ of clusters}} \right\rangle_{\text{b}}\,, \tag{6.3}$$

where c and d are, respectively, the numbers of contracted and deleted bonds and $p = 1 - \exp(-\beta J)$. In the limit $q \to 1$, one can derive the generating function of bond percolation. In bond percolation, an edge of a graph is occupied with probability p and vacant with probability $1 - p$. Equation (6.3) constitutes a fundamental

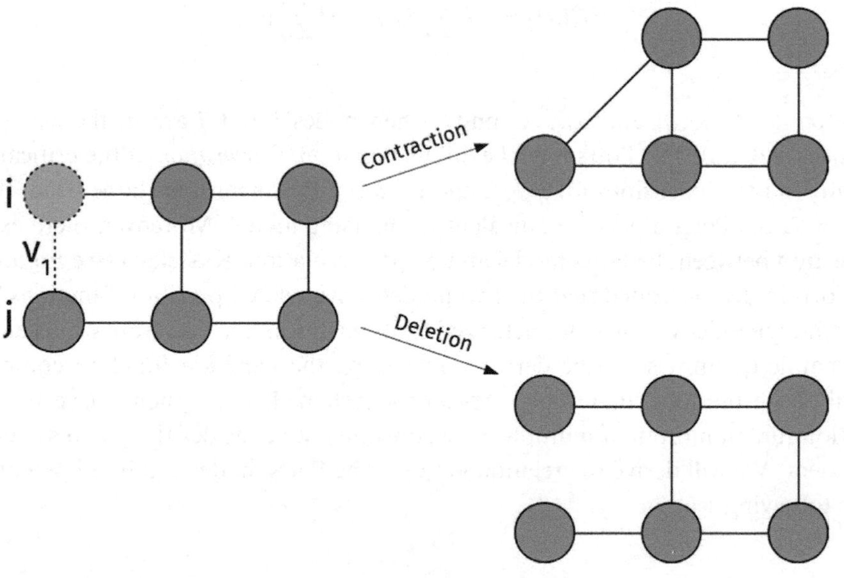

Figure 6.2 Contraction and deletion of bond ν_1 on a graph.

relation between a purely geometrical model (bond percolation) and a magnetic model described by a Hamiltonian (Potts model) [148]. Interestingly, we can now choose non-integer values for q. This was meaningless in the original definition of the Potts model, because q was introduced as the state of each site.

Additional Information: A Proof for the Theorem of Kasteleyn and Fortuin

The partition function is the sum over all possible configurations weighted by the Boltzmann factor and thus given by

$$Z = \sum_X e^{-\beta E(X)} \stackrel{(6.2)}{=} \sum_X e^{-\beta J \sum_\nu \epsilon_\nu} = \sum_X \prod_\nu e^{-\beta J \epsilon_\nu} . \tag{6.4}$$

We now consider a graph where bond ν_1 connects two nodes i and j with states σ_i and σ_j, respectively. If we delete bond ν_1, the partition function is

$$Z_D = \sum_X \prod_{\nu \neq \nu_1} e^{-\beta J \epsilon_\nu} . \tag{6.5}$$

We can thus rewrite eq. (6.4) as

$$\begin{aligned} Z &= \sum_X e^{-\beta J \epsilon_{\nu_1}} \prod_{\nu \neq \nu_1} e^{-\beta J \epsilon_\nu} \\ &= \sum_{X:\sigma_i = \sigma_j} \prod_{\nu \neq \nu_1} e^{-\beta J \epsilon_\nu} + e^{-\beta J} \sum_{X:\sigma_i \neq \sigma_j} \prod_{\nu \neq \nu_1} e^{-\beta J \epsilon_\nu} , \end{aligned}$$

where the first part is the partition function of the contracted graph Z_C and the second part is given by the identity

$$\sum_{X:\sigma_i \neq \sigma_j} \prod_{\nu \neq \nu_1} e^{-\beta J \epsilon_\nu} = \sum_X \prod_{\nu \neq \nu_1} e^{-\beta J \epsilon_\nu} - \sum_{X:\sigma_i = \sigma_j} \prod_{\nu \neq \nu_1} e^{-\beta J \epsilon_\nu} = Z_D - Z_C . \tag{6.6}$$

We thus obtain

$$Z = Z_C + e^{-\beta J} (Z_D - Z_C) = p Z_C + (1-p) Z_D , \tag{6.7}$$

where $p = 1 - e^{-\beta J}$. To be more precise, we expressed the partition function Z as the contracted and deleted partition functions at bond ν_1. We apply this procedure to another bond ν_2 and find

$$Z = p^2 Z_{C_{\nu_1}, C_{\nu_2}} + p(1-p) Z_{C_{\nu_1}, D_{\nu_2}} + (1-p) p Z_{D_{\nu_1}, C_{\nu_2}} + (1-p)^2 Z_{D_{\nu_1}, D_{\nu_2}} . \tag{6.8}$$

After applying deletion and contraction operations to all bonds, we obtain eq. (6.3).

6.3 Coniglio–Klein Clusters

The first who described the dilution of clusters of equal states with probability $p = 1 - e^{-\beta J}$ in the Potts model were Antonio Coniglio and William Klein (see Figure 6.3). Kurt Binder nicely summarized the historical developments around the theorem of

Figure 6.3 Antonio Coniglio (Università di Napoli Federico II) and William Klein (Boston University) introduced the concept of cluster dilution with probability $p = 1 - e^{-\beta J}$ [150].

Kastelyn and Fortuin, Coniglio–Klein Clusters, and cluster algorithms in Ref. [149]. We should bear in mind that the equivalence between bond percolation and the Potts model only concerns their thermodynamic functions since there is no direct relation between bond configurations and spin configurations. The fact that the actual spin values are absent in eq. (6.3) forms the basis for cluster algorithms. The probability of a given cluster C to be in a certain state σ_0 is independent of the state itself:

$$p(C,\sigma_0) = p^{c_C}(1-p)^{d_C} \sum_{\substack{\text{bond percolation} \\ \text{without cluster C}}} q^{\#\text{ of clusters}} p^c (1-p)^d . \tag{6.9}$$

This implies that changing the state of this particular cluster has no effect on the partition function (and therefore the energy) so that it is possible to accept the flip with probability one. This can be seen by looking at the detailed balance condition of the system

$$p(C,\sigma_1) W[(C,\sigma_1) \to (C,\sigma_2)] = p(C,\sigma_2) W[(C,\sigma_2) \to (C,\sigma_1)] , \tag{6.10}$$

and using $p(C,\sigma_1) = p(C,\sigma_2)$. We then obtain for the acceptance probabilities

$$\begin{aligned} A[(C,\sigma_2) \to (C,\sigma_1)] &= \min\left[1, \frac{p(C,\sigma_2)}{p(C,\sigma_1)}\right] = 1 \quad \text{for M(RT)}^2, \\ A[(C,\sigma_2) \to (C,\sigma_1)] &= \frac{p(C,\sigma_2)}{p(C,\sigma_1) + p(C,\sigma_2)} = \frac{1}{2} \quad \text{for Glauber.} \end{aligned} \tag{6.11}$$

Based on these insights, we can focus on cluster algorithms which are much faster than single-spin flip algorithms and less prone to the problem of critical slowing down.

6.4 Swendsen–Wang Algorithm

The Swendsen–Wang algorithm [151, 152] is a refined Monte Carlo technique that uses the advantage of updating whole clusters of spins. It was developed by Robert Swendsen and Jian-Sheng Wang (see Figures 8.6 and 7.3).

For a certain configuration, we iterate over all bonds connecting spins. Whenever two bonded sites are in the same state, the two sites belong to the same cluster with probability $p = 1 - e^{-\beta J}$. Once the clusters are determined, they can be flipped using any of the updating schemes mentioned before. The basic procedure is as follows:

Swendsen–Wang Algorithm

- Occupy the bonds with probability $p = 1 - e^{-\beta J}$ if sites are in the same state.
- Identify the clusters with the Hoshen–Kopelman algorithm (see Section 2.3.4).
- Flip the clusters with probability $1/2$ for Ising or always randomly choose a new state for $q > 2$.
- Repeat the procedure.

Besides moving faster through phase space, the Swendsen–Wang algorithm also reduces critical slowing down considerably since the dynamical critical exponents defined in eq. (5.9) become $z = 0.35(1)$ (2D) and $z = 0.75(1)$ (3D) [151].

6.5 Wolff Algorithm

The Wolff algorithm [153] just considers one cluster instead of all clusters of a configuration and is based on the following steps:

Wolff Algorithm

- Choose a site randomly.
- If the neighboring sites are in the same state, add them to the cluster with probability $p = 1 - e^{-\beta J}$.
- Repeat this for every site on the boundary of the cluster until all bonds of the cluster have been visited exactly once.
- Choose a new state for the cluster.
- Repeat the procedure.

Exercise: Cluster Algorithms

Cluster algorithms can be used to reduce the critical slowing down substantially. In this task you can either pick the Swendsen–Wang or the Wolff algorithm. Both algorithms use the same probability for connecting sites of the same state:

$$p = 1 - \exp(-\beta J)\,.$$

The difference between the algorithms is that in Swendsen–Wang simulations all clusters will be determined and flipped with probability $1/2$ (Ising model), or randomly get a new value (Potts model). In the Wolff algorithm only one cluster is determined by randomly choosing a spin on the lattice. In this case, the cluster will be flipped (Ising model) or gets a new state that is different from the old one (Potts model).

Task

1. Implement either the Swendsen–Wang or the Wolff algorithm (or both).
2. Check your code by plotting $E(T)$ and $|M|(T)$.
3. Show that critical slowing down is substantially reduced by plotting the time dependence of relaxation of E (or $|M|$) around/at T_c into equilibrium in comparison to the single spin-flip algorithm.
4. Measure the nonlinear correlation time for different system sizes at T_c and thus the exponent z and compare it to the values for the single spin-flip algorithms and/or other cluster algorithms.

Exercise: Cluster Algorithms (*cont.*)

Please consider the following points:

- Note that the probability $p = 1 - \exp(-\beta J)$ is derived for the Potts model. This, of course, includes the Ising model, however, using an energy notation (terms $\sigma_i \sigma_j$ in the energy sum take values 0 or 1) different from the one originally used for the Ising model (terms $\sigma_i \sigma_j$ in the energy sum take values ± 1). In the latter case, the probability will be $p = 1 - \exp(-2\beta J)$.
- For the Swendsen–Wang algorithm you will need a working Hoshen–Kopelman algorithm (see Section 2.3.4).
- Using these cluster algorithms for temperatures below T_c, the magnetization of your system will not be "frozen" in one ground state as it was for the single spin-flip algorithms.

6.6 Continuous Degrees of Freedom: The n-Vector Model

Figure 6.4 Harry Eugene ("Gene") Stanley from Boston University (Boltzmann Medal 2004) has made many contributions to statistical and computational physics.

Before focusing on further simulation methods, we briefly discuss other generalizations of the Ising model. The Potts model [144, 145] is only one example. In fact, there exists a large number of models that are modified versions of the Ising model and used to describe related physical phenomena such as antiferromagnetism, spin glasses, and metamagnetism.

One of the possible generalizations of the Ising model is the so called n-vector model [154]. It describes spins as vectors with n components. This model has applications in modeling magnetism and the Higgs mechanism. Its Hamiltonian

$$\mathcal{H} = -J \sum_{\langle i,j \rangle} \mathbf{S}_i \cdot \mathbf{S}_j + \mathbf{H} \sum_i \mathbf{S}_i \,, \tag{6.12}$$

resembles the one of the Potts model in the sense that it favors spin alignment. Here, $\mathbf{S}_i = \left(S_i^1, S_i^2, \ldots, S_i^n\right)$ is an n-dimensional vector and $|\mathbf{S}_i| = 1$. For $n = 1$, we obtain the Ising model, the case $n = 2$ corresponds to the *XY-model*, for $n = 3$ we find the so-called *Heisenberg model* and finally, for $n = \infty$ one obtains the *spherical model* which was solved exactly by Gene Stanley (see Figure 6.4). For $n > 1$, the degrees of freedom of the n-vector model are continuous and as a consequence the XY-model and the Heisenberg model do not exhibit a phase transition from an ordered to a disordered state in two dimensions at a nonzero critical temperature.[2] The proof of this statement can be found in Ref. [156], and it is a consequence of the Mermin–Wagner theorem which states that in dimensions $d \leq 2$ continuous symmetries cannot be spontaneously broken at finite temperature in systems with sufficiently short-range interactions. In three dimensions, however, both the XY and the Heisenberg model exhibit a phase transition at a nonzero critical temperature T_c. The scaling behavior of these models

[2] The two-dimensional XY model does, however, exhibit a transition to a low-temperature phase with power law decaying correlation functions, the Kosterlitz–Thouless transition (Nobel Prize 2016) [155].

close to the critical temperature is very similar to the behavior of the Ising model only with different critical exponents.

For Monte Carlo simulations with vector-valued spins we have to adapt our simulation methods to continuous degrees of freedom. The classical strategy is to flip spins by modifying the spin locally through adding a small $\Delta\mathbf{S}$ such that $\mathbf{S}_i' = \mathbf{S}_i + \Delta\mathbf{S}$ and $\Delta\mathbf{S} \perp \mathbf{S}_i$. The classical Metropolis algorithm can then be used in the same fashion as in the Ising model.

In order to use cluster methods, one can project a group of spins onto a plane, and then reflect the projection of the spins with respect to the plane. In addition, one has to find a method to identify equal spins. The probability to find equal spins in a vector-valued model vanishes and one therefore considers them to be in the same state when the difference between their components is less than a certain tolerance.

Exercise: Heisenberg Model

The Heisenberg model can be seen as a generalization of the Ising model, where a more realistic description of classical magnetic dipoles is represented by the Hamiltonian

$$\mathcal{H} = -J \sum_{\langle i,j \rangle} \mathbf{S_i} \cdot \mathbf{S_j} ,$$

where $\mathbf{S_i} \in \mathbb{S}^2 \subset \mathbb{R}^3$ points to the surface of a three-dimensional sphere of radius one.
While for the XY model (spins in $\mathbb{S}^1$) it is advantageous to use polar coordinates, it is unclear if in the 3D spin case the added complexity introduced by spherical coordinates justifies the memory savings.

1. Implement a MC simulation of the Heisenberg model with a M(RT)^2 algorithm. Remember to normalize your vectors if you use a Cartesian representation.
   - Compute the critical temperature for $J = 1$. You can use either Binder cumulants, or the magnetic susceptibility. You should find $T_c \approx 1.443$.
   - Compute the autocorrelation time, either for E or $\mathbf{M}^2$, at T_c and find the critical dynamical exponent given by the relation $\tau \propto L^{z_c}$.
2. The cluster algorithm can be extended to systems with spins of arbitrary dimensions, with the trick of considering reflections around a random plane at each MC step. For each step select a random unit vector $\vec{r}$ which is chosen to be orthogonal to the reflection plane. Decide if you want to implement Swendsen–Wang or Wolff algorithm, grow a cluster with bond probability

$$p_{i,j} = 1 - \exp[-2\beta(\mathbf{S_i} \cdot \mathbf{r})(\mathbf{S_j} \cdot \mathbf{r})], \tag{6.13}$$

   and flip the spins according to

$$\mathbf{S_i}' = \mathbf{S_i} - 2\mathbf{r}\,(\mathbf{S_i} \cdot \mathbf{r}). \tag{6.14}$$

   - Repeat the computation of T_c and z_c.

7 Histogram Methods

For computing the thermal average as defined by eq. (3.15), we need to sample different configurations at different temperatures. One possibility would be to determine an average at a certain temperature T_0 and extrapolate to another temperature T. For the canonical ensemble, an extrapolation can be achieved by reweighing histograms of energies $p_{T_0}(E)$ with the Boltzmann factor $\exp(E/T - E/T_0)$. Such histogram methods have first been described in Ref. [157]. We now reformulate the computation of the thermal average of a quantity Q and of the partition function as a sum over all possible energies instead of over all possible configurations and find

$$Q(T_0) = \frac{1}{Z_{T_0}} \sum_E Q(E)\, p_{T_0}(E) \quad \text{with} \quad Z_{T_0} = \sum_E p_{T_0}(E), \tag{7.1}$$

where $p_{T_0}(E) = g(E)\, e^{-\frac{E}{k_B T_0}}$ and $g(E)$ defines the *degeneracy of states* (i.e., the number of states with energy E). This takes into account the fact that multiple configurations can have the same energy. Our goal is to compute the quantity Q at another temperature T according to

$$Q(T) = \frac{1}{Z_T} \sum_E Q(E)\, p_T(E). \tag{7.2}$$

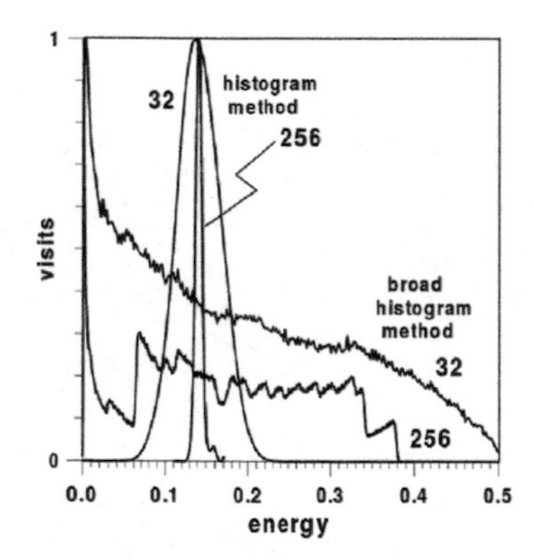

Figure 7.1 An example of the histogram and the broad histogram method for the Ising model on the square lattice with two different system sizes ($L = 32$ and $L = 256$). The figure is taken from Ref. [158].

The degeneracy of states contains all the information needed. Using the definition of $g(E)$ yields

$$p_T(E) = g(E)\, e^{-\frac{E}{k_B T}} = p_{T_0}(E) \exp\left[-\frac{E}{k_B T} + \frac{E}{k_B T_0}\right] \tag{7.3}$$

and with $f_{T_0,T}(E) = \exp\left[-\frac{E}{k_B T} + \frac{E}{k_B T_0}\right]$ we finally obtain

$$Q(T) = \frac{\sum_E Q(E)\, p_{T_0}(E)\, f_{T_0,T}(E)}{\sum_E p_{T_0}(E)\, f_{T_0,T}(E)}. \tag{7.4}$$

With eq. (7.4) we found a way to compute a quantity Q at any temperature T based on a sampling at a certain temperature T_0. The drawback of this method is that the values of $Q(E)$ are sampled around the maximum of $p_T(E)$, which converges to a Dirac delta distribution for large systems (see Figure 7.1). This means that the statistics is very poor if T_0 and T are substantially different, and results become unreliable. One possible solution is to interpolate between several temperatures (multicanonical method) but this involves computations for many temperatures which is also inefficient for large systems. A more efficient solution to this problem, the so-called *broad histogram method*, has been presented in 1996 by Oliveira et al. [159] (see Figure 7.2).

7.1 Broad Histogram Method

The aim of the broad histogram method is to directly compute the degeneracy of states over a broad energy range. We show an illustration of this concept in Figure 7.1. We need to define a Markov process in energy space that is not only exploring regions of certain energies. Let $N_{\text{up}}(E)$ and $N_{\text{down}}(E)$ denote the numbers of processes which, starting from a configuration of energy E, lead to an *increase* and *decrease* in energy, respectively. Furthermore, we have to keep in mind that the degeneracy of states increases exponentially with energy E, because the number of possible configurations increases with energy. To explore all energy regions equally, one imposes a condition equivalent to the one of detailed balance:

$$g(E + \Delta E) N_{\text{down}}(E + \Delta E) = g(E) N_{\text{up}}(E) . \tag{7.5}$$

The motion in phase space toward higher energies can then be penalized with a Metropolis-like dynamics:

Broad Histogram Method

- Choose a new configuration.
- If the new energy is lower, accept the move.
- If the new energy is higher, then accept with probability $\frac{N_{\text{down}}(E+\Delta E)}{N_{\text{up}}(E)}$.

We obtain the function $g(E)$ by taking the logarithm of eq. (7.5) and dividing by ΔE

$$\log[g(E + \Delta E)] - \log[g(E)] = \log\left[N_{\text{up}}(E)\right] - \log[N_{\text{down}}(E + \Delta E)] . \tag{7.6}$$

In the limit of small energy differences, we can approximate this equation by

$$\frac{\partial \log[g(E)]}{\partial E} = \frac{1}{\Delta E} \log\left[\frac{N_{\text{up}}(E)}{N_{\text{down}}(E + \Delta E)}\right], \tag{7.7}$$

which we can numerically integrate over E to obtain $g(E)$. The functions N_{up} and N_{down} can be obtained by checking all possible spin-flips and their corresponding energy change for each new configuration. They become more and more precise as the simulation evolves. In addition, we also need to store the values of the quantity $Q(E)$ for which we wish to compute the thermal average according to

$$Q(T) = \frac{\sum_E Q(E) g(E) e^{-\frac{E}{k_B T}}}{\sum_E g(E) e^{-\frac{E}{k_B T}}} . \tag{7.8}$$

Having the degeneracy of states $g(E)$, we can now compute quantities at any temperature.

Figure 7.2 Paulo Murilo Castro de Oliveira is a professor emeritus at the Universidade Federal Fluminense in Rio de Janeiro.

7.2 Flat Histogram Method

A variant of the broad-histogram method that aims at generating a flat energy histogram was proposed in 2000 by Jian-Sheng Wang (see Figure 7.3) [160]. It is based on the following procedure:

Flat Histogram Method

- Choose a new configuration with energy $E' = E + \Delta E$.
- Flip the spin with probability

$$r(E'|E) = \min\left[1, \frac{N_{\text{down}}(E + \Delta E)}{N_{\text{up}}(E)}\right]. \tag{7.9}$$

- Accumulate statistics for $N_{\text{up}}(E)$.

Figure 7.3 Jian-Sheng Wang (National University of Singapore) made significant contributions to the development of cluster algorithms and histogram methods.

According to eq. (7.5), the spin-flip probability is

$$r(E'|E) = \min\left[1, \frac{g(E)}{g(E + \Delta E)}\right], \tag{7.10}$$

which corresponds to *entropic sampling* Monte Carlo [161] and *multicanonical* Monte Carlo [162, 163]. After the appearance of Ref. [160], another very similar variant was proposed by Wang and Landau [164, 165]:

Wang–Landau Method

- Start with $g(E) = 1$ and set $f = e$.
- Make a Monte Carlo update with $p(E) = 1/g(E)$.
- If the attempt is successful at E: $g(E) \to f \cdot g(E)$.
- Obtain a histogram of energies $H(E)$.
- If $H(E)$ is "flat" (i.e., uniform) enough, then $f \to \sqrt{f}$.
- Stop when f is smaller than some predefined value (e.g., $f \leq 1 + 10^{-8}$).

By setting the Monte Carlo update probability to $1/g(E)$, we obtain smaller transition probabilities for larger energies due to the larger number of possible configurations at the same energy. Therefore, the method tends toward energies with fewer configurations. After a successful update, the values of g and E have to be updated. Once a reasonably flat histogram has been obtained, we can increase the precision by decreasing f. The flatness of the histogram can be measured as the ratio of the minimum to the maximum value.

7.3 Umbrella Sampling

The Umbrella sampling technique was developed and proposed in Ref. [166]. The aim of this method is to overcome the problem of not reaching some configurations for certain energy landscapes. As an example, in the Ising model below T_c, the system could have difficulties in jumping from a positive to a negative magnetization or vice versa if the system is very large. The basic idea is to multiply the probability for configuration X with a function $w(X)$ that is large at the free energy barrier and later on remove this correction in the averaging step:

$$\widetilde{p}(X) = \frac{w(X)\, e^{-\frac{E(X)}{k_B T}}}{\sum_X w(X)\, e^{-\frac{E(X)}{k_B T}}} \quad \text{with} \quad \langle A \rangle = \frac{\langle A/w \rangle_w}{\langle 1/w \rangle_w}, \tag{7.11}$$

where the average $\langle \cdot \rangle_w$ is taken with respect to the distribution $\widetilde{p}(X)$.

Exercise: Broad Histogram Method

- Start from a certain Ising configuration.
- Propose single spin flips for each spin of the configuration (i. e., go sequentially through the whole system) and count the number N_{up} and N_{down} leading to energy increase and energy decrease by ΔE, respectively. *(Do not flip the spins, just check whether energy increases or decreases.)*
- Generate a new configuration by choosing one site randomly and flip the spin if energy is decreased. If energy would be increased, flip the spin with probability $N_{\text{down}}/N_{\text{up}}$ (usually N_{up} is larger than N_{down} which would otherwise preferably lead to increasing energy).
- At each step one accumulates the following values for each energy E for later averaging: $N_{\text{up}}(E)$, $N_{\text{down}}(E)$, $Q(E)$ (the quantity one wants to measure, could be also more than one quantity).

After the simulation you calculate the averages for each energy ("microcanonical average"): $\langle N_{\text{up}}(E) \rangle$, $\langle N_{\text{down}}(E) \rangle$ and $\langle Q(E) \rangle$. From these averages you can determine $g(E)$ via solving eq. (7.7). The thermodynamic average of Q at arbitrary temperatures can then be calculated according to eq. (7.8).

Task

1. Implement the broad histogram algorithm as described above (alternatively you may also try the flat histogram method).
2. Plot the number of times each energy was visited (see Figure 7.1).
3. Measure $g(E)$ and plot it in a semilogarithmic plot.
4. You may plot the energy as a function of temperature (see eq. (7.8)) based on $\langle Q(E) \rangle = E$ (i. e., measuring $Q(E)$ during the simulation is not necessary).
5. Measure the specific heat and susceptibility (see Eqs. (3.26) and (3.27)).
6. Discuss your results.

Exercise: Broad Histogram Method (*cont.*)

Please consider the following points:

- In three dimensions, single spin flips lead to three possible values for ΔE. You can store values for one value of ΔE only (preferably the lowest value) or you store $N_{\text{up}}(E)$ and $N_{\text{down}}(E)$ for each possible value ΔE, resulting in three curves for $g(E)$, which should show the same behavior. Alternatively you can store $(N_{\text{up}}(E))^{1/\Delta E}$ and $(N_{\text{down}}(E))^{1/\Delta E}$.
- As initial configuration, start with a random configuration thermalized at the critical temperature T_c for a few lattice sweeps.
- In order to obtain good correlation free results throw away about 10 lattice sweeps of configuration between the successive values for averaging.
- To accelerate this procedure, you can thermalize the system using Metropolis algorithm at a temperature corresponding to the energy. This may be necessary to avoid blocking of states at low temperatures.

8 Renormalization Group

In this chapter, we discuss renormalization group methods and the importance of symmetries at criticality to improve our simulations. In particular, we focus on the main concepts of renormalization theory and present Monte Carlo renormalization group approaches to determine critical exponents of Ising-like systems. For more details, we refer the interested reader to Ref. [167].

Usually, the more information there is available about a system, the better certain quantities can be computed. Close to critical points, changes in the scale of the system can be used to better extrapolate the values to an infinite system. Furthermore, it is possible to develop scale-invariant theories to describe the properties of models in the vicinity of their critical points.

For self-similar patterns (see Section 2.4), the invariance under scale transformations is obvious. A curve described by the function $f(x)$ is said to be scale-invariant if $f(\lambda x) = \lambda^{\Delta} f(x)$, where λ is a scale factor and Δ is a corresponding exponent. Here, we want to generalize this intuitive treatment of scale changes to concepts from statistical physics. One possibility is to look at the free energy density and its invariance under scale transformations. To renormalize a system means to change its scale by a factor l such that the renormalized length becomes $\tilde{L} = L/l$. This can be done either in position, or in momentum space.

8.1 Real Space Renormalization

If a system is invariant under a certain transformation, we can apply this transformation infinitely often without changing the observables of the system. At the critical point, there exist no finite correlation length scales and the properties of the system are invariant under scale changes. We thus regard criticality as a *fixed point* under such renormalization transformations [169]. In order to put the concept of renormalization into a mathematical framework, we discuss the examples of free energy renormalization and decimation of the one-dimensional Ising model. We then generalize the concept in Section 8.5 and present the implementation of renormalization within Monte Carlo procedures in Section 8.6.

8.2 Renormalization and Free Energy

We first focus on a scale transformation of our system's characteristic length $L \to \tilde{L} = L/l$ and consider the partition function of an Ising system as defined by the

Figure 8.1 Leo P. Kadanoff (1937–2015) contributed to the development of renormalization group theory. He and H. J. Maris describe in "Teaching the Renormalization Group" [168] that "communicating exciting new developments in physics to undergraduate students is of great importance. There is a natural tendency for students to believe that they are a long way from the frontiers of science where discoveries are still being made." Photograph courtesy Thomas A. Witten.

Hamiltonian given in eq. (3.20). At the critical point, a scale transformation with $\tilde{L} = L/l$ leaves the partition function

$$Z = \sum_{\{\sigma\}} e^{-\beta H}, \tag{8.1}$$

and the corresponding free energy invariant [121]. Due to the extensivity of the free energy and in order to keep its density constant, we find

$$f(\epsilon, H) = l^{-d} \tilde{f}\left(\tilde{\epsilon}, \tilde{H}\right), \tag{8.2}$$

where $\epsilon = (T_c - T)/T_c$ is the so-called reduced temperature. The physical law describing the scale invariance at the critical point states that the free energy is a generalized homogeneous function, which means that there exist exponents y_T and y_H such that eq. (8.2) is valid with $\tilde{\epsilon} = l^{y_T}\epsilon$ and $\tilde{H} = l^{y_H} H$ [123], so we obtain

$$\tilde{f}\left(\tilde{\epsilon}, \tilde{H}\right) = \tilde{f}\left(l^{y_T}\epsilon, l^{y_H} H\right). \tag{8.3}$$

The exponents y_T and y_H are known as thermal and magnetic critical exponents. Renormalization also affects the correlation length

$$\xi \propto |T - T_c|^{-\nu} = |\epsilon|^{-\nu}. \tag{8.4}$$

The renormalized correlation length $\tilde{\xi} = \xi/l$ scales as

$$\tilde{\xi} \propto \tilde{\epsilon}^{-\nu}. \tag{8.5}$$

Due to

$$l^{y_T}\epsilon = \tilde{\epsilon} \propto \tilde{\xi}^{-\frac{1}{\nu}} = \left(\frac{\xi}{l}\right)^{-\frac{1}{\nu}} \propto \epsilon l^{\frac{1}{\nu}}, \tag{8.6}$$

we find $y_T = 1/\nu$. For the magnetic critical exponent, one obtains

$$y_H = \frac{d + 2 - \eta}{2} = \frac{d + \gamma/\nu}{2}. \tag{8.7}$$

The critical point is a fixed point of the renormalization transformation since $\epsilon = 0$ at T_c and thus ϵ does not change for any value of the scaling factor.

8.3 Majority Rule

A straightforward example for a renormalization of spin systems in real space is the *majority rule*. Instead of considering all spins in a certain cell separately, we just take the direction of the net magnetization within this cell as a new spin value:

$$\tilde{\sigma}_i = \text{sign}\left(\sum_{\text{region}} \sigma_i\right). \tag{8.8}$$

We have to be careful when applying this transformation. It is not applicable to cells with an even number of spins since this may lead to renormalized spin values with zero net magnetization. The fact that one numerically deals with systems of finite size is also something that we have to take into account. We can only renormalize up to a certain scale before finite size effects become important.

8.4 Decimation of the One-Dimensional Ising Model

Another possible rule is *decimation*, which eliminates certain spins, generally in a regular pattern. As a practical example, we consider the one-dimensional Ising model. The spins only interact with their nearest neighbors and the coupling constant $K = J/(k_B T)$ is the same for all spins. We show an example of such a spin chain in Figure 8.3. To further analyze this system, we compute its partition function Z by distinguishing between even and odd spins:

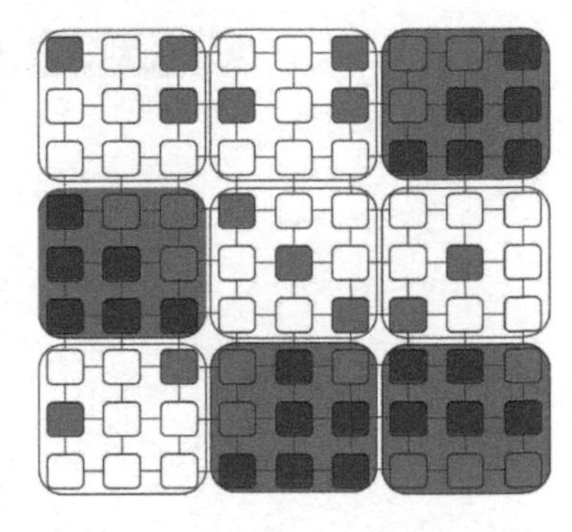

Figure 8.2 An illustration of the majority rule renormalization.

$$
\begin{aligned}
Z = \sum_{\{\sigma\}} e^{K\sum_i \sigma_i} &= \sum_{\sigma_{2i}=\pm 1} \prod_{2i} \left[\sum_{\sigma_{2i+1}=\pm 1} e^{K(\sigma_{2i}\sigma_{2i+1}+\sigma_{2i+1}\sigma_{2i+2})} \right] \\
&= \sum_{\sigma_{2i}=\pm 1} \prod_{2i} \{2\cosh\left[K\left(\sigma_{2i}+\sigma_{2i+2}\right)\right]\},
\end{aligned} \tag{8.9}
$$

where we used that $e^{K(\sigma_{2i}+\sigma_{2i+2})} + e^{-K(\sigma_{2i}+\sigma_{2i+2})} = 2\cosh\left[K\left(\sigma_{2i}+\sigma_{2i+2}\right)\right]$ in the last step. Instead of summing over all configurations explicitly, we first evaluate the contribution of odd spins and then account for the interactions of even spins. This procedure is called *decimation*. In the next step, we rewrite the derived partition function in terms of a renormalized coupling constant $\tilde{K}$:

$$
\begin{aligned}
Z &= \sum_{\sigma_{2i}=\pm 1} \prod_{2i} z(K)\, e^{\tilde{K}\sigma_{2i}\sigma_{2i+2}} \\
&= [z(K)]^{\frac{N}{2}} \sum_{\sigma_{2i}=\pm 1} \prod_{2i} e^{\tilde{K}\sigma_{2i}\sigma_{2i+2}},
\end{aligned} \tag{8.10}
$$

where the function $z(K)$ is a spin-independent multiplicative part of the partition function that does not influence the physics, since it just corresponds to an energy shift. The term $\sum_{\sigma_{2i}=\pm 1} \prod_{2i} e^{\tilde{K}\sigma_{2i}\sigma_{2i+2}}$ in eq. (8.10) describes a partition function with coupling constant $\tilde{K}$ and $N/2$ spins. We thus obtain the following recursion relation:

$$
Z(K,N) = [z(K)]^{\frac{N}{2}}\, Z(\tilde{K}, N/2). \tag{8.11}
$$

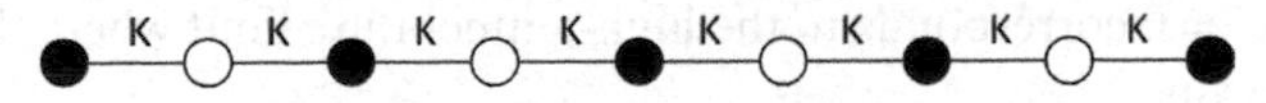

Figure 8.3 Decimation of a one-dimensional Ising chain.

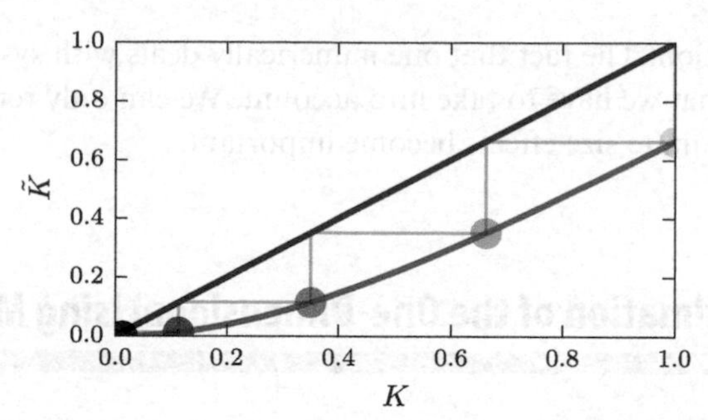

Figure 8.4 An illustration of the fixed point iteration defined by eq. (8.14). The red line corresponds to the function $\tilde{K} = \frac{1}{2} \ln [\cosh (2K)]$.

To connect Eqs. (8.9) and (8.10), we compute the relation $z(K)\, e^{\tilde{K} s_{2i} s_{2i+2}} = 2\cosh [K(s_{2i} + s_{2i+2})]$ explicitly and find

$$z(K)\, e^{\tilde{K} s_{2i} s_{2i+2}} = \begin{cases} 2\cosh(2K) & \text{if } s_{2i} = s_{2i+2}\,, \\ 2 & \text{otherwise}\,. \end{cases} \tag{8.12}$$

We divide and multiply the two expressions of eq. (8.12) with each other and obtain

$$e^{2\tilde{K}} = \cosh(2K) \quad \text{and} \quad z^2(K) = 4\cosh(2K)\,. \tag{8.13}$$

The renormalized coupling constant $\tilde{K}$ in terms of K is therefore

$$\tilde{K} = \frac{1}{2} \ln [\cosh(2K)]\,. \tag{8.14}$$

We illustrate the fixed point iteration of eq. (8.14) in Figure 8.4. For any chosen initial condition, the fixed point iteration approaches the stable fixed point $K^* = 0$. We have thus obtained a rule that describes the change of the coupling constant under renormalization. For a given partition function, we now compute the free energy according to $F = -k_B T N f(K) = -k_B T \ln(Z)$, where $f(K)$ is the dimensionless free energy density. Taking the logarithm of eq. (8.11) yields

$$\ln [Z(K,N)] = N f(K) = \frac{1}{2} N \ln [z(K)] + \frac{1}{2} N f\left(\tilde{K}\right)\,. \tag{8.15}$$

Based on this equation, we can derive the following recursive relation for the free energy density using eq. (8.13):

$$f\left(\tilde{K}\right) = 2f(K) - \ln\left[2\cosh(2K)^{1/2}\right]\,. \tag{8.16}$$

There exists one stable fixed at $K^* = 0$ and another unstable one at $K^* \to \infty$. At every fixed point, eq. (8.16) can be rewritten as

$$f(K^*) = \ln\left[2\cosh(2K^*)^{1/2}\right]\,. \tag{8.17}$$

The case of $K^* = 0$ corresponds to the high-temperature limit where the free energy approaches the value

$$F = -N k_B T f(K^*) = -N k_B T \ln(2)\,. \tag{8.18}$$

In this case, entropy dominates the free energy. For $K^* \to \infty$, the system approaches the low temperature limit and the free energy is

$$F = -Nk_BTf(K^*) = -Nk_BTK = -NJ\,. \tag{8.19}$$

That is, the free energy is dominated by the internal energy.

8.5 Generalization

For the decimation process that we outlined for the one-dimensional Ising system, we found that one renormalized coupling constant was sufficient to perform a renormalization iteration because it was possible to find a solution of eq. (8.12) satisfying all spin values. In general, more coupling constants are necessary to satisfy the equivalent of eq. (8.12) (e. g., in the two-dimensional Ising model). We consider a generalized Hamiltonian of the form

$$\mathcal{H}(\sigma) = \sum_{\alpha=1}^{M} K_\alpha O_\alpha \text{ with } O_\alpha = \sum_i \prod_{k\in c_\alpha} \sigma_{i+k}\,, \tag{8.20}$$

where c_α is the (spin) subset that belongs to operator $\tilde{O}_\alpha$ and M is the total number of operators we want to consider. An example is the Ising Hamiltonian of eq. (3.20) or a three-spin coupling

$$K_3 \sum_{\langle i,j,k\rangle} \sigma_i \sigma_j \sigma_k\,. \tag{8.21}$$

We define the renormalized Hamiltonian $\tilde{\mathcal{H}}$ obtained from a renormalization transformation $P(\tilde{\sigma},\sigma)$ according to [167]

$$e^{G+\tilde{\mathcal{H}}(\tilde{\sigma})} = \sum_\sigma P(\tilde{\sigma},\sigma) e^{\mathcal{H}(\sigma)} \text{ with } \sum_{\tilde{\sigma}} P(\tilde{\sigma},\sigma) = 1\,. \tag{8.22}$$

To satisfy the conservation of the free-energy density, we set

$$\sum_{\tilde{\sigma}} e^{G+\tilde{\mathcal{H}}(\tilde{\sigma})} = \sum_\sigma e^{\mathcal{H}(\sigma)}\,. \tag{8.23}$$

The renormalized Hamiltonian should have the same structure as eq. (8.20):

$$\tilde{\mathcal{H}}(\tilde{\sigma}) = \sum_{\alpha=1}^{M} \tilde{K}_\alpha \tilde{O}_\alpha \text{ with } \tilde{O}_\alpha = \sum_i \prod_{k\in c_\alpha} \tilde{\sigma}_{i+k}\,. \tag{8.24}$$

Inserting Eqs. (8.20) and (8.24) into Eqs. (8.22) and (8.23) and requiring that these equations be valid for all spin configurations, we obtain a set of equations for the renormalized coupling constants as function of the unrenormalized ones, giving the renormalization equation:

$$\tilde{K}_\alpha(K_1,\ldots,K_M) \quad \text{with} \quad \alpha \in \{1,\ldots,M\}\,. \tag{8.25}$$

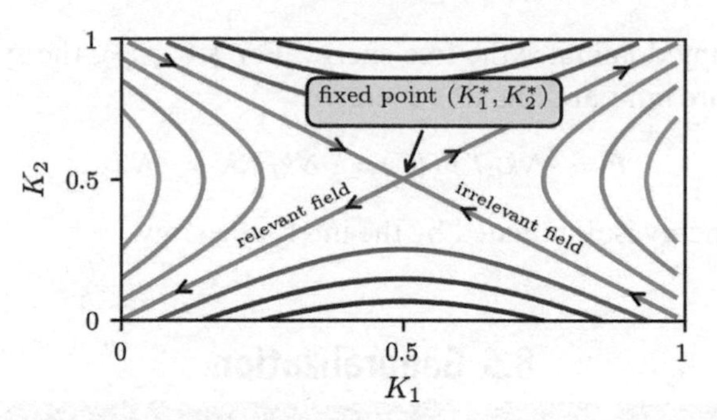

Figure 8.5 Schematic renormalization flow diagram. The point (K_1^*, K_2^*) is the fixed point of the flow.

It is important to note that using only M interaction terms instead of an infinite number is a truncation, leading to systematic errors. In the presence of a critical point there exists a nontrivial fixed point $K_\alpha^* = \tilde{K}_\alpha\left(K_1^*, \ldots, K_M^*\right)$. Solving the generally nonlinear system of equations (8.25), we can now construct a flow diagram of the coupling constant (see Figure 8.5) and obtain values for $\tilde{K}$ for each vector $K = (K_1, \ldots, K_M)$.

A convenient procedure to obtain the corresponding critical exponents is the linearization of eq. (8.25) around this fixed point. Thus, we compute the Jacobian $T_{\alpha\beta} = \frac{\partial \tilde{K}_\alpha}{\partial K_\beta}$ and obtain

$$\tilde{K}_\alpha - K_\alpha^* = \sum_\beta T_{\alpha\beta}\big|_{K^*}\left(K_\beta - K_\beta^*\right). \tag{8.26}$$

To analyze the behavior of the system close to criticality, we consider the eigenvalues $\lambda_1, \ldots, \lambda_M$ and eigenvectors $\phi_1, \ldots, \phi_M$ of the linearized transformation defined by eq. (8.26). The eigenvectors satisfy $\tilde{\phi}_\alpha = \lambda_\alpha \phi_\alpha$ and the fixed point is unstable if there exist eigenvalues $\lambda_\alpha > 1$. Such eigenvalues are called *relevant* and the relevant eigenvalue that corresponds to operators with even numbers of spins (temperature sector) is called λ_T. We can identify the scaling field $\tilde{\epsilon} = l^{y_T}\epsilon$ with the eigenvector of the transformation, and the scaling factor with the eigenvalue $\lambda_T = l^{y_T}$. Then, we can compute the exponent ν according to

$$\nu = \frac{1}{y_T} = \frac{\ln(l)}{\ln(\lambda_T)}. \tag{8.27}$$

Similarly, one obtains y_H from the operators with an odd number of spins like the field term in eq. (3.20) or the three spin coupling of eq. (8.21). Based on eq. (8.27) and using $y_H = (d + 2 - \eta)/2$ (see eq. (8.7)), we obtain

$$d + 2 - \eta = 2y_H = 2\frac{\ln(\lambda_H)}{\ln(l)} \tag{8.28}$$

and can now compute all the other critical exponents using Eqs. (3.34)–(3.37).

8.6 Monte Carlo Renormalization Group

The implementation of real space renormalization with Monte Carlo techniques was first proposed by Shang-keng Ma [170] and then made operational by Bob Swendsen (see Figure 8.6) [171]. Since we are dealing with generalized Hamiltonians with many interaction terms, we compute the thermal average of operator O_α according to

$$\langle O_\alpha \rangle = \frac{\sum_{\{\sigma\}} O_\alpha e^{\sum_\beta K_\beta O_\beta}}{\sum_{\{\sigma\}} e^{\sum_\beta K_\beta O_\beta}} = \frac{\partial F}{\partial K_\alpha}, \tag{8.29}$$

Figure 8.6 Robert Swendsen (Carnegie Mellon University) developed the Swendsen–Wang algorithm together with Jian-Sheng Wang and made important contributions to the Monte Carlo Renormalization Group.

where F is the free energy and $\alpha = 1, \dots, M$. Based on the fluctuation-dissipation theorem (see Section 3.2.2), we can also numerically compute the response functions

$$\chi_{\alpha,\beta} = \frac{\partial \langle O_\alpha \rangle}{\partial K_\beta} = \langle O_\alpha O_\beta \rangle - \langle O_\alpha \rangle \langle O_\beta \rangle,$$

$$\tilde{\chi}_{\alpha,\beta} = \frac{\partial \langle \tilde{O}_\alpha \rangle}{\partial K_\beta} = \langle \tilde{O}_\alpha O_\beta \rangle - \langle \tilde{O}_\alpha \rangle \langle O_\beta \rangle.$$

We find with eq. (8.29) the following set of M^2 equations:

$$\tilde{\chi}^{(n)}_{\alpha,\beta} = \frac{\partial \langle \tilde{O}^{(n)}_\alpha \rangle}{\partial K_\beta} = \sum_\gamma \frac{\partial \tilde{K}_\gamma}{\partial K_\beta} \frac{\partial \langle \tilde{O}^{(n)}_\alpha \rangle}{\partial K_\gamma} = \sum_\gamma T_{\gamma,\beta} \chi^{(n)}_{\alpha,\gamma}. \tag{8.30}$$

It is thus possible to calculate all M^2 components of the matrix $T_{\gamma,\beta}$ from the correlation functions by solving eq. (8.30). At the fixed point $K = K^*$, we can compute the eigenvalues of the matrix and determine critical exponents via eq. (8.27) and (8.28). The fixed point can be localized by iterating the renormalization transformation backward. The accuracy of this method depends on the number of iterations that we want to take into account. There are many error sources in this technique, that originate from the fact that we are using a combination of several approximations to obtain our results:

- statistical errors,
- truncation of the Hamiltonian to M operators,
- finite number of renormalization iterations,
- finite size effects,
- no precise knowledge of K^*.

Nevertheless, it is one of the most accurate numerical methods to obtain critical exponents.

9 Learning and Optimizing

Figure 9.1 Geoffrey Hinton contributed significantly to the development of artificial neural networks. He is currently professor for computer science at the University of Toronto.

There is a growing interest in machine learning and artificial intelligence methods due to the growing amount of training data, refined algorithms, and the increase in computing power [172, 173]. Many algorithms that had been around for decades have been developed further and new methods emerged during the past few years.

The general idea behind machine learning is the development of computational methods to perform a certain task with a performance that can be improved with more training data. In more abstract terms, we could think of the learning problem as a function that depends on multiple parameters and maps a given input to an output. We distinguish between *supervised* and *unsupervised* learning tasks. In supervised learning, we are given input–output pairs and adjust the parameters of the function such that we obtain good performance in mapping given inputs to desired outputs. In the case of unsupervised learning, we aim at extracting the underlying probability distribution of the sample data.

Many machine-learning techniques lack a solid theoretical basis and are therefore not yet very well understood. Attempts have been made to characterize certain learning algorithms with methods from statistical physics such as renormalization group approaches [174]. In this chapter, we focus on one particular example of unsupervised learning, so-called *Boltzmann machine* learning, whose origins lie in statistical physics [175]. This allows us to transfer some of our acquired knowledge on computational sampling techniques in statistical physics to the field of machine learning. After providing an overview of the necessary theoretical concepts, we will discuss the application of Boltzmann machines as generative neural networks, which can be used to learn distributions and generate corresponding samples. Boltzmann machine learning is closely linked to the name Geoffrey Hinton (see Figure 9.1) who made important contributions to this field [176–179].

9.1 Hopfield Network

We begin our excursion to Boltzmann machines with a network consisting of N McCulloch–Pitts neurons [180] that are fully connected (i. e., every single neuron is connected to all other neurons). A McCulloch–Pitts neuron (henceforth neuron) represents a node in a neural network and is nothing but a function of a bias term b and N inputs $\{x_i\}_{i\in\{1,\dots,N\}}$ that are weighted by $\{w_i\}_{i\in\{1,\dots,N\}}$ to compute and output y. We show a single neuron in Figure 9.2 and an example of a Hopfield network with 8 neurons in Figure 9.3.

In a Hopfield network (named after John Hopfield [181], see Figure 9.4), we use discrete inputs $x_i \in \{-1, 1\}$ and weights that satisfy $w_{ij} = w_{ji}$ and $w_{ii} = 0$. The activation of neuron i is described by

$$a_i = \sum_j w_{ij} x_j + b_i \,, \tag{9.1}$$

where b_i is the bias term of neuron i. Similarly to the Ising model, the associated energy is

$$E = -\frac{1}{2} \sum_{i,j} w_{ij} x_i x_j - \sum_i b_i x_i \,. \tag{9.2}$$

The dynamics of a Hopfield network evolves in discrete time steps Δt according to

$$x_i(t + \Delta t) = \begin{cases} 1\,, & \text{if } a_i(t) \geq 0\,, \\ -1\,, & \text{otherwise}\,, \end{cases} \tag{9.3}$$

where $a_i(t)$ is the activation of neuron i at time t (see eq. (9.1)).

A closer look at the update in eq. (9.3) and the energy as defined by eq. (9.2) shows that the energy in a Hopfield network is never increasing for asynchronous updates: If $a_i = \sum_j w_{ij} x_j + b_i \geq 0$, we set $x_i(a_i) = 1$ and if $a_i < 0$ we set $x_i(a_i) = -1$. In both cases, we obtain a negative energy contribution in eq. (9.2). The energy difference ΔE_i between neuron i in states -1 and $+1$ is

$$\Delta E_i = E(x_i = -1) - E(x_i - 1) = 2\left(b_i + \sum_j w_{ij} x_j\right) . \tag{9.4}$$

We thus showed that the activation defined by eq. (9.1) equals one half of the energy difference ΔE_i. We can absorb the bias b_i in the weights w_{ij} by associating an extra active unit to every node in the network. In this way, the energy difference is

$$\Delta E_i = 2 \sum_j w_{ij} x_j \,. \tag{9.5}$$

One possible application of Hopfield networks is to store certain information in local minima by adapting the weights. For example, we could be interested in storing n binary (black and white) patterns in a Hopfield network. These n patterns are described by the set $\{p_i^\nu = \pm 1 | 1 \leq i \leq N\}$ with $\nu \in \{1, \ldots, n\}$. To learn the weights that describe our binary patterns, we can employ the *Hebbian learning* rule [182].

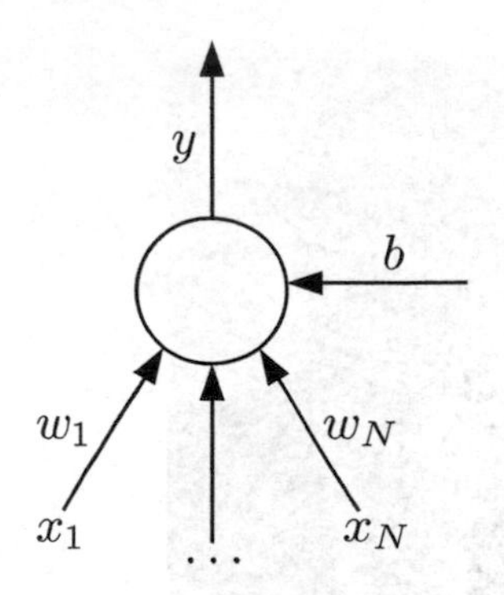

Figure 9.2 An illustration of a single neuron with output y, inputs $\{x_i\}_{i \in \{1,\ldots,N\}}$, weights $\{w_i\}_{i \in \{1,\ldots,N\}}$, and bias b.

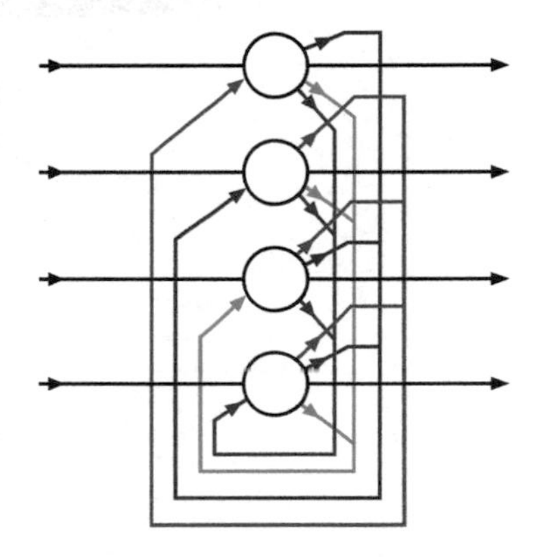

Figure 9.3 An example of a Hopfield network with four neurons. Each neuron is connected to all other neurons through colored edges. Black lines indicate inputs and outputs.

Hebbian Learning

- Set the weights to

$$w_{ij} = w_{ji} = \frac{1}{n} \sum_\mu p_i^\mu p_j^\mu \,. \tag{9.6}$$

Figure 9.4 John Hopfield is a professor emeritus at Princeton University.

For Hopfield networks with a large number of neurons (thermodynamic limit), the number of random patterns that can be stored is approximately 14 percent of the number of neurons [183]. After employing the Hebbian learning rule, we may consider certain initial configurations of x_i and let them evolve according to eq. (9.3). A Hopfield network is said to correctly represent a pattern $\{p_i^\nu = \pm 1 | 1 \leq i \leq N\}$ if $x_i(t+\Delta t) = x_i(t) = p_i^\nu(t)$ for all i (i.e., if p_i^ν is a fixed point of the system). Depending on the number of stored patterns and the chosen initial configuration, the Hopfield network may approach a local minimum that does not correspond to the desired pattern. To avoid getting trapped in local minima, it is possible to employ a stochastic update rule instead of the deterministic one of eq. (9.3) as will be discussed in the next section.

9.2 Boltzmann Machine Learning

In Hopfield networks one can take an initial configuration that is sufficiently close to the desired local energy minimum and retrieve the corresponding pattern. For some applications, finding a local minimum based on the deterministic update rule defined by eq. (9.3) might not be sufficient. In general constraint satisfaction tasks, such as learning a certain distribution, our goal is that the underlying system is able to escape from local minima to find a global minimum given a certain input [176]. We therefore consider Boltzmann machines that are similar to Hopfield networks but use the update probability

$$\sigma_i = \sigma(\Delta E_i/T) = \sigma(2a_i/T) = \frac{1}{1+\exp(-\Delta E_i/T)} \tag{9.7}$$

to set neuron i to unity (i.e., activate it) independently of its state [176]. Here, $\sigma(x) = 1/(1+\exp(-x))$ denotes the sigmoid function (see Figure 9.5). Such a probabilistic update rule is similar to the Monte Carlo methods that we utilized to study Ising-like systems in Chapter 4. As defined in eq. (9.4), ΔE_i is the energy difference between an inactive neuron i and an active one. The parameter T acts as temperature equivalent.[1]

A closer look at Eqs. (9.2) and (9.7) tells us that we are simulating a Hamiltonian system with Glauber dynamics (see Section 4.6). Due to the satisfied detailed balance condition, we reach thermal equilibrium and find for the probabilities of the system to be in state X or Y[2]

$$\frac{p_{\text{eq}}(Y)}{p_{\text{eq}}(X)} = \exp\left(-\frac{E(Y)-E(X)}{T}\right). \tag{9.8}$$

Independent of the initial configuration, this stochastic update procedure always leads to a thermal equilibrium configuration that is fully determined by its energy.

[1] For $T \to 0$, we recover deterministic dynamics as described by eq. (9.3).
[2] Here we set $k_B = 1$.

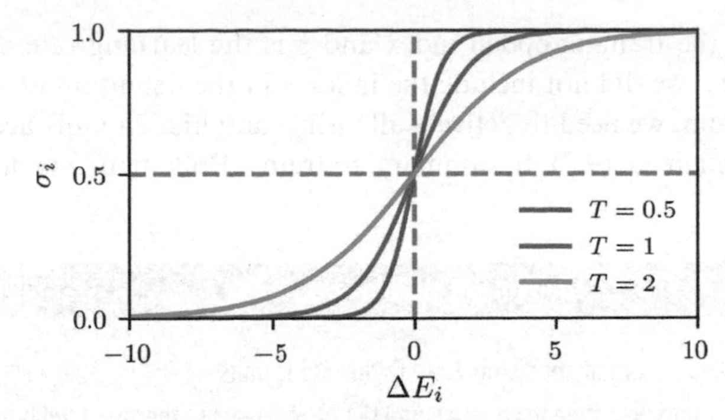

Figure 9.5 The sigmoid function σ_i (eq. (9.7)) for different temperatures T.

Boltzmann Machines

For Boltzmann machines (BMs), we distinguish between two types of nodes: *visible units* and *hidden units* (see Figure 9.6) that are represented by the nonempty set V and the possibly empty set H, respectively. While training a BM, visible units are set by the environment (e. g., an empirical distribution), whereas hidden units may be used to account for constraints involving more than two states in the input data.

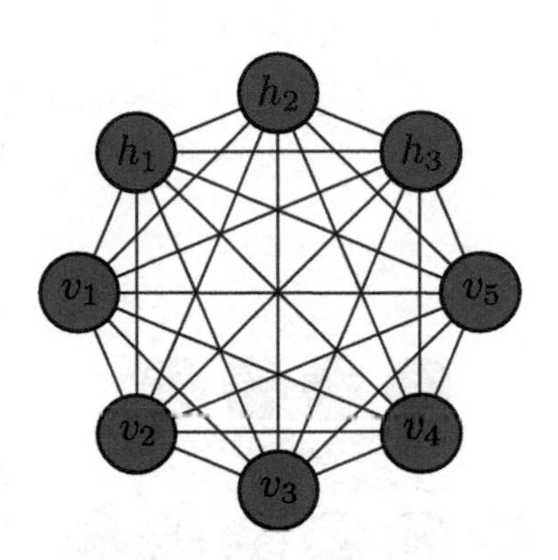

Figure 9.6 Boltzmann machines are composed of visible units (blue) and hidden units (red). In the shown example, there are five visible units $\{v_i\}_{i\in\{1,\ldots,5\}}$ and three hidden units $\{h_j\}_{j\in\{1,\ldots,3\}}$. The network underlying a Boltzmann machine is complete.

Let $P'(\{v\})$ be the probability distribution of all configurations of visible units $\{v\}$ in a freely running network. The term *freely running* just describes that no external inputs are clamped/connected to the visible units. We obtain the distribution $P'(\{v\})$ by summing (i. e., marginalizing) over the corresponding joint probability distribution:

$$P'(\{v\}) = \sum_{\{h\}} P'(\{v\},\{h\}) , \tag{9.9}$$

where we sum over all possible configurations of hidden units $\{h\}$. The goal is to come up with a method such that $P'(\{v\})$ approaches the unknown environment (i. e., data) distribution $P(\{v\})$. We measure the difference between $P'(\{v\})$ and $P(\{v\})$ in terms of the Kullback-Leibler (KL) divergence (or relative entropy)

$$G(P,P') = \sum_{\{v\}} P(\{v\}) \ln\left[\frac{P(\{v\})}{P'(\{v\})}\right] . \tag{9.10}$$

To minimize G, we perform (as outlined in the gray box on pages 117–118) a gradient descent according to [176]

$$\frac{\partial G}{\partial w_{ij}} = -\frac{1}{T}\left(p_{ij} - p'_{ij}\right) , \tag{9.11}$$

where p_{ij} is the probability that two units i and j are active if the environment is determining the states of the visible units and p'_{ij} is the corresponding probability in a freely running network without a coupling to the environment. Both probabilities are measured at thermal equilibrium. The weights w_{ij} of the network are then updated according to [176]

$$w_{ij}^{(n+1)} = w_{ij}^{(n)} - \Delta w_{ij}^{(n)} \quad \text{with} \quad \Delta w_{ij} = \epsilon\left(p_{ij} - p'_{ij}\right) , \tag{9.12}$$

where n denotes the training epoch index and ϵ is the learning rate. Note that, for the sake of brevity, we did not include the index n in the definition of Δw_{ij}. To reach thermal equilibrium, we need to activate all visible and hidden units according to the probability defined in eq. (9.7). In summary, to train a Boltzmann machine we have to do the following:

Boltzmann Machine Learning

1. Clamp the input data (environment distribution) to the visible units.
2. Activate all hidden units according to eq. (9.7) until the system reaches thermal equilibrium.
3. Update all weights according to eq. (9.12) and return to step 2 or stop if the weight updates are sufficiently small.

After training a machine, we can unclamp the visible units from the environment and generate samples to assess the quality of the samples that our machine generates. We therefore consider different initial configurations and activate neurons according to eq. (9.7) until we reach thermal equilibrium. If the learning of our BM was successful, the distribution of states of the unclamped visible units should follow the environment distribution.

Restricted Boltzmann Machines

Boltzmann machines are impractical for general learning tasks, because for large system sizes the underlying complete network structure requires an enormous computational effort to reach thermal equilibrium.

So-called restricted Boltzmann machines (RBMs) are not taking into account connections within the set of hidden and visible units (see Figure 9.7), and turned out to be more suitable for learning tasks. The resulting network structure is called *bipartite*. For RBMs, the updates of visible and hidden units are done in an alternating way. Due to the missing intralayer connections, we can first update all visible units in parallel. In particular, a visible unit i that we denote by v_i is activated with the conditional probability

$$p(v_i = 1|\boldsymbol{h}) = \sigma\left(\sum_j w_{ij}h_j + b_i\right), \tag{9.13}$$

where b_i is the bias of visible unit v_i and $\boldsymbol{h}$ a given configuration of hidden units. In the second step, we activate all hidden units in parallel according to the conditional probability

$$p(h_j = 1|\boldsymbol{v}) = \sigma\left(\sum_i w_{ij}v_i + c_j\right), \tag{9.14}$$

where h_j denotes hidden unit j, c_j its corresponding bias, and $\boldsymbol{v}$ a given configuration of visible units. This sampling technique is known as *block Gibbs sampling*.

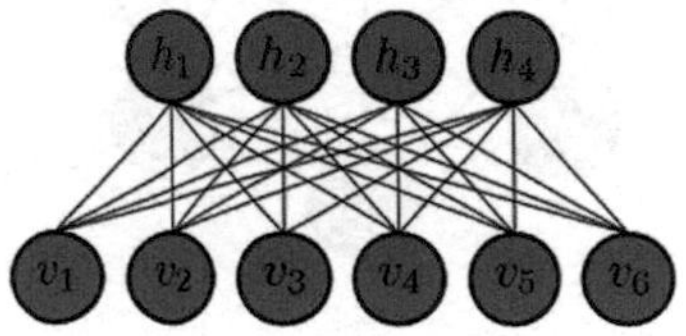

Figure 9.7 Restricted Boltzmann machines are composed of a visible layer (blue) and a hidden layer (red). In the shown example, the respective layers consist of six visible units $\{v_i\}_{i\in\{1,\ldots,6\}}$ and four hidden units $\{h_i\}_{j\in\{1,\ldots,4\}}$. The network structure underlying a restricted Boltzmann machine is bipartite and undirected.

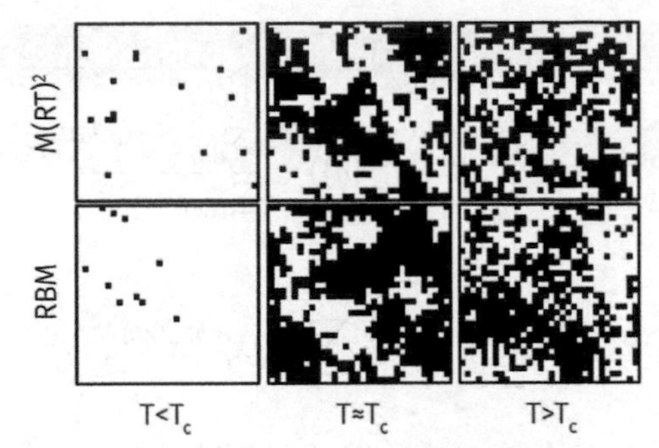

Figure 9.9 Snapshots of 32 × 32 Ising configurations for $T \in \{1.5, 2.5, 4\}$. The configurations are based on M(RT)^2 and RBM samples, respectively.

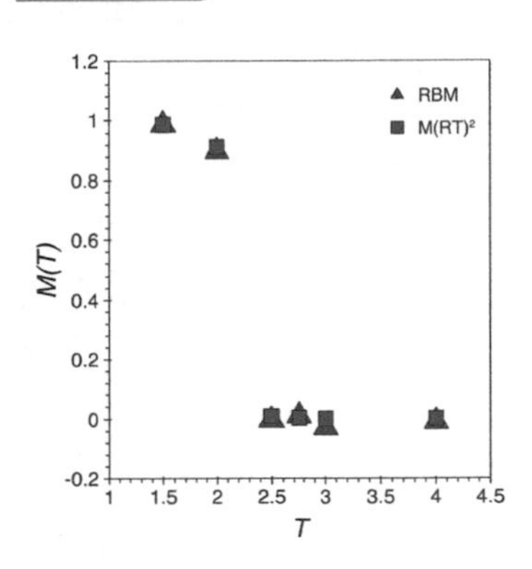

Figure 9.8 Temperature dependence of the magnetization of the Ising model on a square lattice with 32 × 32 lattice sites. Blue triangles correspond to the mean magnetization of 20×10^4 M(RT)^2 samples. Orange squares indicate samples that we generated with RBMs that were trained on the shown M(RT)^2 samples. Each RBM was trained for a distinct temperature $T \in \{1.5, 2, 2.5, 2.75, 3, 4\}$ [185].

Training an RBM is similar to training a BM. The major difference is that we have to account for the bipartite network structure in eq. (9.12) and thus obtain for the weight update

$$\Delta w_{ij} = \epsilon \left(\langle v_i h_j \rangle_{\text{data}} - \langle v_i h_j \rangle_{\text{model}} \right) . \tag{9.15}$$

Instead of sampling the configurations for computing $\langle v_i h_j \rangle_{\text{data}}$ and $\langle v_i h_j \rangle_{\text{model}}$ at thermal equilibrium, we could also just consider a few relaxation steps. This method is called *contrastive divergence* [184] and defined by the following update rule:

$$\Delta w_{ij}^{\text{CD}} = \epsilon \left(\langle v_i h_j \rangle_{\text{data}}^0 - \langle v_i h_j \rangle_{\text{model}}^k \right) , \tag{9.16}$$

where the superscript index k indicates the number of updates.

Restricted Boltzmann machines have been applied to study phase transitions [186, 187] and represent wave functions [188, 189] and other physical features [190]. In Figure 9.8, we show the temperature dependence of the magnetization of RBM-generated Ising samples. Three snapshots of M(RT)^2 and corresponding RBM-generated Ising configurations with 32 × 32 spins are shown in Figure 9.9. The RBM has been trained with 20×10^4 realizations of Ising configurations at different temperatures [185].

Additional Information: Gradient Descent in RBMs

To derive eq. (9.11) by minimizing the KL divergence $G(P, P')$, we proceed according to Ref. [176]. We first note that the environment (or clamped) distribution $P(\{v\})$ is independent of w_{ij} and thus

$$\frac{\partial G}{\partial w_{ij}} = - \sum_{\{v\}} \frac{P(\{v\})}{P'(\{v\})} \frac{\partial P'(\{v\})}{\partial w_{ij}} . \tag{9.17}$$

Next, we have to calculate the gradient $\partial P'(\{v\}) / \partial w_{ij}$. For a freely running network, the probability distribution of the visible units at equilibrium is

Additional Information: Gradient Descent in RBMs (*cont.*)

$$P'(\{v\}) = \sum_{\{h\}} P'(\{v\},\{h\}) = \frac{\sum_{\{h\}} e^{-E(\{v,h\})/T}}{\sum_{\{v,h\}} e^{-E(\{v,h\})/T}}, \tag{9.18}$$

with the energy

$$E(\{v,h\}) = -\frac{1}{2}\sum_{i,j} w_{ij} x_i^{\{v,h\}} x_j^{\{v,h\}}, \tag{9.19}$$

where $x_i^{\{v,h\}}$ is the state of neuron i for a system in state $\{v,h\}$. Using

$$\frac{\partial e^{-E(\{v,h\})/T}}{\partial w_{ij}} = \frac{1}{T} x_i^{\{v,h\}} x_j^{\{v,h\}} e^{-E(\{v,h\})/T} \tag{9.20}$$

leads to

$$\begin{aligned}\frac{\partial P'(\{v\})}{\partial w_{ij}} &= \frac{\frac{1}{T}\sum_{\{h\}} x_i^{\{v,h\}} x_j^{\{v,h\}} e^{-E(\{v,h\})/T}}{\sum_{\{v,h\}} e^{-E(\{v,h\})/T}} - \frac{\sum_{\{h\}} e^{-E(\{v,h\})/T}\frac{1}{T}\sum_{\{v,h\}} x_i^{\{v,h\}} x_j^{\{v,h\}} e^{-E(\{v,h\})/T}}{\left(\sum_{\{v,h\}} e^{-E(\{v,h\})/T}\right)^2} \\ &= \frac{1}{T}\left[\sum_{\{h\}} x_i^{\{v,h\}} x_j^{\{v,h\}} P'(\{v\},\{h\}) - P'(\{v\})\sum_{\{v,h\}} x_i^{\{v,h\}} x_j^{\{v,h\}} P'(\{v\},\{h\})\right].\end{aligned} \tag{9.21}$$

We now plug this result into eq. (9.17) and find

$$\begin{aligned}\frac{\partial G}{\partial w_{ij}} &= -\sum_{\{v\}} \frac{P(\{v\})}{P'(\{v\})}\frac{1}{T}\left[\sum_{\{h\}} x_i^{\{v,h\}} x_j^{\{v,h\}} P'(\{v\},\{h\}) - P'(\{v\})\sum_{\{v,h\}} x_i^{\{v,h\}} x_j^{\{v,h\}} P'(\{v\},\{h\})\right] \\ &= -\frac{1}{T}\left[\sum_{\{v,h\}} x_i^{\{v,h\}} x_j^{\{v,h\}} P(\{v\},\{h\}) - \sum_{\{v,h\}} x_i^{\{v,h\}} x_j^{\{v,h\}} P'(\{v\},\{h\})\right],\end{aligned} \tag{9.22}$$

where we used that $\sum_{\{v\}} P(\{v\}) = 1$ and $\frac{P(\{v\})}{P'(\{v\})} P'(\{v\},\{h\}) = P(\{v\},\{h\})$. This equation follows from the definition of joint probability distributions

$$P(\{v\},\{h\}) = P(\{h\}|\{v\})\,P(\{v\}) \quad \text{and} \quad P'(\{v\},\{h\}) = P'(\{h\}|\{v\})\,P'(\{v\}) \tag{9.23}$$

and $P(\{h\}|\{v\}) = P'(\{h\}|\{v\})$. The distributions $P(\{h\}|\{v\})$ and $P'(\{h\}|\{v\})$ are identical, because in an equilibrated system the probability of observing a certain hidden state h given a visible one v is independent of the origin of v. That is, for the conditional probability it is irrelevant if v is given by the environment (clamped) or by a freely running machine. With the definitions

$$p_{ij} := \sum_{\{v,h\}} x_i^{\{v,h\}} x_j^{\{v,h\}} P(\{v\},\{h\}) \quad \text{and} \quad p'_{ij} := \sum_{\{v,h\}} x_i^{\{v,h\}} x_j^{\{v,h\}} P'(\{v\},\{h\}), \tag{9.24}$$

we finally obtain eq. (9.11).

Exercise: Learning the Ising Model with RBMs

In this exercise, we are going to use an RBM to generate two-dimensional Ising configurations with $L = 32$ at a certain temperature T. Therefore, we choose the number of visible nodes to be $|V| = 32 \times 32$.
Before samples can be drawn, the machine has to be *trained*. By training we mean updating the weights and biases according to some training data. First, we have to generate about 5000 two-dimensional Ising configurations on a square lattice of linear size $L = 32$ using the M(RT)^2 algorithm. We then train the machine via *contrastive divergence*. The update rule for the weights is given by

$$w_{ij} \to w_{ij} - \epsilon\Big(\langle v_i h_j\rangle^0_{\text{data}} - \langle v_i h_j\rangle^k_{model}\Big) ,$$

where ϵ is the *learning rate*. For completeness, the update rules for the biases (visible and hidden) are given by

$$b_i \to b_i - \epsilon\Big(\langle v_i\rangle^0_{\text{data}} - \langle v_i\rangle^k_{\text{model}}\Big) ,$$
$$c_j \to c_j - \epsilon\Big(\langle h_j\rangle^0_{\text{data}} - \langle h_j\rangle^k_{\text{model}}\Big) .$$

The expectation values are understood to be averages over the whole set of training data. Note that this procedure is in general very slow, because the averages are always computed over the whole training data. Instead of performing this kind of optimization for the weights and biases, it is beneficial to divide the whole set of training data into subsets and compute the averages only over these subsets. The quantity $(v_i h_j)_{\text{data}}$ is computed by taking a vector $\boldsymbol{v}$ from the training data and determining the corresponding vector $\boldsymbol{h}$ as described above. For the quantity $(v_i h_j)^k_{\text{model}}$ one has to take a vector from the training data, compute the corresponding vector $\boldsymbol{h}$, compute the new $\boldsymbol{v}$, and perform k more back-and-forth operations.
Use your training data to find the optimal weights and biases for your RBM.
Once the machine is trained, it can be used to generate new samples. We do this in the following way:

1. Set the nodes in the visible layer to random values (either 0 or 1).
2. Let the machine evolve. That is, go several times back and forth between the visible and hidden layer.
3. Read out the nodes in the visible layer and store this RBM-based Ising sample.

Use the RBM to obtain new Ising configurations for a certain temperature and store these samples in a separate file. You can examine the quality of the RBM samples by computing their magnetization, energy, and correlation function.

9.3 Simulated Annealing

In Sections 9.1 and 9.2, we discussed how Hopfield networks and Boltzmann machines can be used to learn distributions and generate corresponding samples. The underlying optimization problem (i.e., approximating a distribution based on a given set of samples) was solved with tailored learning rules. For the approximate solution of general optimization problems, we may use simulated annealing, or SA for short,

Figure 9.10 Scott Kirkpatrick is a computer scientist and professor at the Hebrew University, Jerusalem.

which is a stochastic optimization technique that was introduced by Scott Kirkpatrick (see Figure 9.10) and collaborators in 1983 [191]. Consider, for instance, a given set S of possible solutions and a cost function $F : S \to \mathbb{R}$. We seek its global minimum:

$$s^* \in S^* := \{s \in S : F(s) \leq F(t)\ \forall t \in S\}.$$

Finding the solution will become increasingly difficult for large S, particularly when, for instance, $|S| = n!$. The name of this optimization technique stems from annealing in metallurgy (a technique that involves heating and controlled cooling of a material to increase the size of its crystals and reduce their defects).

In each step of the SA algorithm, we replace the current solution by a random "nearby" solution that is chosen and accepted with a given probability which depends on the difference between the corresponding values of the cost function and a global parameter T (temperature) which is gradually decreased during the process. The current solution changes almost randomly for high temperatures but comes closer to the minimum as we let T approach zero. We permit nonetheless "uphill" movements in order to avoid that the algorithm gets stuck in a local minimum (this will all be explained in the following example).

Traveling Salesman

An optimization task that can be tackled with simulated annealing is the so-called traveling salesman problem. To illustrate this problem, we consider n cities σ_i and denote the traveling cost from city σ_i to city σ_j by $c(\sigma_i, \sigma_j)$.

We now look for the cheapest trip through all these cities in order to minimize our costs. The set S of solutions is

$$S = \{\text{permutations of } \{1, \ldots, n\}\} \tag{9.25}$$

and the cost function (for a given order, characterized by the permutation π) is

$$F(\pi) = \sum_{i=1}^{n-1} c(\sigma_{\pi_i}, \sigma_{\pi_{i+1}}) \quad \text{for} \quad \pi \in S\,. \tag{9.26}$$

A quick word about permutations: Let us say we have eight cities; then one possible permutation (the trivial one) is $\pi_1 = \{1, 2, 3, 4, 5, 6, 7, 8\}$, and a second one may be $\pi_2 = \{1, 8, 3, 4, 5, 6, 7, 2\}$ (numbers describe the order of the cities on the trip). For the second permutation (see Figure 9.11), the first three summands in $F(\pi_2)$ will be $c(\sigma_1, \sigma_8) + c(\sigma_8, \sigma_3) + c(\sigma_3, \sigma_4)$. Finding the best trajectory is a NP-complex problem. That is, the time that is necessary to solve the problem grows faster than any polynomial of degree n.

We can now apply the simulated annealing algorithm to find the order of cities that minimizes the cost function $F(\sigma)$, which is here the total length of the path. We start from a certain initial configuration of cities σ, generate a permutation σ', and apply the following procedure:

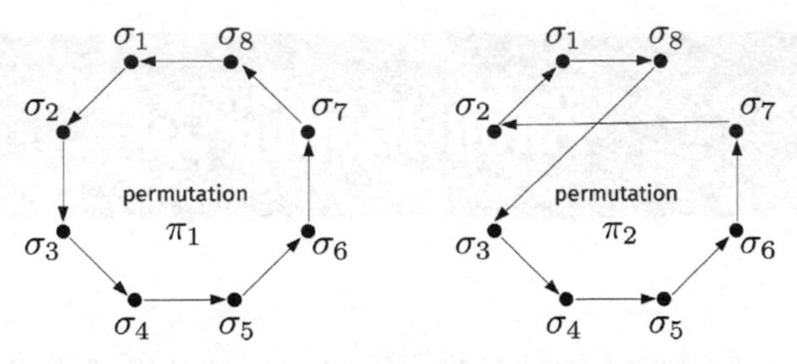

Figure 9.11 Two possible permutations of cities in the traveling salesman problem.

Simulated Annealing

- If $F(\sigma') \leq F(\sigma)$: replace σ by σ'.
- If $F(\sigma') > F(\sigma)$: replace σ by σ' with probability $\exp(-\Delta F/T)$ (where $\Delta F = F(\sigma') - F(\sigma) > 0$).

The T appearing in the probability is an artificial parameter of SA (temperature). In the course of the algorithm we slowly let T go to zero in order to find the global minimum.

Exercise: Traveling Salesman

Apply the simulated annealing algorithm to find paths that minimize the cost function of the traveling salesman problem.

Consider a configuration of n cities (e. g., $n = 10$) and assign corresponding costs $c(\sigma_i, \sigma_j)$ between city σ_i and σ_j. You can use uniformly distributed random numbers as cost values.

After initializing the sets of cities and costs, start from an initial permutation of cities π_1 and apply the simulated annealing algorithm to find permutations of cities that minimize the total cost.

Please consider the following points:

- It might be helpful to try out different ways (slower/faster) of decreasing the temperature T in your simulated annealing implementation.
- You can plot the time until you reach a certain total cost value against T to determine temperatures that are useful for your specific optimization task.

10 Parallelization

After having discussed different algorithms for the simulation of equilibrium systems, we now discuss parallelization techniques that can help making the necessary computations faster. Computation times of parallel codes can be orders magnitudes faster compared to purely serial simulations! The easiest way to parallelize and obtain better statistics is *farming* (i.e., executing the same program on different computers or processors with different initial conditions). This technique is useful for Ising Monte Carlo and related simulations for which one has to repeat the exact same sequence of operations as often as possible. Farming is therefore just a method to improve statistics by generating more samples than one could obtain from a single machine. More sophisticated parallelization techniques are presented in the subsequent sections.

10.1 Multispin Coding

The idea behind *multispin coding* [192] is based on the idea to efficiently use all available bits (e.g., 64 bits if you are using a 64-bit system) for computations. Multispin coding is useful when one deals with only a few integer variables. For the Ising model, we are simulating interactions of spins $\sigma_i \in \{\pm 1\}$. Storing such spin values in 64 bit integers is not only a waste of memory but also leads to a waste of computation time since most of the bits are not carrying any information. On a cubic lattice, the local field $h_i \in \{0, \pm 2, \pm 4, \pm 6\}$ exhibits only seven distinct values. In binary representation, three bits are sufficient to store up to eight different numbers. Therefore, three bits would be enough to represent all possible local field and energy values of any site σ_i while 61 bits remain unused. Spin configurations can be also stored more efficiently using the sequences (000) and (001) which correspond to spin down and spin up, respectively. Applying the XOR function ($\oplus$) to two spin sequences in the aforementioned representation yields (000) whenever two spins are parallel and (001) for antiparallel spins. For the truth table of the XOR function, see Table 10.1.

We can now store 21 spin values $\sigma_i, \ldots, \sigma_{i+20}$ in a single 64-bit integer word:

$$N = (\delta_1, \underbrace{\delta_2, \delta_3, \delta_4}_{\sigma_{i+20}}, \ldots, \underbrace{\delta_{62}, \delta_{63}, \delta_{64}}_{\sigma_i}), \tag{10.1}$$

where $\delta_{(\cdot)} \in \{0, 1\}$. Thus, 63 bits of each integer are used and only one bit remains unused. Similarly, we also store the six neighbors of each spin in N in integer words

Table 10.1 XOR truth table

a	b	$a \oplus b$
0	0	0
0	1	1
1	0	1
1	1	0

$K_1, \ldots, K_6$. No spins in N shall be neighbors to avoid memory conflicts and one has to be very careful with the organization of storing spins. The flipping probabilities for all 21 spins are determined by the energy

$$E = N \oplus K_1 + N \oplus K_2 + N \oplus K_3 + N \oplus K_4 + N \oplus K_5 + N \oplus K_6 . \qquad (10.2)$$

Here $\oplus$ is the bitwise XOR, which realizes the XOR function for each bit in parallel, executing 64 logical operations simultaneously. To determine the flipping probabilities, each value has to be extracted and compared with a random number. We apply the mask $7 = (0, \ldots, 0, 1, 1, 1)$ and the bitwise AND (&) function to eq. (10.1) and illustrate how to extract information of one lattice site:

$$7\&N_j = (0, \ldots, 0, 1, 1, 1) \wedge (\delta_1, \ldots, \delta_{62}, \delta_{63}, \delta_{64}) = (0, \ldots, 0, \tilde{\delta}_{62}, \tilde{\delta}_{63}, \tilde{\delta}_{64}) . \qquad (10.3)$$

Here $\tilde{\delta}_{(\cdot)}$ are the last three bit values of the integer word defined by eq. (10.1). With the bitwise shift operator to the right or to the left we can also access the remaining bits of eq. (10.1). A possible implementation of Monte Carlo updates of the Ising model using the multispin coding algorithm would be the following:

```
cw=0;
for(i=1;i<=21;i++)
{
        z=ranf();
        if(z<P(E&7)) cw=(cw|1);
        cw=ror(cw,3);
        E=ror(E,3);
}
cw=ror(cw,1);
N=(N^cw);
```

In the above pseudocode, we use the abbreviation "cw" to denote the "changer word," which is 1 if the spin is flipped and 0 if the spin is not flipped. Furthermore, we use "&," "|," and "∧" to denote bitwise AND, OR, and XOR operations, respectively. The circular-right-shift operation is abbreviated by "ror." With multispin coding, we can update 21 sites simultaneously and reduce the memory requirement by a factor 21.

According to Ref. [193], "*multi spin coding makes a program more complicated and error-prone but may save a lot of memory and computer time for large systems. The more complicated the interaction is, the less useful is multispin coding, and for continuous degrees of freedom it does not seem to work at all.*"

10.2 Vectorization

Vectorization is a technique which allows to perform multiple operations at the same time like on a car assembly line. It only works in the innermost loops of a program. As an example, we consider the following loop:

```
i_max = 10000;
for(i=1; i<=i_max; i++)
{
        A(i) = B(i) * (C(i) + D(i));
}
```

With the help of vectorizing compilers, this loop can be made more efficient for simulations by executing multiple operations simultaneously and the longer the loop the better. In the case of more complicated routines, this may have to be implemented by hand. The following situations are problematic for vectorization:

- multiple short loops,
- conditional branchings like if-statements,
- indirect addressing.

For optimal performance, the instructions must be repetitive without interruptions or exceptions such as "if" statements. However, there are ways to handle some of such cases. An example would be replacing

```
if(P(i)>z)
{
        s = -s;
}
```

by

```
s = s*sign(z-P(i));
```

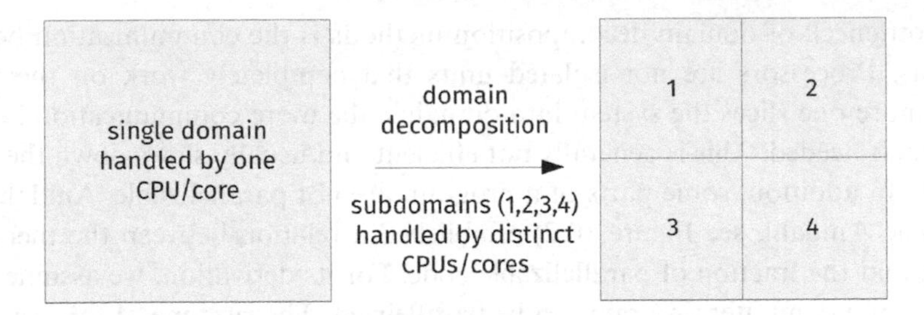

Figure 10.1 Domain decomposition procedure.

Moreover, one has to make sure that the inner loop is the longest and loops that cannot be vectorized should be split up. It is also recommended to make use of vectorized random number generators. Today most compilers automatically try to vectorize programs in order to fill the pipelines of the processors.

10.3 Domain Decomposition

If one needs to simulate very large systems with local interactions, one can use domain decomposition methods on several processors (see Figure 10.1). Instead of using one CPU/core to simulate a certain dynamics (e. g., spin interactions) on the whole domain, one can decompose the simulation domain into subdomains and parallelize the computations. This method has the advantage that a large proportion of interactions can be treated in parallel. Communication between processing units is only necessary to appropriately describe interactions between boundary regions. Monte Carlo simulations of systems with local interactions on regular lattices are well suited for parallelization with domain decomposition.

Depending on the given domain, certain decompositions may have an advantage over others. In general, one has to find a good trade-off between the speed-up gain resulting from a larger number of subdomains and the corresponding communication between CPUs that may slow down the simulations substantially. It is also possible to dynamically change the domain sizes and interface positions depending on the load that each processor is just handling (*dynamic load sharing*).

To parallelize a routine one has to use programming languages created specifically for such tasks or embed special libraries in which the parallelization has been implemented in such a way that it can be summoned in standard programming languages such as C++. For programs that are intended to run on a graphics processing unit (GPU), one can use CUDA (Compute Unified Device Architecture), a parallel computing platform and programming model created by NVIDIA and implemented in their GPUs. Moreover, MPI (Message Passing Interface) is a standardized and portable message-passing system. This means that instructions can be passed between processors within programs written in languages such as Java or C++.

Figure 10.2 Gene M. Amdahl (1922–2015) contributed to the development of mainframe computers at IBM and later founded the company Amdahl Corporation.

The bottleneck of domain decomposition methods is the communication between processors. Processors are not isolated units that completely work on their own, and the more one slices the system into domains, the more communication between processors is needed. This is generally not efficient and usually slows down the parallelization. In addition, some parts of a program are not parallelizable. Amdahl's law (after Gene Amdahl, see Figure 10.2) established a relation between the theoretical speed-up and the fraction of parallelizable code. For its derivation, we assume that a fraction p of a computer program can be parallelized. The runtime of this fraction p will be distributed among n processors, whereas the runtime of the non-parallelizable part of the program stays the same. The total runtime is thus

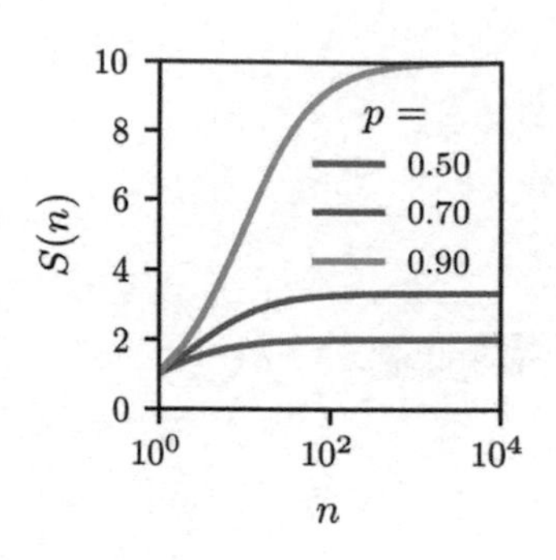

Figure 10.3 Amdahl's law (see eq. (10.5)) for different fractions of parallelizable code p and numbers of processors n.

$$T(n) = (1 - p)T_0 + \frac{p}{n}T_0, \tag{10.4}$$

where T_0 is the runtime before parallelization. The resulting speedup is simply

$$S(n) = \frac{T_0}{T(n)} = \frac{1}{(1 - p) + \frac{p}{n}}. \tag{10.5}$$

We show some characteristic speed-up curves in Figure 10.3.

Nonequilibrium Systems 11

11.1 Directed Percolation and Gillespie Algorithms

Systems that approach thermal equilibrium obey reversibility. That is, their underlying probability flows cancel out each other. The Ising model is one important example of an equilibrium model in statistical physics. In Section 6.2, we outlined that the Ising model is closely related to isotropic percolation (see Section 2.1), which consequently can be also considered an equilibrium model. In addition to isotropic percolation, one can define directed percolation, where sites can be only occupied (made "active") along a certain direction. The dynamics thus has an irreversible character. For this reason, directed percolation is a nonequilibrium process with universal properties different from those of isotropic percolation [121, 194]. Both processes are very fundamental in the sense that they are simple to define, and include features that are observed in a variety of contexts. For instance, models that describe certain hadronic interactions [195] and the spreading of opinions [196–200] or diseases [201–206] have been mapped to nonequilibrium (directed percolation) processes [19, 121]. Isotropic percolation (see Section 2.1) is an equilibrium process in which a site or bond is occupied with a certain probability p and thus no time is needed to make a configuration – all sites can be occupied in parallel. In the case of directed percolation, an additional constraint is defined, namely, at each time step occupied sites or bonds occupy their neighbors with probability p only along a given direction [207]. Time is therefore a relevant parameter in directed-percolation simulations. Both models, directed and isotropic percolation, have in common that there exists a critical probability p_c at which a percolating phase occurs. We illustrate this behavior in Figure 11.1.

In the vicinity of p_c, the percolation order parameter $P(p)$, the density of active sites (directed percolation), takes the form $P(p) \propto (p - p_c)^{\beta}$. In directed percolation and related nonequilibrium systems, we define an absorbing phase ($p \leq p_c$) in which the "spreading" stops and a fluctuating phase ($p > p_c$) that is characterized by $P(p) > 0$.

Although the free energy is not defined for nonequilibrium processes, we can distinguish between first-order and continuous phase transitions based on the behavior of the order parameter (see Section 3.1.7).

Another formulation of directed percolation is given by the *contact process* (CP). On a d-dimensional lattice with active ($s_i(t) = 1$) and inactive ($s_i(t) = 0$) sites, we sequentially update the system according to the following dynamics: Let the number of active neighbors neighbors be $n_i(t) = \sum_{\langle i,j \rangle} s_j(t)$. A new value $s_i(t + \mathrm{d}t) \in \{0, 1\}$ is obtained according to the transition rates.

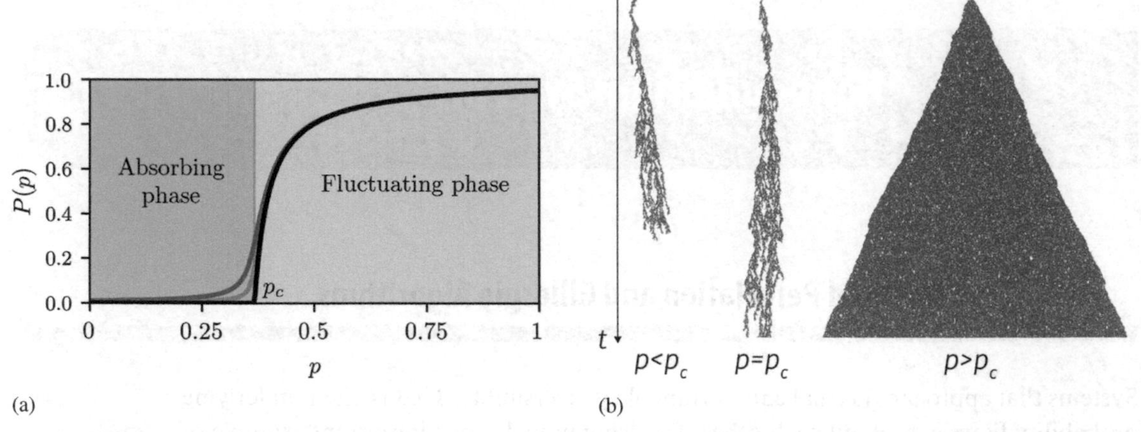

Figure 11.1 The order parameter of the directed-percolation second-order phase transition (a). The gray lines illustrate the effect of a field-like term. Three clusters of directed bond percolation on a tilted square lattice (i. e., the square lattice is tilted by 45 degrees, such that all bonds are equivalent) with different occupation probabilities p (b). At around $p_c = 0.6447$, a second-order phase transition occurs [121].

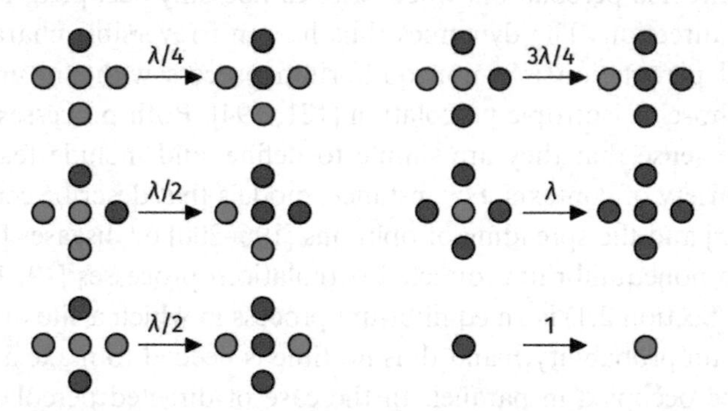

Figure 11.2 Update rules of the contact process on a two-dimensional lattice (spatial rotation invariance of the reactions is assumed). Red (green) nodes represent active (inactive) states.

$$w[0 \to 1, n_i] = (\lambda n_i)/(2d) \quad \text{and} \quad w[1 \to 0, n_i] = 1\,, \tag{11.1}$$

where λ is the spreading rate.

We show an illustration of the CP updates in Figure 11.2. If $\lambda \leq \lambda_c$, the "epidemic" will die out. However, for $\lambda > \lambda_c$, we observe an epidemic (or fluctuating) phase. That is, we observe a mean density of active sites

$$n(t) = \sum_i s_i(t)\,, \tag{11.2}$$

which is larger than zero for $\lambda > \lambda_c$. Starting with every site active at λ_c, the mean density of active sites scales as $n(t) \propto t^{-\delta}$ (see Figure 11.3). In Table 11.1, we summarize the critical exponents β and δ in different dimensions, and the critical values λ_c that are observed in CP simulations on d-dimensional lattices. Note that the contact process is also often referred to as (asynchronous) *susceptible-infected-susceptible* (SIS) model.

In the SIS model (see Figure 11.4), susceptible nodes become infected and infected nodes recover as for the contact process with rates as defined by eq. (11.1). The

Table 11.1 The critical infection rate λ_c and the critical exponents β and δ of the contact process on a square lattice for different dimensions d

	(1+1)D	(2+1)D	(3+1)D	MF
λ_c	3.2978(2)	1.64872(3)	1.31683(2)	–
β	0.2767(3)	0.580(4)	0.818(4)	1
δ	0.15944(2)	0.4510(4)	0.7398(10)	1

Note: At and above the critical dimension $d_c = 4$, the critical exponents are described by mean-field (MF) theory [121]. The values of λ_c are taken from Ref. [208], and the values of β and δ are taken from Ref. [209]. Note that λ_c is dimension dependent, even in the MF case [208].

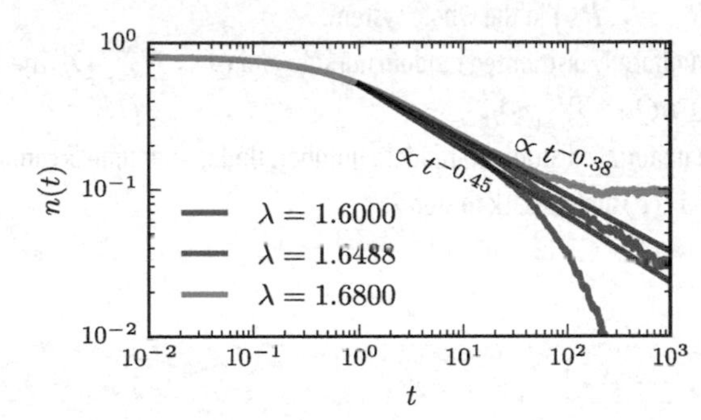

Figure 11.3 We show the relaxation of the density $n(t)$. For $\lambda = 1.6488 \approx \lambda_c$, the density decays as $n(t) \propto t^{-\delta}$ with $\delta = 0.4510(4)$ [209]. The upper guide-to-the-eye (black solid line) has slope $\propto t^{-0.38}$, and the one below has slope $\propto t^{-0.45}$.

SIS model is used to model epidemic diseases without (long-lasting) immunization. Examples include bacterial infections, common cold, and influenza.

The contact process has the same critical exponents as directed percolation. This implies that they both belong to the same universality class. According to a conjecture by Grassberger [210] and Janssen [211], the directed percolation universality class requires [121]:

1. a continuous phase transition from a fluctuating phase to a unique absorbing state,
2. the transition having a one-component order parameter,
3. local process dynamics,
4. no additional attributes, such as conservation laws or special symmetries.

Nonequilibrium systems are, in general, not described by a Hamiltonian and we cannot use the same algorithms that we employed for simulating Ising-like systems or other equilibrium models. Time is now a physical parameter and not a mere virtual property of a Markov chain.

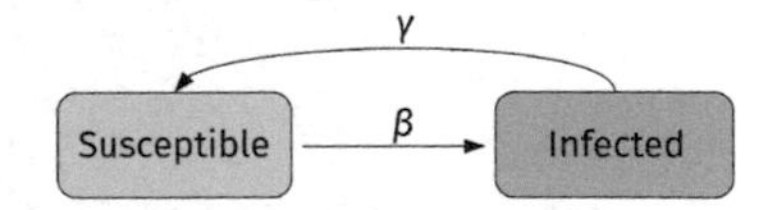

Figure 11.4 Schematic of the suscpetible-infected-suscpetible (SIS) model. Susceptible individuals become infected at rate β and recover at rate γ. The use of β and γ as transition rate symbols is common in the epidemic-modeling literature. Note that the transition rate β should not be confused with a critical exponent. In terms of a contact process description and according to eq. (11.1), we associate β with $\lambda n_i/(2d)$ and set $\gamma = 1$.

A standard algorithm for simulating nonequilibrium dynamics is *kinetic Monte Carlo* that is also known under the name *Gillespie algorithm* (after Daniel T. Gillespie [212, 213]).

As an example, we apply the kinetic Monte Carlo algorithm to the contact process on a square lattice. According to eq. (11.1), recovery and infection define $N = 2$ processes with corresponding rates $R_1 = 1$ and $R_2 = \lambda n_i/4$. At time t, the total recovery rate is given by $Q_1(t) = \sum_i s_i(t)$. On a square lattice only nearest-neighbor interactions are considered and the total rate of the second process (infection) is $Q_2(t) = \sum_i \lambda n_i/4$ (see Figure 11.2).

The following steps summarize a general kinetic Monte Carlo update [201]:[1]

Kinetic Monte Carlo (Gillespie) Algorithm

1. For all N processes, identify the corresponding rates (per node) $\{R_1, R_2, \ldots, R_N\}$.
2. Determine the overall rates (all nodes) $\{Q_1, Q_2, \ldots, Q_N\}$ by summing up all the individual rate contributions $\{R_1, R_2, \ldots, R_N\}$ in the whole system.
3. Let $\eta \in [0, 1)$ be a uniformly distributed random number and $Q = \sum_{i=1}^{N} Q_i$. The process with rate R_j occurs if $\sum_{i=1}^{j-1} Q_i \leq \eta Q < \sum_{i=1}^{j} Q_i$.
4. Let $\epsilon \in [0, 1)$ be a uniformly distributed random number. Update the time according to $t \to t + \Delta t$, where $\Delta t = -Q^{-1} \ln(\epsilon)$ and go back to step 2.

Additional Information: Kinetic Monte Carlo (Gillespie) Time Step and Process Selection

The kinetic Monte Carlo (Gillespie) algorithm uses a specific time update schema $t \to t + \Delta t$, where $\Delta t = -Q^{-1} \ln(1 - \epsilon)$ and $\epsilon \in [0, 1)$. The attentive reader may realize that this update rule is related to the discussion of transforming a uniform distribution to an exponential distribution (see Section 1.6). To better understand the origin of the specific form of the time step, we now present a derivation that is based on Refs. [212, 213].

Our goal is to transform a given uniformly distributed random number to a random number that is distributed according to a distribution which determines the time evolution of the underlying model. We use $P(\tau, i)\,\mathrm{d}\tau$ to denote the probability that the next process with rate Q_i will occur in the time interval $(t+\tau, t+\tau+\mathrm{d}\tau)$. We rewrite the reaction probability density function $P(\tau, i)$ according to

$$P(\tau, i)\,\mathrm{d}\tau = P_0(\tau) Q_i\,\mathrm{d}\tau\,, \tag{11.3}$$

where $P_0(\tau)$ is the probability that no reaction occurred in $(t, t+\tau)$, and $Q_i\,\mathrm{d}\tau$ is the probability that a Q_i process will occur in $(t+\tau, t+\tau+\mathrm{d}\tau)$. To determine the functional form of $P_0(\tau)$, we use

[1] To guarantee efficient simulations, one usually stores the site identifiers/numbers that belong to a certain reaction in corresponding lists. After each update, one should only update the list entries of the sites that were involved in a certain reaction and not recompute all lists.

Additional Information: Kinetic Monte Carlo (Gillespie) Time Step and Process Selection *(cont.)*

$$P_0(\tau' + d\tau') = P_0(\tau')(1 - Q d\tau'), \tag{11.4}$$

where $(1 - Q\,d\tau')$ is the probability that no process occurs in $d\tau'$. For sufficiently small $d\tau'$, the solution of eq. (11.4) is

$$P_0(\tau) = \exp(-Q\tau). \tag{11.5}$$

Inserting eq. (11.5) into eq. (11.3) yields

$$P(\tau, i) = Q_i \exp(-Q\tau). \tag{11.6}$$

We now use $P(\tau, i) = P(i|\tau)P(\tau)$, where $P(\tau)$ is the probability that a process occurs in $(t + \tau, t + \tau + d\tau)$ and $P(i|\tau)$ is the corresponding conditional probability that the process with rate Q_i is selected. Moreover, we obtain $P(\tau) = \sum_i P(\tau, i) = Q\exp(-Q\tau)$ and consequently $P(i|\tau) = Q_i/Q$. Finally, we apply the transformation method (see Section 1.6) to $P(\tau)$ and find

$$\epsilon = \int_0^{\Delta t} P(\tau)\, d\tau = 1 - \exp(-Q\Delta t), \tag{11.7}$$

where ϵ is a uniformly distributed random number between 0 and 1. Inverting eq. (11.7) yields the time step $\Delta t = -Q^{-1}\ln(1 - \epsilon) = -Q^{-1}ln(\epsilon)$ since $\epsilon \in [0, 1)$.

The outlined algorithm is based on the assumption that the underlying dynamics is governed by independent Poisson processes. Thus, the corresponding interevent times are exponentially distributed. However, the interevent time distributions of many natural processes are nonexponential. Examples of such processes include earthquakes [214], firing of neurons [215], and social dynamics [216, 217]. Deviations from a Poisson description occur, for instance, because of correlation and memory (non-Markovian) effects [218].

Figure 11.5 Marian Boguñá works at the Universitat de Barcelona.

To be able to simulate processes with nonexponential interevent time distributions, Marian Boguñá (see Figure 11.5) and collaborators introduced the non-Markovian Gillespie algorithm (nMGA) [219]:

Non-Markovian Gillespie Algorithm

1. Let $\psi_i(\tau)$ be the probability density function of interevent times τ of process $i \in \{1, 2, \ldots, N\}$) and $\Psi_i(\tau)$ the corresponding survival probability.
2. Initialize the elapsed times $\{t_j\}$ ($j \in \{1, 2, \ldots, N\}$) for all processes (e.g., $t_j = 0$).
3. Determine the time until the next event, Δt, according to $\Phi(\Delta t|\{t_j\}) = \epsilon$, where $\epsilon \in [0, 1)$ is a uniformly distributed random number and $\Phi(\Delta t|\{t_j\}) = \prod_{j=1}^{n} \frac{\Psi_j(t_j+\Delta t)}{\Psi_j(t_j)}$.
4. Let $\eta \sim [0, 1)$ be a uniformly distributed random number and $Q(\Delta t|\{t_j\}) = \sum_{j=1}^{N} Q_j(t_j + \Delta t)$ with $Q_j(t_j + \Delta t) = \psi_j(t_j + \Delta t)/\Psi_j(t_j + \Delta t)$. Process i occurs if $\sum_{i=1}^{j-1} Q_i(t_i + \Delta t) \leq \eta Q(\Delta t|\{t_j\}) < \sum_{i=1}^{j} Q_i(t_i + \Delta t)$.

Non-Markovian Gillespie Algorithm (*cont.*)

5. Update the times $\{t_j\}$ according to

$$t_j \to t_j + \Delta t \,\forall j \neq i \quad \text{and} \quad t_i = 0\,. \tag{11.8}$$

6. Go back to step 2.

We describe the details of this algorithm in the gray box below. Potential drawbacks of the nMGA are that it (i) is exact only for a sufficiently large number of processes N and (ii) requires one to update the rates $Q_j(t_j)$ of *all* processes after the occurrence of an event. Algorithms that overcome these problems are the shortest-path kinetic Monte Carlo algorithm [220] and the Laplace Gillespie algorithm [221].

Additional Information: Non-Markovian Gillespie Algorithm

Let $\psi_i(\tau)$ be the probability density function of interevent times τ of process $i \in \{1, 2, \ldots, N\}$). Furthermore, let t_i be the time elapsed since the last occurrence of process i. The waiting time τ until the next event is distributed according to

$$\psi_i(\tau|t_i) = \frac{\psi_i(\tau + t_i)}{\Psi_i(t_i)}\,, \tag{11.9}$$

where $\Psi_i(t_i) = \int_{t_i}^{\infty} \psi_i(\tau')\,\mathrm{d}\tau'$ is the survival probability of process i. That is, $\Psi_i(t_i)$ is the probability that the interevent time is larger than t_i. We have to bear in mind that process i coexists with $N-1$ other processes. The probability density that process i and not another process generates the next event after time Δt is

$$\phi(\Delta t, i|\{t_j\}) = \psi_i(\Delta t|t_i) \prod_{j \neq i} \Psi_j(\Delta t|t_j)\,, \tag{11.10}$$

where

$$\Psi_j(\Delta t|t_j) = \int_{\Delta t}^{\infty} \psi_j(\tau|t_j)\,\mathrm{d}\tau' = \frac{\Psi_j(t_j + \Delta t)}{\Psi_j(t_j)} \tag{11.11}$$

is the conditional survival probability of process j (i.e., the probability that process j occurs with an interevent time larger than Δt, given that process j occurred at time t_j before). We thus obtain

$$\phi(\Delta t, i|\{t_j\}) = \frac{\psi_i(\tau + t_i)}{\Psi_i(t_i)} \Phi(\Delta t|\{t_j\})\,, \tag{11.12}$$

where

$$\Phi(\Delta t|\{t_j\}) = \prod_{j=1}^{n} \frac{\Psi_j(t_j + \Delta t)}{\Psi_j(t_j)} \tag{11.13}$$

is the probability that no event occurs for time Δt. We therefore have to generate ϵ, a uniformly distributed random number between 0 and 1, and solve $\epsilon = \Phi(\Delta t|\{t_j\})$ to determine the time until the next event, Δt. Now we approximate $\Phi(\Delta t|\{t_j\})$ for the limit of a large number of processes N and thus small time steps Δt according to

Additional Information: Non-Markovian Gillespie Algorithm (*cont.*)

$$\begin{aligned}\Phi(\Delta t|\{t_j\}) &= \exp\left[\sum_{j=1}^{N}\ln\left(\frac{\Psi_j(t_j+\Delta t)}{\Psi_j(t_j)}\right)\right] \\ &= \exp\left[\sum_{j=1}^{N}\ln\left(\frac{\Psi_j(t_j)-\psi_j(t_j)\Delta t+\mathcal{O}\left(\Delta t^2\right)}{\Psi_j(t_j)}\right)\right] \\ &\approx \exp\left[-\Delta t\left(\sum_{j=1}^{N}Q_j(t_j)\right)\right] = \exp\left[-\Delta t\, Q(t_j)\right],\end{aligned} \tag{11.14}$$

where we used that $Q_j(t_j) = \psi_j(t_j)/\Psi_j(t_j)$. In contrast to the standard Gillespie algorithm, the rates $Q_j(t_j)$ depend on the times t_j elapsed since the last event. Finally, we solve $\Phi(\Delta t|\{t_j\}) = \epsilon$ to obtain $\Delta t = -Q(t_j)^{-1}\ln(\epsilon)$. In the Markovian case where $Q_j(t_j) = Q_j$, we recover the standard Gillespie algorithm with $\Delta t = -Q^{-1}\ln(\epsilon)$.

Exercise: Kinetic Monte Carlo Simulation of the Contact Process

Write a program to simulate the contact process on a 2D square lattice with the kinetic Monte Carlo method. First, consider a lattice with $N = 128 \times 128$ sites and study the evolution of $n(t)$ at $\lambda = \lambda_c \approx 1.6488$. Can you observe that $n(t) \propto t^{-\delta}$ with $\delta \approx 0.45$? Your results will improve, if you consider larger lattices (e.g., $N = 1024 \times 1024$ sites).

Next, we focus on the stationary behavior for different values of λ. Wait until the system reaches its stationary state $n_{st}(\lambda)$. For $\lambda \leq \lambda_c \approx 1.6488$ and large system sizes, we observe $n_{st} = 0$ for $\lambda < \lambda_c$ and $n_{st} > 0$ for $\lambda > \lambda_c$. After reaching a stationary state, measure $n_{st}(\lambda)$. We have to measure $n_{st}(\lambda)$ a certain number of times to compute the average density $\langle n_{st}(\lambda)\rangle$. Compute $\langle n_{st}(\lambda)\rangle$ for different values of λ, plot $\langle n_{st}(\lambda)\rangle$ against λ, and determine the exponent β.

Consider the following points:

1. Define an initial seed (e.g., a circular infected region or a small fraction of randomly infected nodes).
2. Let the system evolve for sufficiently long times. After a certain number of updates, the system will reach a stationary state or fluctuate around a stationary state. You can monitor the equilibration process by keeping track of the variance within a certain time interval.

11.2 Cellular Automata

Cellular automata (CA) were introduced by Stanislaw Ulam and John von Neumann. A cellular automaton is a model that consists of a grid of cells and each one of the cells can be in a finite number of states (e.g., ± 1). Cellular automata use the configuration of cells at time t to compute the subsequent configuration at time $t+1$ based on deterministic rules. To summarize:

- Cellular automata are defined on a grid.
- The variables of each grid cell have discrete values (e.g., Boolean).
- At each finite time step, we deterministically compute the configuration of cells at the next step $t+1$ based on the configuration of the last time step t.

If we consider a binary variable σ_i at a certain lattice site i, the next time step $t+1$ is determined by

$$\sigma_i(t+1) = f_i(\sigma_1(t), \ldots, \sigma_k(t)), \tag{11.15}$$

where k is the number of inputs that are assumed to be fixed in time for every site. In this example, the binary function f_i determines the value of $\sigma_i(t+1)$ based on the values of $\sigma_1(t), \ldots, \sigma_k(t)$. We see that there are thus 2^{2^k} possible rules! We distinguish between *sequential* and *parallel* updates. For sequential updates, we update one variable σ_i per time step, whereas for parallel updates all variables are updated at once.

In the following sections, we will discuss only a few properties of cellular automata and note that much more information on this topic can be found in Refs. [222, 223].

11.2.1 Classification of Cellular Automata

We consider $k = 3$ (3 inputs) as an example. There are $2^3 = 8$ possible binary entries ($n = 111, 110, 101, 100, 011, 010, 001, 000$) and the rule needs to define an output for each of these entries. Let us consider the following rule:

entries:	111	110	101	100	011	010	001	000
$f(n)$:	0	1	1	0	0	1	0	1

Furthermore, we define

$$c = \sum_{n=0}^{2^k-1} 2^n f(n),$$

which for the mentioned rule yields

$$c = 0 \cdot 2^7 + 1 \cdot 2^6 + 1 \cdot 2^5 + 0 \cdot 2^4 + 0 \cdot 2^3 + 1 \cdot 2^2 + 0 \cdot 2^1 + 1 \cdot 2^0 = 64 + 32 + 4 + 1 = 101\,.$$

We can thus identify a rule by its c number. We consider a few rules to see how this works:

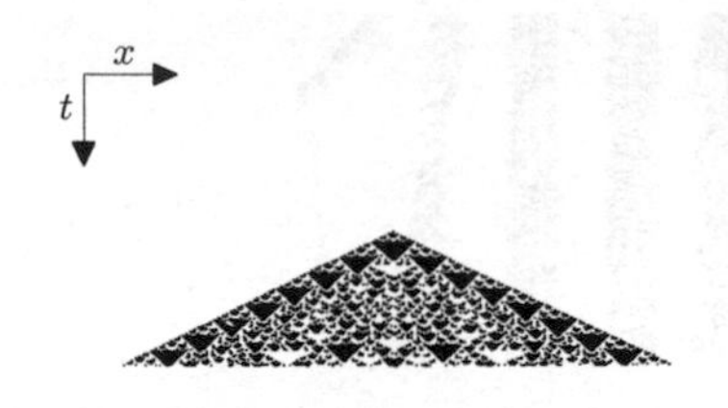

(a) Evolution according to rule 50 for $k = 5$ (class 3).

(b) Evolution following rule 20 for $k = 5$ (class 4).

Figure 11.6 Evolution for two different CA rules. The horizontal describes a one-dimensional chain of cells, time goes from top to bottom, black is 1, and white is 0. The figure is taken from Ref. [222].

Rule number c	Entries n:	111	110	101	100	011	010	001	000
4	$f_4(n)$:	0	0	0	0	0	1	0	0
8	$f_8(n)$:	0	0	0	0	1	0	0	0
20	$f_{20}(n)$:	0	0	0	1	0	1	0	0
28	$f_{28}(n)$:	0	0	0	1	1	1	0	0
90	$f_{90}(n)$:	0	1	0	1	1	0	1	0

11.2.2 Time Evolution

We can study the time evolution of a given rule. The evolution patterns differ significantly from rule to rule. In fact, in some cases special patterns start to appear while in other cases every site ends up in the same state. Two examples starting with a random initial configuration are illustrated in Figure 11.6.

11.2.3 Classes of Automata

Stephen Wolfram (see Figure 11.7) divided the rules into four groups according to their dynamical characteristics [224]:

- Class 1: Initial patterns evolve quickly into a stable, homogeneous state, any trace of the initial patterns disappears.
- Class 2: Almost all initial patterns evolve quickly into stable striped or oscillating structures. Some traces of initial patterns remain. Local changes to the initial pattern tend to remain local.
- Class 3: Nearly all initial patterns evolve in a pseudo-random or chaotic manner. Any stable structures that appear are quickly destroyed by the surrounding noise. Local changes to the initial pattern tend to spread indefinitely.

Figure 11.7 Stephen Wolfram investigated the patterns formed by cellular automata and developed the scientific computing software Mathematica.

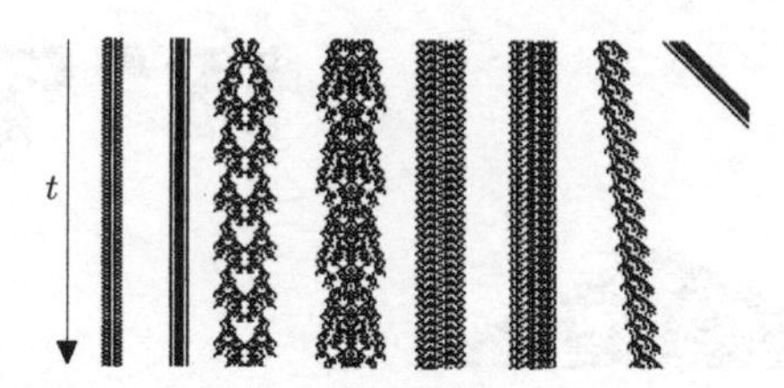

Figure 11.8 Examples of persistent structures of a class 4 (rule 20 for $k = 5$) CA. The figure is taken from Ref. [222].

- Class 4: Nearly all initial patterns evolve into structures that interact in complex and interesting ways. These structures collide and sometimes annihilate each other slowly disappearing and eventually exhibiting class 2 behavior; the time necessary to reach that point is very large and diverges with system size.

We show some persistent structures of a class 4 automaton in Figure 11.8.

11.2.4 The Game of Life

Figure 11.9 John Horton Conway (1937–2020) was a British mathematician.

One example of a class 4 cellular automaton is the Game of Life, which was invented by Conway (see Figure 11.9) in 1970. Let us consider a square lattice, and let n be the number of nearest and next-nearest neighbors that are in state "1." We use the following rule for a certain lattice site:

- if $n < 2$: set the site to 0,
- if $n = 2$: stay as before,
- if $n = 3$: set the site to 1,
- if $n > 3$: set the site to 0.

This rule is known under the name *Game of Life*. Cellular automata that only depend on the number n of 1s in a given neighborhood like the Game of Life are called *totalistic*. During its temporal evolution, the Game of Life keeps producing "gliders" from a "gun" (see Figure 11.10).

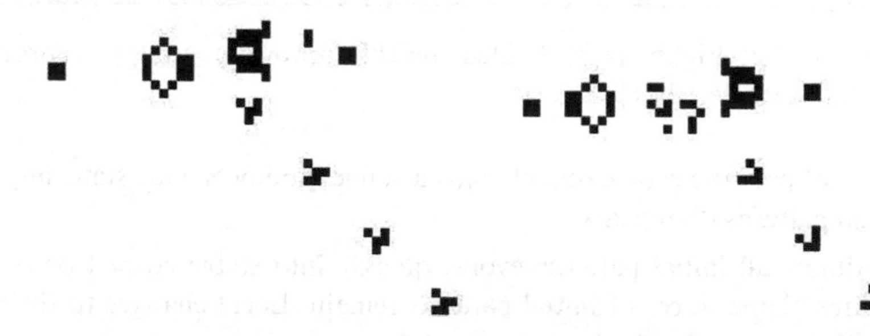

Figure 11.10 Glider gun of ones (black) within zeros (white). Steps 0 and 16 (in a period of 30 time steps) of the Game of Life [225].

Exercise: Game of Life

Simulate Conway's Game of Life (see Section 11.2.4) on a square lattice with periodic boundary conditions. Implement different initial configurations and study their evolution. Can you observe the glider behavior that we show in Figure 11.10?

Tasks

1. How many types of different gliders do you observe in your simulations for different initial conditions?
2. Study the density of 1s for different system sizes as a function of time.

11.2.5 Q2R

For $E_{\max} \to 0$, the Creutz algorithm (see Section 4.8) for microcanonical simulations of the Ising model on the square lattice becomes a cellular automaton called *Q2R* [226]. The update rules on a square lattice for spins $\sigma_{ij} \in \{0, 1\}$ are

$$\sigma_{ij}(\tau + 1) = f(x_{ij}) \oplus \sigma_{ij}(\tau), \tag{11.16}$$

where $\oplus$ is the binary XOR operation and

$$x_{ij} = \sigma_{i-1j} + \sigma_{i+1j} + \sigma_{ij-1} + \sigma_{ij+1} \quad \text{and} \quad f(x) = \begin{cases} 1 & \text{if } x = 2 \\ 0 & \text{if } x \neq 2 \end{cases}. \tag{11.17}$$

For the outlined Q2R dynamics, spins are flipped if and only if the change in energy is zero. This can be implemented in a very efficient way using multi spin coding (see Section 10.1). The update rule of eq. (11.16) can be expressed in a very elegant way using logical functions:

$$\sigma(\tau + 1) = \sigma(\tau) \oplus \{[(\sigma_1 \oplus \sigma_2) \wedge (\sigma_3 \oplus \sigma_4)] \vee [(\sigma_1 \oplus \sigma_3) \wedge (\sigma_2 \oplus \sigma_4)]\}, \tag{11.18}$$

where $\wedge$ and $\vee$ are the logical AND and logical OR operators, respectively. This logical expression can be computed in roughly 12 cycles that last typically around 10 ns for 64 sites in parallel for bitwise logical functions. The method is extremely fast, deterministic, and reversible. It represents the fastest possible simulation for an Ising model [227]. The problem is that it is not ergodic and that it strongly depends on the initial spin configuration.

11.3 Irreversible Growth

For many applications, we cannot assume that the underlying system is in thermal equilibrium. Many natural phenomena are essentially irreversible. Examples include

Figure 11.11 Fereydoon Family (Emory University) (top) and Tamás Vicsek (Eötvös Loránd University) (bottom) introduced the concept of dynamic scaling.

- deposition and aggregation,
- fluid invasion,
- electric breakdown,
- biological morphogenesis,
- fracture and fragmentation.

In the following sections, we will get to know some methods that help us understand irreversible growth. More details can be found in Ref. [228].

11.3.1 Self-Affine Scaling

Let us consider a growing interface, like the front of an invading fluid or the surface of a tumor. We can analyze the dependence of the interface width W (see eq. (4.47)) on the variables L (lattice length) and t (time). The following relation is called Family–Vicsek scaling [230]

$$W(L,t) = L^{\xi} f\left(\frac{t}{L^z}\right), \tag{11.19}$$

where ξ is the roughening exponent and z the dynamic exponent. This scaling relation was introduced by Fereydoon Family and Tamás Vicsek (see Figure 11.11). We can now consider two limits: (i) $t \to \infty$ and (ii) $L \to \infty$. To simplify the expression, we substitute the argument of f by $u = t/L^z$ and find

- $t \to \infty : W \propto L^{\xi} \Longrightarrow f(u \to \infty) = \text{const},$
- $L \to \infty : W \propto t^{\beta} \Longrightarrow f(u \to 0) \propto u^{\beta},$

where we have introduced the so-called growth exponent $\beta = \frac{\xi}{z}$ (see Figure 11.12). We can numerically verify these laws by observing a data collapse.

11.3.2 Random Deposition

The simplest model for irreversible growth models is random deposition. It is based on only one rule:

Random Deposition

- Pick a column uniformly at random and add a particle of unit height to that column (see Figure 11.13).

The height is directly proportional to time (i.e., number of depositions), $h \propto t$ and the width W is proportional to $t^{\frac{1}{2}}$. The roughening exponent ξ and the growth exponent β are both $\frac{1}{2}$, so the Family–Vicsek scaling law (see eq. (11.19)) is

$$W(L,t) = \sqrt{L} f\left(\frac{t}{L}\right). \tag{11.20}$$

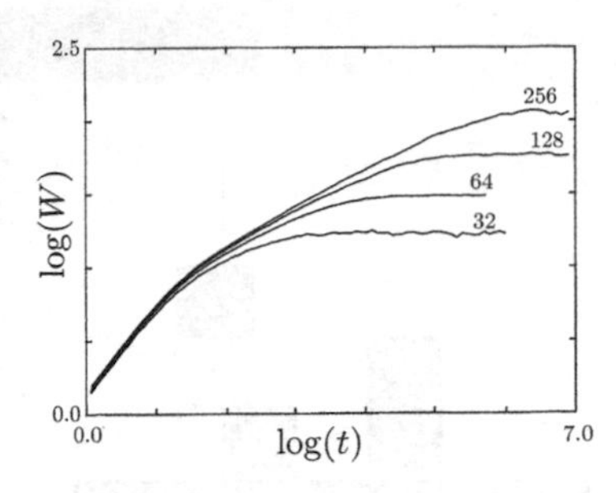

Figure 11.12 Interface width W versus time t for ballistic deposition and four different linear lattice sizes L. The figure is taken from Ref. [229].

We show a typical random-deposition pattern in Figure 11.13. One immediately notices the sharp spikes which are completely independent from one another and that the height varies wildly between columns.

11.3.3 Random Deposition with Surface Diffusion

For many applications, the random deposition model is oversimplified. If we would like to get something more realistic/less spiky, we need to give the particles a bit more freedom. This is done by permitting the particles to move a short distance to find a more stable configuration.

The following procedure accounts for such short-distance movements:

Random Deposition with Surface Diffusion

- Pick a random column i.
- Compare the heights $h(i)$ and $h(i+1)$.
- The particle is added to the lower column. For equal heights, the new particle is added to column i or $i+1$ with equal probability.

We illustrate this procedure (along with a typical growth pattern) in Figure 11.14. Using this growth model, the average height now increases with $\sqrt{t}$ while the width W increases with $t^{\frac{1}{4}}$. The roughening exponent ξ is $\frac{1}{2}$ and the growth exponent β is $\frac{1}{4}$, so the Family–Vicsek scaling law is

$$W(L,t) = \sqrt{L} f\left(\frac{t}{L^2}\right). \tag{11.21}$$

In Figure 11.14, we observe that the surface is much "smoother" and not as jagged anymore as without diffusion.

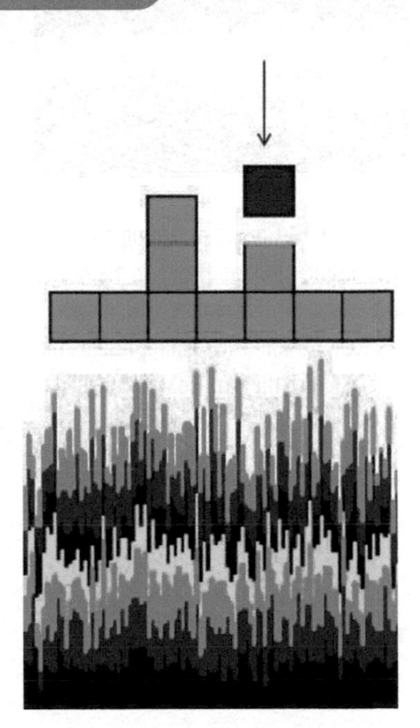

Figure 11.13 Growth by adding one particle and result after adding many particles using the random deposition algorithm. The colors represent the times at which certain particles were added.

11.3.4 Restricted Solid on Solid Model

One can make the surfaces even smoother by requiring that neighboring sites may not have a height difference larger than one unit. The procedure then needs to reflect this restriction with an additional condition (which can lead to the rejection of a particle addition, if it does not meet the condition). The recipe incorporating this idea defines the restricted solid on solid model (RSOS):

Restricted Solid on Solid Model

- Pick a random column i.
- Add a particle if $|h(i+1) - h(i-1)| \leq 1$.
- If the particle could not be placed, pick a different column.

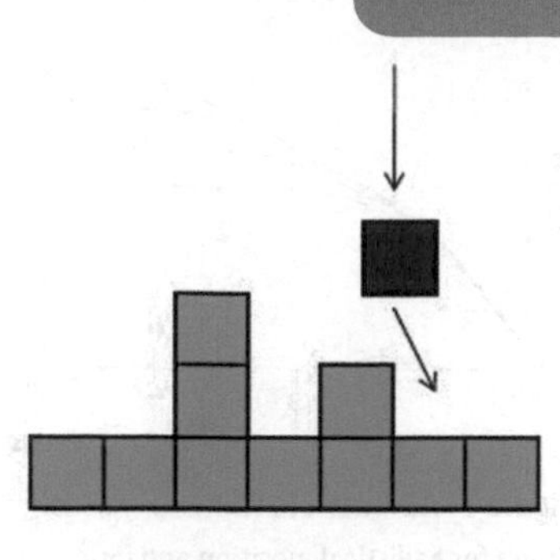

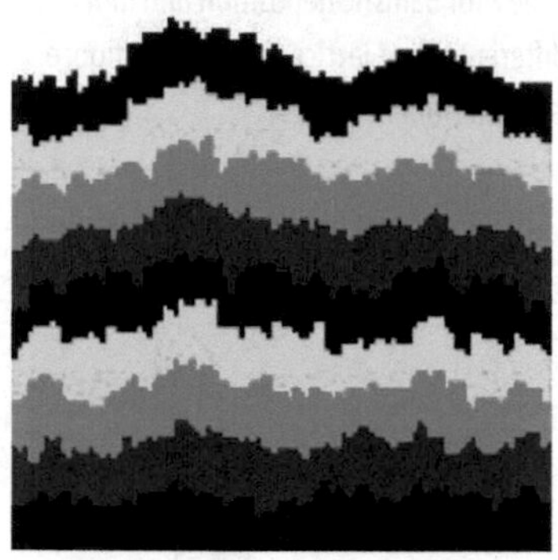

Figure 11.14 Growth by adding one particle and result after adding many particles using random deposition with surface diffusion. The color represents the time at which the particle was attached.

We illustrate the procedure, again along with a typical growth pattern, in Figure 11.15. Because of the rather stringent requirement introduced in the RSOS, the height now varies less than for the previous models and the interfaces become smoother. In Figure 11.12, we see how for the RSOS model the width increases with time for two different system sizes. Again, we may ask for the roughening exponent ξ which is $\frac{1}{2}$ and the growth exponent β which is $\frac{1}{3}$. The Family–Vicsek scaling law for RSOS is thus

$$W(L,t) = \sqrt{L} f\left(\frac{t}{L^{\frac{3}{2}}}\right).$$

11.3.5 Eden Model

The Eden model was introduced 1961 by Murray Eden [231] to describe tumor growth and epidemic spreading. In this model, each neighbor of an occupied site has the same probability of being occupied. The growth algorithm is as follows:

Eden Model

- Make a list of all empty nearest neighbors (red in Figure 11.16).
- Pick one empty nearest neighbor at random from the list. Add it to the cluster (green in Figure 11.16).
- Add the empty neighbors of this site to the list, if they are not already there.

There may be more than one growth site in a column, so there can be overhangs and there can also be lakes of empty sites inside the occupied zone. We illustrate the Eden model together with a typical realization in Figure 11.16. The roughening exponent ξ is $\frac{1}{2}$ again while the growth exponent β is $\frac{1}{3}$. The Family–Vicsek scaling is the same as for RSOS.

We can grow a cluster from a point-like seed according to the Eden model (a so-called Eden cluster) and use it as a simple model for simulating tumor growth.

We show a typical growth pattern in Figure 11.17 and see that it is essentially just a big blob with rough edges.

11.3.6 Ballistic Deposition

In yet another model, we drop particles from above onto the surface; they stick on the surface either when they touch a particle below or a neighbor on one side. The algorithm is as follows:

Ballistic Deposition

- Pick a column i.
- Let a particle fall vertically until it touches a neighbor (either below or on either side).

We show the growth sites (red) of ballistic deposition and a typical realization in Figure 11.18. Considering only the external surface (not the internal holes), the roughening and growth exponents are again $\xi = \frac{1}{2}$ and $\beta = \frac{1}{3}$, respectively. The size of the largest hole in fact grows logarithmically with time. The model is often generalized to particles impacting with a certain angle instead of falling down vertically.

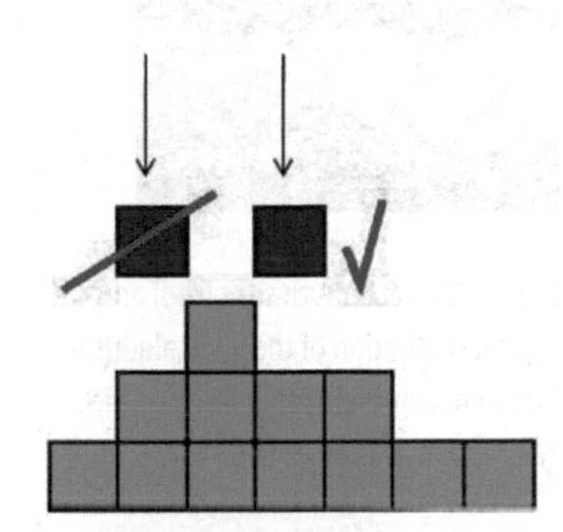

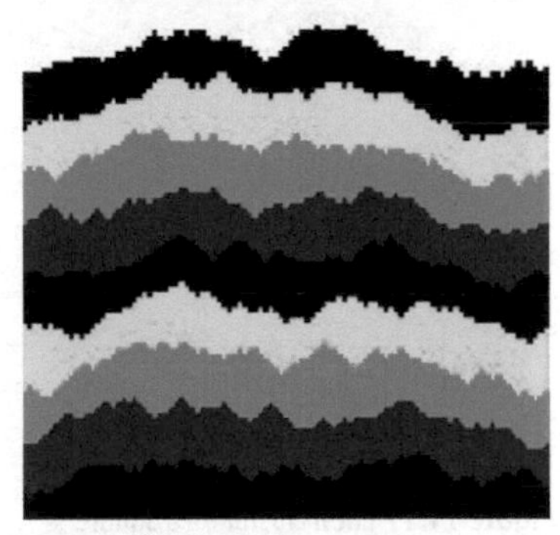

Figure 11.15 Growth rule and typical simulation of the RSOS algorithm. The color represents the time at which the particle was attached.

11.3.7 Continuum Description

We have seen a handful of models to grow irreversible patterns according to a given set of rules. Let us now formulate an equation for the surface height $h(x, t)$ based on some symmetry arguments:

1. Invariance under translation in time $t \to t + \delta t$, where δt is a constant.
2. Translation invariance along the growth direction: $h \to h + \delta h$, where δh is a constant.
3. Translation invariance along the substrate: $x \to x + \delta x$, where δx is a constant.
4. Isotropic growth dynamics (e. g., $x \to -x$).
5. Conserved relaxation dynamics: The deterministic part can be written as the gradient of a current.

Of course, not all growth algorithms are going to respect every single one of these symmetries (nor would we want them to!). Depending on the growth process we envision, a given set of symmetries will need to be respected.

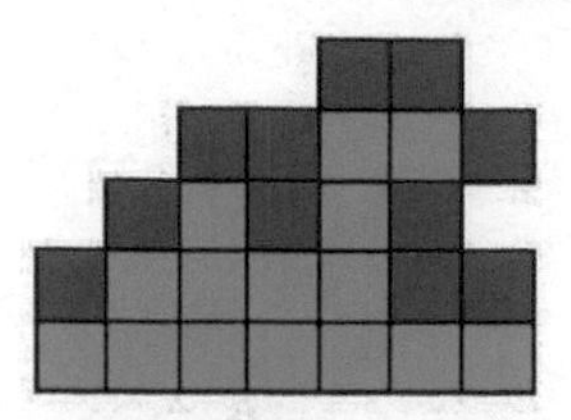

Figure 11.16 Growth sites (red) and a typical realization of the Eden algorithm. The color represents the time at which the particle was attached.

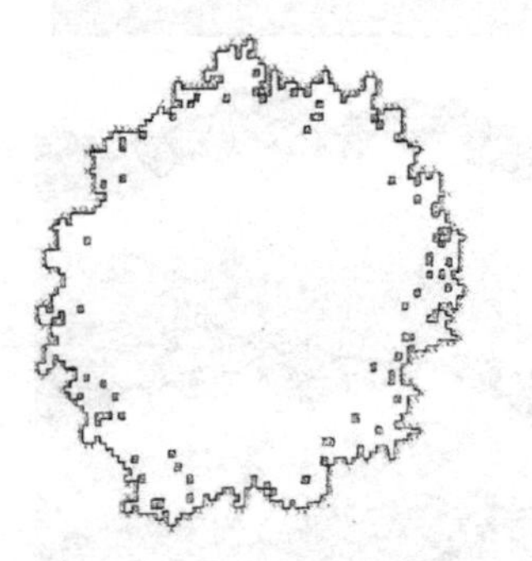

Figure 11.17 Eden cluster on a square lattice grown from a single seed. The figure was taken from Ref. [228].

An equation for a growth process respecting all five symmetries is the Edwards–Wilkinson (EW) equation [232]

$$\frac{\partial h(x,t)}{\partial t} = \nu \Delta h + \eta(x,t), \tag{11.22}$$

where ν is the surface tension and η is Gaussian white noise with average

$$\langle \eta(x,t) \rangle = 0 \tag{11.23}$$

and second moment

$$\langle \eta(x,t)\eta(x',t') \rangle = 2D\delta(x-x')\delta(t-t'). \tag{11.24}$$

The resulting time evolution of $h(t)$ follows Family–Vicsek scaling with exponents $\xi = \frac{1}{2}$ and $\beta = \frac{1}{4}$. Random deposition with surface diffusion belongs to this universality class.

When we include the next leading order in the gradient (which is nonlinear), we drop the fifth requirement (conserved relaxation dynamics) and obtain the Kardar–Parisi–Zhang (KPZ) equation [233]

$$\frac{\partial h}{\partial t} = \nu \Delta h + \frac{\lambda}{2}(\nabla h)^2 + \eta(x,t). \tag{11.25}$$

This equation was first proposed by M. Kardar, G. Parisi, and Y.-C. Zhang (in 1986). Its scaling exponents are $\xi = \frac{1}{2}$ and $\beta = \frac{1}{3}$. Thus, the RSOS model, the Eden model, and ballistic deposition belong all to the universality class of KPZ.

The EW and KPZ equations both describe the evolution of the height h. We see that one phenomenon that is common to both equations is noise, parametrized as $\eta(x,t)$, which means that these are stochastic equations that require specific numerical methods [234]. Note that only the KPZ equation takes into account differences in height along the cluster (via the term $\propto (\nabla h)^2$), which describes lateral growth.

11.3.8 Diffusion-Limited Aggregation

Another irreversible growth model is diffusion-limited aggregation (DLA), which has been introduced by Witten and Sander [235] (see Figure 11.19). It describes the aggregation of diffusing particles to a cluster. The algorithm is as follows:

Diffusion-Limited Aggregation

1. Consider a particle far away from the surface of the cluster.
2. Let the particle diffuse (i.e., perform a random walk). If the particle moves too far away, go back to step 1.
3. Attach the particle to the growth cluster if it touches it. Then go back to step 1.

Exercise: Diffusion-Limited Aggregation

In diffusion-limited aggregation (DLA), we start with a seed point. In each of the following steps a particle is released from the outside and performs a random walk until it reaches the perimeter of the structure, where it is attached to the structure (or seed surface). This can be implemented using continuous space or on a lattice (to simulate snow flakes, a triangular lattice is used).
For simplicity, generate DLA clusters on a two-dimensional lattice (square lattice, or/and triangular lattice). As a seed, you may use a site or a surface (in 2D a line). For a seed site, it is practical to define a start radius, from where the random walker starts, and a maximal radius. When the walker reaches the maximal radius, the random walk is stopped and a new walker starts at the start radius (for a seed surface it can be implemented similarly).

- Color sites according to their aggregation time. Which sites of the perimeter are occupied? What does it mean for the probability of occupying perimeter sites (e.g., as compared to the Eden model where each perimeter site is occupied with equal probability)?
- Implement a method of "noise reduction" such that a perimeter site has to be visited m times before it is occupied (i.e., store number of visits at each perimeter site). What is the qualitative difference when varying m?
- Determine the fractal dimension with the methods you know already (sandbox, box counting, and ensemble method).
- Generate a DLA cluster in continuous space (i.e., no lattice).

Please consider the following points:

- On the lattice, a list of all perimeter sites can be helpful to determine if a site belongs to the perimeter (this would also allow you to easily change your implementation to the Eden model). Another (simpler) possibility is to check if the neighboring sites are occupied.
- For efficiency, the start radius and the maximal radius can be adjusted to the actual radius of the cluster at each time step (e.g., start radius two sites larger than current cluster radius, maximal radius twice the current cluster radius).

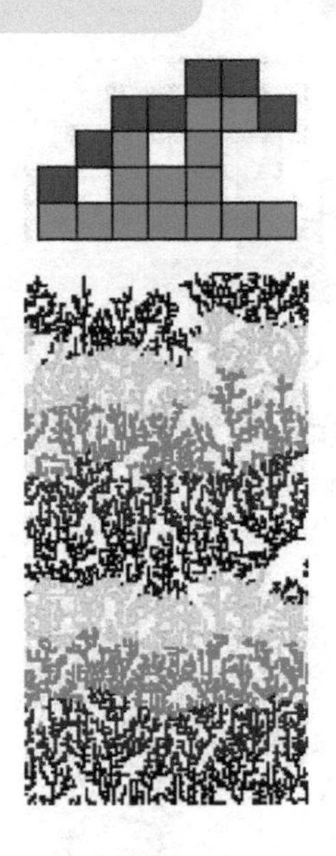

Figure 11.18 Ballistic deposition growth sites (red) and typical simulation. The color represents the time at which the particle was attached.

Figure 11.19 Thomas Witten is a professor at the University of Chicago (top), and Leonard M. Sander is a professor emeritus at the University of Michigan (bottom).

We show an example of a DLA growth process in Figure 11.20. DLA clusters are fractal with fractal dimensions of $d \approx 1.71$ in two and $d_f \approx 2.53$ in three dimensions. Applications of DLA include electro- and mineral deposition, viscous fingering, and the formation of ice dendrites.

11.3.9 Dielectric Breakdown Model (DBM)

A very interesting generalization of DLA is a model originally designed to describe dielectric breakdown [236].

Figure 11.20 Growth rule and a typical realization of diffusion-limited aggregation. The color represents the time at which the particle was attached.

We start out from the Laplace equation

$$\Delta \phi = 0 . \tag{11.26}$$

In the center of the system, we place the seed for the growing cluster on which we impose the boundary condition $\phi = 0$, while on the external boundary of the system, we impose $\phi = 1$. After finding the solution, one adds a site on the surface of the cluster with probability

$$p \propto (\nabla \phi)^{\eta} , \tag{11.27}$$

where the gradient is taken perpendicular to the surface and η is a parameter of the model. If one replaces the electric field ϕ by pressure, the above problem describes viscous fingering, if one identifies ϕ with temperature, one describes crystal growth in an undercooled melt, and if ϕ is chosen to be the probability to find a random walker, we model diffusion, like in Section 11.3.8. We recover two previous models for special values of η:

- For $\eta = 1$, we recover DLA.
- For $\eta = 0$, growth becomes independent of ϕ and DBM produces Eden clusters.

We show DBM configurations for different values of η in Figure 11.21. For large η, the clusters become more anisotropic resembling electric discharges in a random medium, which gives the model its name.

Figure 11.21 Clusters obtained by the dielectric breakdown model for different values of η. The field ϕ is shown in gray.

11.3.10 Random Fuse Model and Fracture

The random fuse model was introduced by Lucilla de Arcangelis et al. [237] (see Figure 11.22) to describe brittle fracture of disordered solids. This model is based on a network of fuses that are defined as electrical resistors which follow Ohm's law until they reach a randomly chosen breaking current at which they fail (see Figure 11.23). If one applies an electrical potential drop U between two sides of the system that one slowly increases, one by one, the fuses will break each time the current flowing through them reaches their threshold. When all conducting paths are disrupted, the system fractures in two pieces and no current can flow anymore. In order to simulate this model, one must calculate the currents passing through each resistor after each breaking. This is done by imposing Kirchhoff's law at each point of the grid, which produces a system of coupled linear equations. These can be solved with LU decomposition since the matrix is sparse.

Figure 11.22 Lucilla de Arcangelis is a professor at the University of Campania "Luigi Vanvitelli" in Naples, Italy.

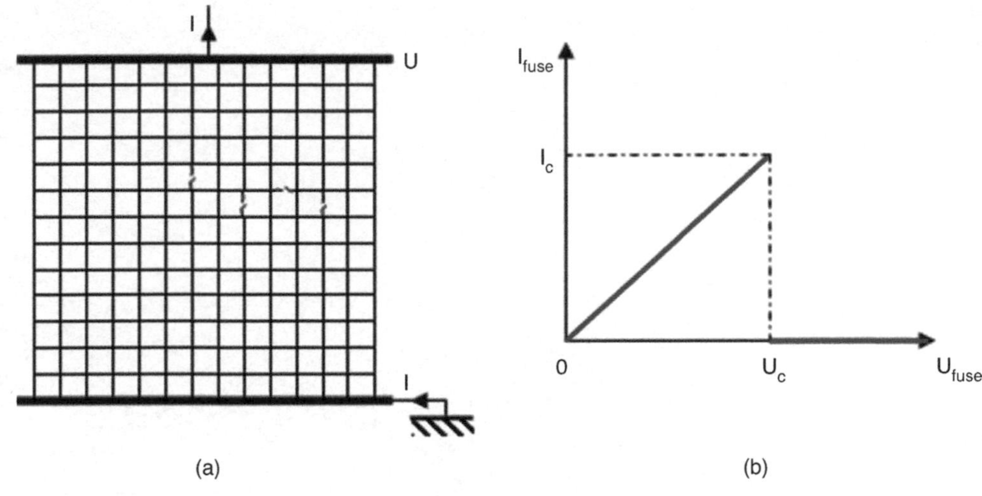

Figure 11.23 (a) Resistor network with a potential difference U between the top and bottom parts. Note that some resistors are broken (indicated by broken bonds). (b) If the current exceeds a threshold I_c, the fuse breaks, and current can no longer flow through it.

PART II

MOLECULAR DYNAMICS

Basic Molecular Dynamics 12

12.1 Introduction

In Part I, we introduced different stochastic methods to describe the properties of a large number of interacting units, for example particles, spins, and other entities. Instead of a statistical description, we now focus on modeling microscopic particle interactions by solving the corresponding equations of motion. This technique is also known as *molecular dynamics* (MD). There are several excellent books dedicated to this method [238]. We start with some straightforward and intuitive methods such as the so-called Verlet and leapfrog schemes for solving Newton's equations of motion. In what follows, we discuss Lagrange multipliers and long-range potential methods as tools to simulate composed and long-range interacting particle systems. For hard-sphere interactions and long-lasting contacts, we give an overview of event-driven MD and contact dynamics. We conclude this second part of our book with a discussion of different quantum-mechanical (ab initio) MD approaches.

In classical MD simulations, we have to solve Newton's equations of motion. The mathematical framework of classical mechanics dates back to the work of Isaac Newton in the seventeenth century [239]. But only with the rise of computers in the second half of the twentieth century, it became possible to perform first MD simulations. Many of the techniques contained in this chapter form the basis of modern commercial softwares that are frequently applied to many engineering and industrial problems. Some examples of fields where MD simulations are relevant include

- simulation of atoms and molecules,
- gravitational interactions in astrophysics,

Additional Information: History of Numerical Methods

Numerical methods are almost as old as mathematics itself. Simple calculations such as linear interpolation and square root approximations were developed a few thousand years ago (see, e.g., the Babylonian clay tablet). More refined methods in differential analysis started to become of great interest with the rise of physical sciences in the seventeenth century. As an example, Euler's method was described by Leonhard Euler in 1768 but was certainly known for a long time before. *Runge–Kutta* methods, which are often used today for very precise calculations (e.g., in celestial mechanics), were developed around 1900. For Hamiltonian problems, there exist different tailored numerical methods [240].

- fluid dynamics (including traffic and pedestrians [241]),
- biopolymers,
- granular materials,
- dislocations, voids, quasi-particles,
- electrons (Car–Parrinello [242]),
- virtual reality simulations.

One of the pioneers of this field is Berni Alder (see Figure 12.1). He was one of the first who developed MD methods to computationally study particle interactions [243].

12.2 Equations of Motion

To model interacting particle systems, we use generalized coordinates

$$\mathbf{q}_i = \left(q_i^1, \ldots, q_i^d\right) \quad \text{and} \quad \mathbf{p}_i = \left(p_i^1, \ldots, p_i^d\right) \tag{12.1}$$

Figure 12.1 Berni Alder (1925–2020) was a professor at the University of California, Davis. He won the Boltzmann Medal in 2001. The Berni Alder Prize is the most prestigious European prize for computational physics.

in a system where each particle i has d degrees of freedom. We then describe the system consisting of N particles by

$$Q = (\mathbf{q}_1, \ldots, \mathbf{q}_N) \quad \text{and} \quad P = (\mathbf{p}_1, \ldots, \mathbf{p}_N)\,, \tag{12.2}$$

and the Hamiltonian

$$\mathcal{H}(P, Q) = K(P) + V(Q) \tag{12.3}$$

with $K(P) = \sum_{i,k} \frac{(p_i^k)^2}{2m_i}$ being the kinetic energy, m_i the mass of the ith particle, and $V(Q)$ the potential energy. The sum over $k \in \{1, \ldots, d\}$ accounts for the d degrees of freedom. The potential (e.g., an attractive or repulsive electromagnetic potential) determines particle interactions, and therefore their dynamics. The potential energy can be decomposed in the following way:

$$V(Q) = \sum_i v_1(q_i) + \sum_i \sum_{j>i} v_2\left(q_i, q_j\right) + \sum_i \sum_{j>i} \sum_{k>j>i} v_3\left(q_i, q_j, q_k\right) + \ldots \,. \tag{12.4}$$

The used summation ($j > i$, $k > j > i$) avoids to sum over any pair twice. The first term $v_1(\cdot)$ describes interactions of particles with the environment (e.g., gravity and boundary conditions for particle–wall interactions). The second term $v_2(\cdot)$ accounts for pairwise interactions, and $v_3(\cdot)$ for three-body interactions between particles. Examples of potentials involving three particles are the Stillinger–Weber potential [244] and the Axilrod–Teller potential [245]. Usually, three or more body interactions are, however, neglected and their effect is considered in an effective two-body interaction described by

$$v_2^{\text{eff}}\left(q_i, q_j\right) = v^{\text{attr}}(r) + v^{\text{rep}}(r) \quad \text{with} \quad r = \left|\mathbf{q}_i - \mathbf{q}_j\right|\,, \tag{12.5}$$

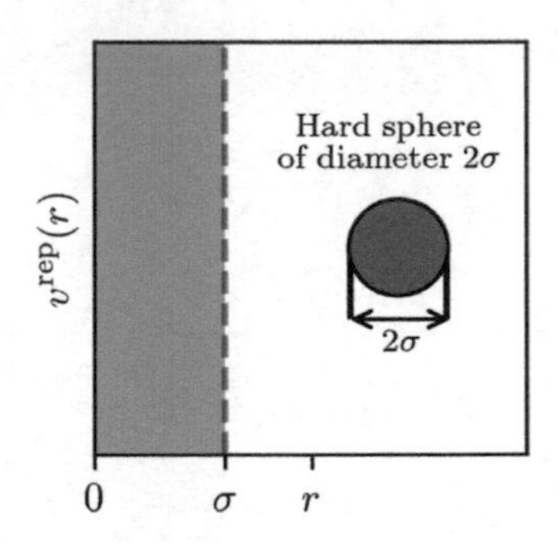

Figure 12.2 An example of a hard-sphere potential. In the blue shaded region the potential is infinite.

where $v^{\mathrm{attr}}(r)$ and $v^{\mathrm{rep}}(r)$ represent attractive and repulsive parts of the potential, respectively. The simplest potential is the hard-sphere interaction potential

$$v^{\mathrm{rep}}(r) = \begin{cases} \infty, & \text{if } r < \sigma, \\ 0, & \text{if } r \geq \sigma. \end{cases} \tag{12.6}$$

An illustration of the hard-sphere potential is shown in Figure 12.2. This potential is not well suited for numerical simulations due to the fact that the force $\mathbf{f} = -\nabla v^{\mathrm{rep}}(r)$ becomes infinite at $r = \sigma$. A softer potential should be used in numerical simulations. One possibility is to use a spring-like repulsion potential

$$v^{\mathrm{rep}}(r) = \begin{cases} \frac{k}{2}(R-r)^2, & \text{if } r < R, \\ 0, & \text{if } r > R, \end{cases} \tag{12.7}$$

where k is the elastic spring constant, and $R = R_1 + R_2$ is the sum of the radii of the individual particles. Such a soft-sphere interaction may lead to unrealistic simulation outcomes (overlapping particles). Another important potential for modeling molecular interactions is the *Lennard–Jones* potential

$$v^{\mathrm{LJ}}(r) = 4\epsilon\left[\left(\frac{\sigma}{r}\right)^{12} - \left(\frac{\sigma}{r}\right)^{6}\right], \tag{12.8}$$

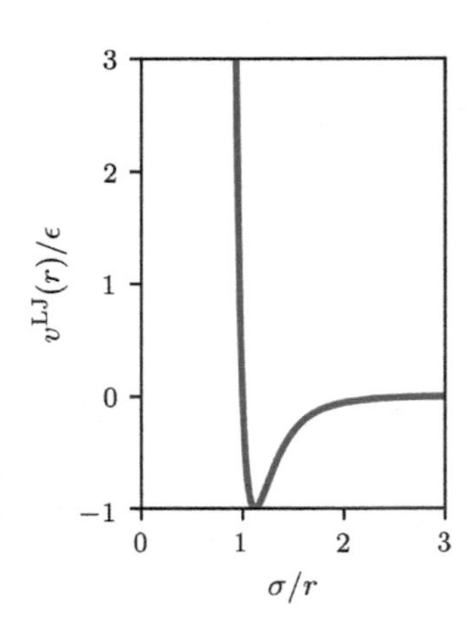

Figure 12.3 The Lennard–Jones potential approaches zero for large distances r and reaches a minimum at $r = 2^{1/6}\sigma$.

where ϵ is the depth of the potential well and σ is the interaction range. It is a mathematically simple model that approximates the spherically symmetric interaction between a pair of neutral atoms or molecules. The first term accounts for Pauli repulsion at short ranges and the second term describes attractive van der Waals forces. An example of a Lennard–Jones potential is shown in Figure 12.3. It is possible to define a cutoff for certain short-range potentials to reduce the computational effort. This is, however, not applicable to long-range potentials that are therefore more challenging to simulate.

Once the interaction potential has been defined, we can easily derive the equations of motion using the Hamilton equations

$$\dot{q}_i^k = \frac{\partial \mathcal{H}}{\partial p_i^k} \quad \text{and} \quad \dot{p}_i^k = -\frac{\partial \mathcal{H}}{\partial q_i^k}, \tag{12.9}$$

where $k \in \{1, \ldots, d\}$ and $i \in \{1, \ldots, N\}$. Let us identify $\mathbf{q}_i$ with the position vector $\mathbf{x}_i$ and $\dot{\mathbf{q}}_i = \dot{\mathbf{x}}_i$ with the velocity vector $\mathbf{v}_i$. Using $\dot{\mathbf{x}}_i = \mathbf{v}_i = \mathbf{p}_i/m_i$ and $\dot{\mathbf{p}}_i = -\nabla_i V(Q) = \mathbf{f}_i$, we obtain Newton's equations of motion

$$m_i\ddot{\mathbf{x}}_i = \dot{\mathbf{p}}_i = \mathbf{f}_i = \sum_j \mathbf{f}_{ij}, \tag{12.10}$$

where $\mathbf{f}_{ij}$ is the force exerted on particle i by particle j. Simulating Hamiltonian dynamics implies that our molecular system exhibits certain conservation laws. First, the total energy is conserved if our considered Hamiltonian $\mathcal{H}(P, Q)$ is not explicitly time dependent (i.e., if $\partial_t \mathcal{H} = 0$). Second, if the system is translational invariant in a certain direction, the corresponding momentum is conserved. And third, if the simulated system is rotational invariant about a certain axis, the corresponding angular momentum component is conserved. For example, a cubic box with periodic boundary conditions leads to a conserved total momentum

$$\mathbf{P} = \sum_i \mathbf{p}_i \,. \tag{12.11}$$

For a spherically symmetric box, the total angular momentum

$$\mathbf{L} = \sum_i \mathbf{x}_i \wedge \mathbf{p}_i \tag{12.12}$$

about the center of symmetry is conserved.

12.3 Contact Time

We now have to solve the equations of motion defined by eq. (12.10) with the help of appropriate numerical integration methods. We are going to numerically compute the motion of particles via an explicit forward integration and it is thus very important to use a sufficiently small time step Δt. If our chosen time step is too large, we may encounter unrealistic overlaps and particles may pass through the interaction range of other particles without interacting with them. We therefore need a measure that estimates the required time step and the corresponding integration error. For potentials of finite range, such a measure is the so-called *contact time* that corresponds to the time span during which a contact lasts. Since we consider an interaction force which only depends on distance, we first analyze our particle dynamics in a one-dimensional setting. The interaction of the particle with the potential is illustrated in Figure 12.4. Using energy conservation

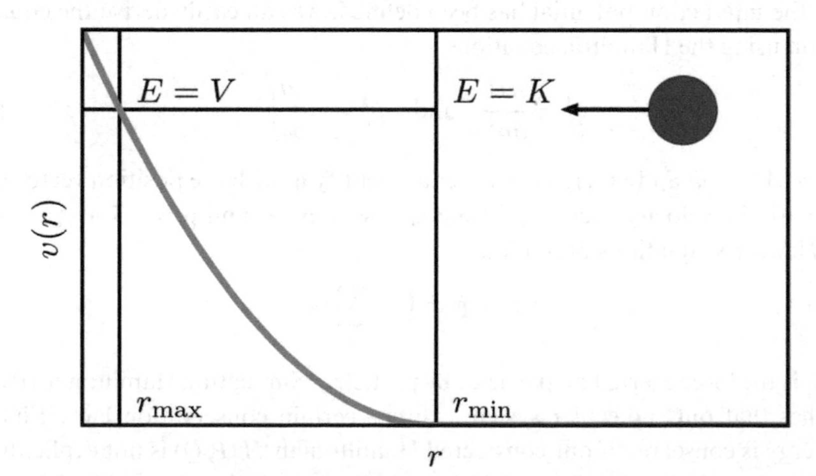

Figure 12.4 Derivation of the contact time. The particle with a kinetic energy K approaches another particle at distance $r_{\min}$ and comes to a halt at a distance $r_{\max}$, when its potential energy equals K. Then it returns and leaves the collision region with kinetic energy K.

$$E = \frac{1}{2} m \dot{r}^2 + V(r) = \text{const.}, \tag{12.13}$$

we obtain the velocity

$$\frac{dr}{dt} = \left[\frac{2}{m} (E - V(r)) \right]^{\frac{1}{2}} \tag{12.14}$$

and the contact time

$$t_c = 2 \int_0^{\frac{1}{2}t_c} dt = 2 \int_{r_{\min}}^{r_{\max}} \frac{dt}{dr} dr = 2 \int_{r_{\min}}^{r_{\max}} \left[\frac{2}{m} (E - V(r)) \right]^{-\frac{1}{2}} dr, \tag{12.15}$$

where $r_{\min}$ and $r_{\max}$ are the range of the potential and the turning point of a colliding particle, respectively. The contact time sets the scale for the appropriate time step of the explicit forward integration of the equations of motion in an MD simulation. We expect accurate results only if the time step is at least ten times smaller than the smallest contact time. The time integration of the equations of motion is then possible using an explicit forward integration method such as

- Euler's method,
- Runge–Kutta methods,
- predictor-corrector methods,
- Verlet method,
- leapfrog method.

In the subsequent sections, we only focus on the last two methods which have been developed specifically for solving Newton's equations.

12.4 Verlet Method

The Verlet integration method was proposed by Loup Verlet (see Figure 12.5) to solve Newton's equations of motion [246]. The procedure is simple and related to forward Euler integration. We begin with a Taylor expansion of the position $\mathbf{x}_i(t \pm \Delta t)$ of the ith particle for sufficiently small time steps Δt:

$$\begin{aligned} \mathbf{x}_i(t + \Delta t) &= \mathbf{x}_i(t) + \Delta t\, \mathbf{v}_i(t) + \frac{1}{2} \Delta t^2\, \dot{\mathbf{v}}_i + \mathcal{O}\left(\Delta t^3\right), \\ \mathbf{x}_i(t - \Delta t) &= \mathbf{x}_i(t) - \Delta t\, \mathbf{v}_i(t) + \frac{1}{2} \Delta t^2\, \dot{\mathbf{v}}_i - \mathcal{O}\left(\Delta t^3\right). \end{aligned} \tag{12.16}$$

Adding these two expressions yields

Verlet Integration

$$\mathbf{x}_i(t + \Delta t) = 2\mathbf{x}_i(t) - \mathbf{x}_i(t - \Delta t) + \Delta t^2\, \ddot{\mathbf{x}}_i(t) + \mathcal{O}\left(\Delta t^4\right). \tag{12.17}$$

Figure 12.5 Loup Verlet (1931–2019) was a research director at CNRS in Orsay. Photograph courtesy Jean-Jaques Banide and the Les Houches School of Physics (reprinted with kind permission of Christiane DeWitt).

Newton's second law enables us to express the acceleration $\mathbf{a}_i(t) = \dot{\mathbf{v}}_i(t) = \ddot{\mathbf{x}}_i(t)$ as

$$\ddot{\mathbf{x}}_i(t) = \frac{1}{m_i} \sum_j \mathbf{f}_{ij}(t) \quad \text{with} \quad \mathbf{f}_{ij}(t) = -\nabla V\left(r_{ij}(t)\right) . \tag{12.18}$$

The particle trajectories are then computed by plugging these results into eq. (12.17). Ideally, we should use a time step of approximately $\Delta t \approx t_c/20$. Some general remarks about the Verlet method:

- Two time steps need to be stored (t and $t - \Delta t$).
- Velocities can be computed with $\mathbf{v}_i(t) = \frac{\mathbf{x}_i(t+\Delta t) - \mathbf{x}_i(t-\Delta t)}{2\Delta t}$.
- The local numerical error is of order $O\left(\Delta t^4\right)$ (i.e., it is globally a third order algorithm).
- The numbers which are added are of order $O\left(\Delta t^0\right)$ and $O\left(\Delta t^2\right)$, posing a challenge to numerical accuracy.
- We can include higher-order terms (very inefficient).
- The method is time reversible, which allows us to estimate the accumulation of numerical errors by reversing the process and comparing it to the initial configuration.

12.5 Leapfrog Method

The leapfrog method was introduced by Hockney in 1970 [247]. For the derivation of the leapfrog method, we start with the velocities of the ith particle at intermediate steps

$$\begin{aligned} \mathbf{v}_i\left(t + \frac{1}{2}\Delta t\right) &= \mathbf{v}_i(t) + \frac{1}{2}\Delta t \dot{\mathbf{v}}_i(t) + O\left(\Delta t^2\right), \\ \mathbf{v}_i\left(t - \frac{1}{2}\Delta t\right) &= \mathbf{v}_i(t) - \frac{1}{2}\Delta t \dot{\mathbf{v}}_i(t) + O\left(\Delta t^2\right). \end{aligned} \tag{12.19}$$

Taking the difference of these two equations yields

$$\mathbf{v}_i\left(t + \frac{1}{2}\Delta t\right) = \mathbf{v}_i\left(t - \frac{1}{2}\Delta t\right) + \Delta t\, \ddot{\mathbf{x}}_i(t) + O\left(\Delta t^3\right), \tag{12.20}$$

where $\ddot{\mathbf{x}}_i(t)$ is given in eq. (12.18). Finally, the position is updated according to

$$\mathbf{x}_i(t + \Delta t) = \mathbf{x}_i(t) + \Delta t\, \mathbf{v}_i\left(t + \frac{1}{2}\Delta t\right) + O\left(\Delta t^4\right). \tag{12.21}$$

The leapfrog method has the same order of accuracy as the Verlet method. However, both methods differ in the way in which the variables are integrated. To summarize, the leapfrog method is based on the following update scheme:

Leapfrog Integration

$$\ddot{\mathbf{x}}_i(t) = \frac{1}{m_i}\sum_j \mathbf{f}_{ij}(t)\,,$$
$$\mathbf{v}_i\left(t+\frac{1}{2}\Delta t\right) = \mathbf{v}_i\left(t-\frac{1}{2}\Delta t\right) + \Delta t\,\ddot{\mathbf{x}}_i(t)\,, \tag{12.22}$$
$$\mathbf{x}_i(t+\Delta t) = \mathbf{x}_i(t) + \Delta t\,\mathbf{v}_i\left(t+\frac{1}{2}\Delta t\right).$$

Additional Information: Comparison of Euler and Leapfrog Methods

If we compare the leapfrog algorithm with a forward Euler scheme

$$\ddot{\mathbf{x}}_i(t) = \frac{1}{m_i}\sum_j \mathbf{f}_{ij}(t)\,,$$
$$\mathbf{v}_i(t+\Delta t) = \mathbf{v}_i(t) + \Delta t\,\ddot{\mathbf{x}}_i(t)\,, \tag{12.23}$$
$$\mathbf{x}_i(t+\Delta t) = \mathbf{x}_i(t) + \Delta t\,\mathbf{v}_i(t)\,,$$

we see that the leapfrog method uses velocity and acceleration at different time steps to compute the new particle positions.

Verlet and leapfrog schemes are microcanonical (i. e., energy conserving). Therefore, for sufficiently small time steps Δt, energy is conserved on average during a simulation and fluctuations are due to roundoff errors. Large fluctuations in energy are usually a hint for the time step being too large or an erroneous code. It is also possible to estimate the accuracy of the method by analyzing the underlying energy fluctuations.

As in the case of the Verlet integration scheme, the leapfrog method is completely time reversible and we can assess the error by looking at the change in the initial configuration after having simulated forward and then backward in time. The difference between the two configurations can be taken as a measure for the error of the method.

Another approach is to take two initial configurations that differ only slightly in the velocity or the position of a particle, and to observe the difference in the time evolution of the two systems. Initially, both systems should behave similarly, whereas the difference increases exponentially with time. For instance, let ΔE be the energy difference between both systems. The slope of $\log \Delta E$ as function of time is an indicator of the precision of the method. The slope is the so-called *Lyapunov exponent*, and describes the divergence of the simulated trajectory from the true one. To reduce this sensitive dependence on initial conditions, other integration methods (e. g., splitting methods) can be used to simulate the long-term evolution of such Hamiltonian systems.

Exercise: Simulating Interacting Particles in a Box

We now simulate a system of N interacting particles in a box using the Verlet scheme.
The programming code of this exercise will be the basis for further improvements and modifications in the following exercises.
Consider a system of N point particles of mass m (in a box of length L with periodic boundary conditions) that interact via the Lennard–Jones potential

$$V(r) = 4\left[\left(\frac{1}{r}\right)^{12} - \left(\frac{1}{r}\right)^{6}\right].$$

The total Hamiltonian of the system is given by

$$H = \frac{1}{2}\mathrm{m}\sum_i \boldsymbol{v}_i^2 + \frac{1}{2}\sum_{i\neq j} V(|\boldsymbol{r}_i - \boldsymbol{r}_j|).$$

Task 1: Simulate the system by using the Verlet scheme. Use sufficiently small time steps and a cutoff length $r_c = 2.5$ for the Lennard–Jones potential.
Task 2: Investigate the influence of the choice of the time step size.
Task 3: Measure the energy of the system as well as its fluctuations. Is the energy conserved?

Visualization of the results: For visualizing the particles you have different options. You can simply use a graphical tool (e. g., gnuplot, Matlab, matplotlib) which reads in the coordinates for a given time step. Another possibility would be to use either *vmd* (www.ks.uiuc.edu/Research/vmd/) or *paraview* (www.paraview.org/) to visualize the dynamics and create animations.

13 Optimizing Molecular Dynamics

Usually, we are interested in performing MD simulations for a large number of particles. This makes it necessary to compute the forces between any pair of all N particles – an operation of complexity $O(N^2)$. However, there exist different ways of optimizing our MD simulations. For example, let us consider a potential that is a function of even powers of the particle distance r. That is, our potential has the form $v(r) \propto r^{-2n}$ with $n \geq 1$. If we now determine the force between two particles i and j

$$\mathbf{f}_{ij} = -\nabla r_{ij}^{-2n} \propto r_{ij}^{2(n-1)} \mathbf{r}_{ij}, \tag{13.1}$$

we realize that it is possible to omit the computation of the square root in

$$r_{ij} = \sqrt{\sum_{k=1}^{d} \left(x_i^k - x_j^k\right)^2}. \tag{13.2}$$

We can simply evaluate

$$r_{ij}^{2(n-1)} = \left[\sqrt{\sum_{k=1}^{d} \left(x_i^k - x_j^k\right)^2}\right]^{2(n-1)} = \left[\sum_{k=1}^{d} \left(x_i^k - x_j^k\right)^2\right]^{(n-1)}. \tag{13.3}$$

If the potential has a complicated form which makes it computationally expensive to evaluate, we may discretize the potential and store its values in a lookup table.

For short range potentials, we define a cutoff r_c and divide the interval $(0, r_c^2)$ into K pieces with subinterval lengths

$$l_j = \frac{j}{K} r_c^2, \tag{13.4}$$

where $j \in \{1, \ldots, K\}$. The force values stored in a lookup table are $f_j = f\left(\sqrt{l_j}\right)$, and the corresponding index j is given by

$$j = \left\lfloor S \sum_{k=1}^{d} \left(x_i^k - x_j^k\right)^2 \right\rfloor + 1, \tag{13.5}$$

where $\lfloor \cdot \rfloor$ denotes the floor function[1] and $S = K/r_c^2$. Interpolation methods are useful to obtain intermediate values. The definition of a cutoff introduces a discontinuity

[1] The floor function $\lfloor x \rfloor$ of a real number x corresponds to the greatest integer that is less than or equal to x.

of the potential at r_c and thus an infinite force at r_c. To avoid this problem, we can introduce a modified potential $\tilde{v}(r)$ according to

$$\tilde{v}(r) = \begin{cases} v(r) - v(r_c) - \frac{\partial v}{\partial r}\Big|_{r=r_c}(r - r_c) & \text{if } r \le r_c, \\ 0 & \text{if } r > r_c, \end{cases} \tag{13.6}$$

where $v(r_c)$ is the value of the original potential at r_c. Subtracting $v(r_c)$ removes the jump which represents a physically irrelevant shift in the energy scale. The subtraction of the first derivative makes the potential differentiable at r_c but slightly modifies the physics. For the Lennard–Jones potential, a common cutoff value is $r_c = 2.5\sigma$. Potentials that decay slower than r^{-d} are called long-range potentials and for them it is not possible to introduce any cutoff, because the forces at large distances are not negligible.

13.1 Verlet Tables

To reduce the amount of computations, we should ignore particle–particle interactions whenever their force contributions are negligible. Therefore, we only consider particles in a certain range $r_l > r_c$. An illustration of this concept is shown in Figure 13.1. For every particle, we store the coordinates of the neighboring particles in a list which is referred to as *Verlet table*. As the particles move, the table has to be updated after

$$n = \frac{r_l - r_c}{\Delta t v_{\max}} \tag{13.7}$$

time steps with $v_{\max}$ being the largest velocity occurring in the system. Updating the whole list is still an operation of complexity $O(N^2)$.

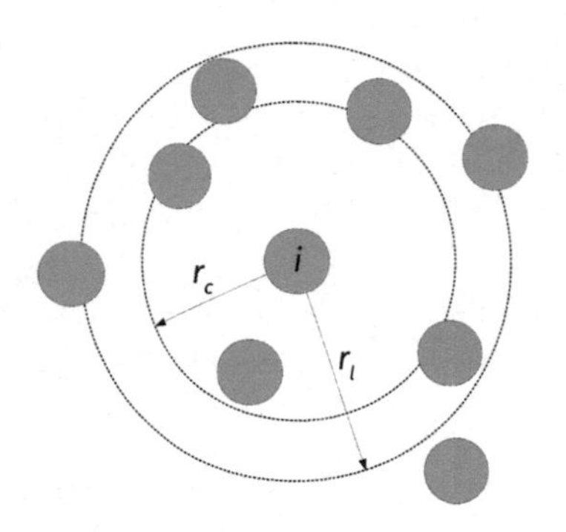

Figure 13.1 An illustration of the Verlet table method. Only particles with their center being within a distance of $r_l > r_c$ from particle i are considered.

The Verlet table LIST is a one-dimensional vector in which all the particles in the neighborhood of particle i are sequentially stored for all particles i, starting with $i = 1$ and ending with $i = N$. A second vector POINT[i] contains the index of the first particle in the neighborhood of particle i in LIST. Therefore, the particles in the neighborhood of particle i are: LIST[POINT[i]], ..., LIST[POINT[$i + 1$]-1]. The force is calculated by going through this list and finding all neighbors of particle i.

13.2 Linked-Cell Method

A more efficient way to identify interactions between neighboring particles in our MD simulations is the *linked-cell method* [248, 249] (see Figure 13.2). In this method, the domain is discretized using a regular grid with a grid spacing of $M > r_c$ where r_c is the range of the potential (or the cutoff range). Each particle is located within a certain grid cell. Due to our choice of M, we only have to account for particle interactions in certain cell neighborhoods. Particles located in more distant cells do not contribute to the force computation. In d dimensions there are 3^d neighboring cells of interest. On average, we thus have to compute the interactions of $N3^dN/N_M$ particles where N_M is the number of grid cells. To keep track of the locations of all particles, we use linked lists which were introduced by Donald E. Knuth (see Figure 13.3) [9]. In a vector (or list) FIRST of length N_M, we store the index of a particle located in cell j in FIRST[j]. If cell j is empty, then FIRST[j]=0. The indices of the remaining particles that are located in the same cell are stored in a second vector LIST of length N. If particle i is the last one in a cell, then LIST[i]=0.

In the following code listing, we show an example of how to extract all particles that are located in cell $i = 2$.

Figure 13.2 An illustration of the linked-cell method. A grid with grid spacing M ($\frac{r_c}{2} < M < r_c$) is placed on top of the system. Only interactions between particles in a certain cell neighborhood have to be considered.

```
i=2;
M[1]=FIRST[i];
j=1;
while(LIST[M[j]]!=0)
{
    M[j+1]=LIST[M[j]];
    j=j+1;
}
```

When a particle changes its cell, FIRST and LIST can be updated locally. The algorithm is thus of order $O(N)$. In addition, this method is well suited for parallelization (domain decomposition).

Exercise: Linked-Cell Method

In this exercise, we are going to make use of the linked-cell method in our MD simulations. Note that the length of the particle lists may alter from one time step to another. If you use C++, you have the choice of storing lists of pointers to a particle, or lists of particle indices.

Task

1. Implement the linked-cell method in your existing MD program for particles with Lennard–Jones potentials (see previous exercise).
2. How many particles can you simulate?
3. Try different optimizations:
   - Work with r^2 rather than r (avoiding calculating square roots).
   - Use a look-up table for the force as a function of discrete values of r^2 and interpolate for other values.

Figure 13.3 Donald E. Knuth is a professor emeritus at Stanford University. He is the inventor of the typesetting system "Tex" and author of the famous textbook *The Art of Computer Programming* [9].

Exercise: Linked-Cell Method (*cont.*)

Please consider the following points:

- Since the particles are in a cubic box, divide the system into smaller cubic regions of minimum size r_c, where $r_c = 2.5\sigma$ is the interaction range (cutoff of the Lennard–Jones potential).
- Note that for small systems (a few hundred particles) checking all possible particle pairs can be faster than using the linked-cell method. For large systems, the saved computation time can be enormous.

14 Dynamics of Composed Particles

Until now, we just considered single-particle interactions with potentials that depend on the distance between particles. In nature, however, there exist many examples of systems in which the interactions also depend on size and shape of the considered particles. Examples of such systems include sand grains and molecules such as the water molecule that we show in Figure 14.1. To describe the time evolution of such composed systems, we have to consider their shapes and constituents.

14.1 Lagrange Multipliers

One possibility to model composed particle systems is to introduce additional constraint forces [250]. Such constraint forces are used to establish rigid bonds or angles between individual particles. The idea is to rewrite the equation of motion for each particle i according to

$$m_i \ddot{\mathbf{x}}_i = \underbrace{\mathbf{f}_i}_{\text{interaction forces}} + \underbrace{\mathbf{g}_i}_{\text{constraint forces}} , \tag{14.1}$$

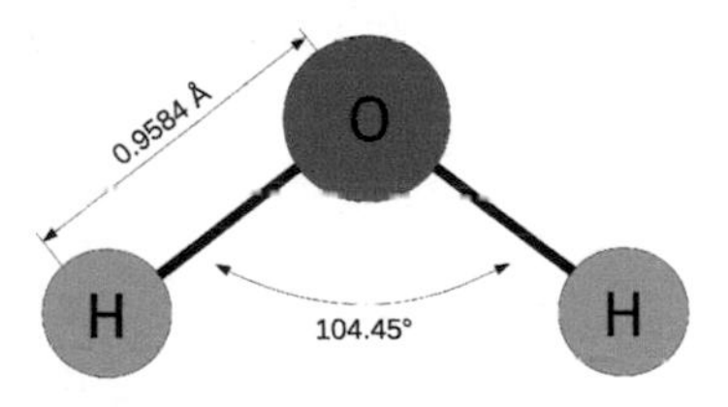

Figure 14.1 A water molecule as a composed particle system consisting of two hydrogen atoms and one oxygen atom.

where the first term accounts for interactions between particles and the second one for constraint forces. We impose the constraints forces to account for the geometric structure of molecules (e. g., certain distances d_{12} and d_{23} between atoms). We therefore define a potential such that the constraint forces $\mathbf{g}_i$ are proportional to the difference between the actual and the desired distance of the particles. As an example, we consider a water molecule that consists of three particles (see Figure 14.1). We assume that distance and angles between atoms are constant as it is the case in most practical applications. The differences between actual and desired distances

$$\chi_{12} = r_{12}^2 - d_{12}^2 \quad \text{and} \quad \chi_{23} = r_{23}^2 - d_{23}^2 \tag{14.2}$$

are zero if the particles have the desired distance, with $r_{ij} = \|\mathbf{r}_{ij}\|$ and $\mathbf{r}_{ij} = \mathbf{x}_i - \mathbf{x}_j$. We define the force

$$\mathbf{g}_k = \frac{\lambda_{12}}{2} \nabla_{\mathbf{x}_k} \chi_{12} + \frac{\lambda_{23}}{2} \nabla_{\mathbf{x}_k} \chi_{23} , \tag{14.3}$$

for $k \in \{1, 2, 3\}$, and λ_{12} and λ_{23} are the yet undetermined Lagrange multipliers. We determine these multipliers by imposing the constraints. Inserting eq. (14.2) into eq. (14.3) yields

$$\mathbf{g}_1 = \lambda_{12} \mathbf{r}_{12} , \quad \mathbf{g}_2 = \lambda_{23} \mathbf{r}_{23} - \lambda_{12} \mathbf{r}_{12} , \quad \mathbf{g}_3 = -\lambda_{23} \mathbf{r}_{23} . \tag{14.4}$$

These equations describe nothing but linear springs with yet to be determined spring constants $\lambda_{(\cdot)}$. To obtain the values of the Lagrange multipliers $\lambda_{(\cdot)}$, the Verlet algorithm is executed in two steps. We first compute the Verlet update without constraint to obtain

$$\tilde{\mathbf{x}}_i\,(t + \Delta t) = 2\mathbf{x}_i - \mathbf{x}_i\,(t - \Delta t) + \Delta t^2 \frac{\mathbf{f}_i}{m_i} \,. \tag{14.5}$$

Next, we correct the value using the constraints according to

$$\mathbf{x}_i\,(t + \Delta t) = \tilde{\mathbf{x}}_i\,(t + \Delta t) + \Delta t^2 \frac{\mathbf{g}_i}{m_i} \,. \tag{14.6}$$

Combining Eqs. (14.6) and (14.4), the updated positions are

$$\mathbf{x}_1\,(t + \Delta t) = \tilde{\mathbf{x}}_1\,(t + \Delta t) + \Delta t^2 \frac{\lambda_{12}}{m_1} \mathbf{r}_{12}\,(t) \,, \tag{14.7}$$

$$\mathbf{x}_2\,(t + \Delta t) = \tilde{\mathbf{x}}_2\,(t + \Delta t) + \Delta t^2 \frac{\lambda_{23}}{m_2} \mathbf{r}_{23}\,(t) - \Delta t^2 \frac{\lambda_{12}}{m_2} \mathbf{r}_{12}\,(t) \,, \tag{14.8}$$

$$\mathbf{x}_3\,(t + \Delta t) = \tilde{\mathbf{x}}_3\,(t + \Delta t) - \Delta t^2 \frac{\lambda_{23}}{m_3} \mathbf{r}_{23}\,(t) \,. \tag{14.9}$$

With these expressions, we now obtain λ_{12} and λ_{23} by inserting Eqs. (14.7)–(14.9) into the constraint conditions:

$$\begin{aligned} |\mathbf{x}_1\,(t + \Delta t) - \mathbf{x}_2\,(t + \Delta t)|^2 &= d_{12}^2 \,, \\ |\mathbf{x}_2\,(t + \Delta t) - \mathbf{x}_3\,(t + \Delta t)|^2 &= d_{23}^2 \,, \end{aligned} \tag{14.10}$$

which gives

$$\begin{aligned} \left| \tilde{\mathbf{r}}_{12}\,(t + \Delta t) + \Delta t^2 \lambda_{12} \left(\frac{1}{m_1} + \frac{1}{m_2} \right) \mathbf{r}_{12}\,(t) - \Delta t^2 \frac{\lambda_{23}}{m_2} \mathbf{r}_{23}\,(t) \right|^2 &= d_{12}^2 \,, \\ \left| \tilde{\mathbf{r}}_{23}\,(t + \Delta t) + \Delta t^2 \lambda_{23} \left(\frac{1}{m_2} + \frac{1}{m_3} \right) \mathbf{r}_{23}\,(t) - \Delta t^2 \frac{\lambda_{12}}{m_2} \mathbf{r}_{12}\,(t) \right|^2 &= d_{23}^2 \,, \end{aligned} \tag{14.11}$$

where $\tilde{\mathbf{r}}_{ij} = \tilde{\mathbf{x}}_i - \tilde{\mathbf{x}}_j$. These coupled quadratic equations are solved for λ_{12} and λ_{23} to compute the next position $\mathbf{x}_i\,(t + \Delta t)$. Depending on the precision needed, we may ignore the higher-order terms of Δt and just solve two coupled quadratic equations.

Exercise: Composed Particle Dynamics

Consider a system of N diatomic molecules (in a box of length L with periodic boundary conditions) that interact with each other. For simplicity, we assume that all atoms are identical. Atoms of different molecules interact via the Lennard–Jones potential

$$V(r) = 4\left[\left(\frac{1}{r}\right)^{12} - \left(\frac{1}{r}\right)^{6} \right] .$$

Exercise: Composed Particle Dynamics (*cont.*)

To ensure that the two particles in a molecule keep their distance r, we have to add the constraint forces

$$g_1 = \lambda r \quad \text{and} \quad g_2 = -\lambda r$$

to the equation(s) of motion.

Task 1: Derive an analytical expression for the Lagrange multiplier λ.

Task 2: Simulate the system using the Verlet scheme.

Please consider the following points:

- The distance between the two atoms in a molecule should be small enough such that other molecules do not cross their bond.
- It might be easier to consider the 2D case and check if the behavior of the molecules is correct before you go to the 3D case.

Task 3: Extend your program to simulate three-component molecules.

14.2 Rigid Bodies

Systems whose n constituents of mass m_i are located at fixed positions $\mathbf{x}_i$ with respect to each other are referred to as rigid bodies. The motion of such objects is described by translations of the center of mass and rotations around it. The center of mass is

$$\mathbf{x}_{\text{cm}} = \frac{1}{M} \sum_{i=1}^{n} \mathbf{x}_i m_i \quad \text{with} \quad M = \sum_{i=1}^{n} m_i \,. \tag{14.12}$$

The equation of motion of the center of mass and the corresponding torque are given by

$$M\ddot{\mathbf{x}}_{\text{cm}} = \sum_{i=1}^{n} \mathbf{f}_i = \mathbf{f}_{\text{cm}} \quad \text{and} \quad \mathbf{T} = \sum_{i=1}^{n} \mathbf{d}_i \wedge \mathbf{f}_i, \tag{14.13}$$

where $\mathbf{d}_i = \mathbf{x}_i - \mathbf{x}_{\text{cm}}$. In two dimensions, the rotation axis always points in the direction of the normal vector on the plane. Therefore, there only exist three degrees of freedom: two translational and one rotational. In three dimensions, there are six degrees of freedom: three translational and three rotational.

We first discuss the treatment of rigid bodies in two dimensions and then generalize it to the three-dimensional case.

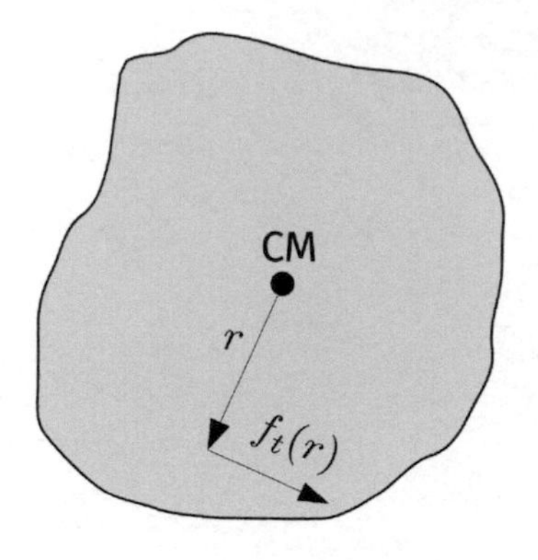

Figure 14.2 An example of a rigid body in two dimensions. The black dot shows the center of mass (CM), and $f_t(r)$ represents the tangential force component at position r.

14.2.1 Two Dimensions

In two dimensions, the moment of inertia and torque are

$$I = \int\int_A r^2 \rho(r)\, \mathrm{d}A \quad \text{and} \quad \mathrm{T} = \int\int_A r f_t(r)\, \mathrm{d}A\,, \tag{14.14}$$

where $\rho(r)$ is the mass density, A the area, and $f_t(r)$ the tangential force (see Figure 14.2). In general, the mass density may be constant or depend on the actual position. The equation of motion is

$$I\dot{\omega} = \mathrm{T}\,, \tag{14.15}$$

where ω is the angular velocity. We now apply the Verlet algorithm to the position $\mathbf{x}$ and rotation angle ϕ and compute the corresponding time evolutions according to

$$\begin{aligned} \phi(t+\Delta t) &= 2\phi(t) - \phi(t-\Delta t) + \Delta t^2 \frac{\mathrm{T}(t)}{I}\,, \\ \mathbf{x}(t+\Delta t) &= 2\mathbf{x}(t) - \mathbf{x}(t-\Delta t) + \Delta t^2 M^{-1} \sum_{j\in A} f_j(t)\,, \end{aligned} \tag{14.16}$$

where the total torque T is the sum over all the torques acting on the rigid body. That is,

$$\mathrm{T}(t) = \sum_{j\in A} \left[f_j^y(t)\, d_j^x(t) - f_j^x(t)\, d_j^y(t) \right]. \tag{14.17}$$

14.2.2 Three Dimensions

In three dimensions, the moment of inertia is a tensor that needs to be diagonalized in order to solve the equation of motion. Only in the body-fixed coordinate system a systematic diagonalization of this tensor can be achieved. To describe the motion of rigid bodies in three dimensions, we consider a lab-fixed and a body-fixed coordinate system with coordinates $\mathbf{x}$ and $\mathbf{y}$, respectively. We assume that the point $\mathbf{y} = 0$ corresponds to $\mathbf{x} = 0$ in the lab-fixed system. The transformation between both systems is

$$\mathbf{x} = R(t)\mathbf{y}\,, \tag{14.18}$$

where $R(t) \in \mathrm{SO}(3)$ denotes a rotation matrix.[1] Furthermore, we define $\Omega = R^T \dot{R}$ and find with $R^T R = 1$ that

$$R^T \dot{R} + \dot{R}^T R = \Omega + \Omega^T = 0\,. \tag{14.19}$$

This equation implies that Ω is skew-symmetric and thus of the form

$$\Omega = \begin{pmatrix} 0 & -\omega_3 & \omega_2 \\ \omega_3 & 0 & -\omega_1 \\ -\omega_2 & \omega_1 & 0 \end{pmatrix} \quad \text{and} \quad \Omega\mathbf{y} = \omega \wedge \mathbf{y}\,, \tag{14.20}$$

[1] SO(3) is the so-called three-dimensional rotation group, or special orthogonal group. All rotation matrices $R \in \mathrm{SO}(3)$ satisfy $R^T R = R R^T = 1$.

where $\omega = (\omega_1, \omega_2, \omega_3)$ is the angular velocity in the body-fixed frame and $\wedge$ denotes the cross product. Therefore, the angular momentum is

$$\mathbf{L} = \sum_{i=1}^{n} m_i \mathbf{x}_i \wedge \dot{\mathbf{x}}_i = \sum_{i=1}^{n} m_i R\mathbf{y}_i \wedge \dot{R}\mathbf{y}_i \,. \tag{14.21}$$

Combining Eqs. (14.20) and (14.21) yields

$$\mathbf{L} = R \sum_{i=1}^{n} m_i \mathbf{y}_i \wedge (\omega \wedge \mathbf{y}_i) = R \sum_{i=1}^{n} m_i \left[\omega \left(\mathbf{y}_i \cdot \mathbf{y}_i\right) - \mathbf{y}_i \left(\omega \cdot \mathbf{y}_i\right)\right] . \tag{14.22}$$

The components of the inertia tensor are

$$I_{jk} = \sum_{i=1}^{n} m_i \left[(\mathbf{y}_i \cdot \mathbf{y}_i)\, \delta_{jk} - \mathbf{y}_i^j \mathbf{y}_i^k \right] \tag{14.23}$$

and thus

$$\mathbf{L} = R\mathbf{S} \quad \text{with} \quad S_j = \sum_{k=1}^{3} I_{jk} \omega_k \,. \tag{14.24}$$

Considering a coordinate system whose axes are parallel to the principal axes of the moment of inertia of the body, the inertia tensor takes the form

$$I = \begin{pmatrix} I_1 & 0 & 0 \\ 0 & I_2 & 0 \\ 0 & 0 & I_3 \end{pmatrix} \quad \text{and} \quad S_j = I_j \omega_j \,. \tag{14.25}$$

We use eq. (14.24) to obtain the equations of motion

$$\dot{\mathbf{L}} = \dot{R}\mathbf{S} + R\dot{\mathbf{S}} = \widetilde{\mathbf{T}} \,, \tag{14.26}$$

where $\widetilde{\mathbf{T}} = R\mathbf{T}$ represents the torque in the lab-fixed coordinate system. Multiplying the prior equation with R^T, we find *Euler's equations* in the principal axes coordinate system

$$\begin{aligned}
\dot{\omega}_1 &= \frac{T_1}{I_1} + \left(\frac{I_2 - I_3}{I_1}\right) \omega_2 \omega_3 \,, \\
\dot{\omega}_2 &= \frac{T_2}{I_2} + \left(\frac{I_3 - I_1}{I_2}\right) \omega_3 \omega_1 \,, \\
\dot{\omega}_3 &= \frac{T_3}{I_3} + \left(\frac{I_1 - I_2}{I_3}\right) \omega_1 \omega_2 \,.
\end{aligned} \tag{14.27}$$

We then integrate the angular velocities according to

$$\begin{aligned}
\omega_1 (t + \Delta t) &= \omega_1 (t) + \Delta t \frac{T_1 (t)}{I_1} + \Delta t \left(\frac{I_2 - I_3}{I_1}\right) \omega_2 \omega_3 \,, \\
\omega_2 (t + \Delta t) &= \omega_2 (t) + \Delta t \frac{T_2 (t)}{I_2} + \Delta t \left(\frac{I_3 - I_1}{I_2}\right) \omega_3 \omega_1 \,, \\
\omega_3 (t + \Delta t) &= \omega_3 (t) + \Delta t \frac{T_3 (t)}{I_3} + \Delta t \left(\frac{I_1 - I_2}{I_3}\right) \omega_1 \omega_2 \,.
\end{aligned} \tag{14.28}$$

From these expressions, we obtain the angular velocity in the laboratory frame

$$\tilde{\omega}(t + \Delta t) = R\omega(t + \Delta t) \,. \tag{14.29}$$

Since the particles are moving all the time, the rotation matrix is not constant. We have to find an efficient way to determine and update R at every step in our simulation. In the following section, we therefore discuss three-dimensional rotations introducing Euler angles and quaternions.

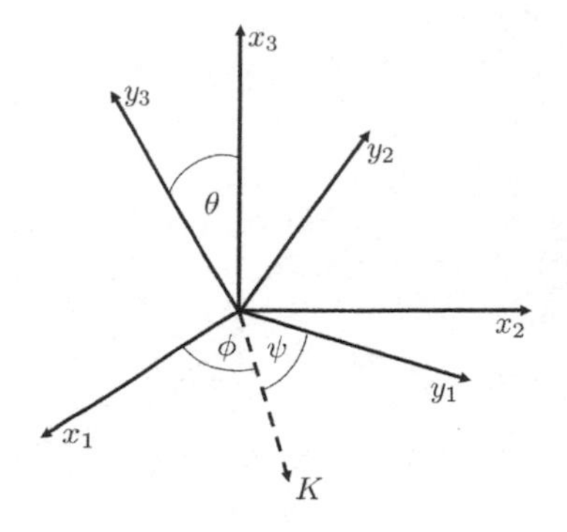

Figure 14.3 To transform the lab-fixed x-system into the body-fixed y-system, we first rotate the lab-fixed system around the x_3-axis at an angle ϕ. Next, we rotate the system around the K-axis at an angle θ and finally apply a rotation around the y_3-axis at an angle ψ.

14.2.3 Euler Angles

One possible parameterization of the rotation matrix R is given by the Euler angles (ϕ, θ, ψ). These angles correspond to three rotations that transform the lab-fixed system into a body-fixed system (see Figure 14.3). First, we rotate the lab-fixed system at an angle ϕ around the x_3-axis and refer to the rotated x_1-axis as K-axis. In the second step, we rotate the system around the K-axis at an angle θ and finally apply a rotation around the y_3-axis at an angle ψ. Mathematically, this means

$$\begin{aligned} R &= R(\phi, \theta, \psi) \\ &= \begin{pmatrix} \cos\phi & -\sin\phi & 0 \\ \sin\phi & \cos\phi & 0 \\ 0 & 0 & 1 \end{pmatrix} \begin{pmatrix} 1 & 0 & 0 \\ 0 & \cos\theta & -\sin\theta \\ 0 & \sin\theta & \cos\theta \end{pmatrix} \begin{pmatrix} \cos\psi & -\sin\psi & 0 \\ \sin\psi & \cos\psi & 0 \\ 0 & 0 & 1 \end{pmatrix} . \end{aligned} \tag{14.30}$$

As a consequence of the occurrence of products of multiple trigonometric functions, this parameterization is not well suited for efficient computations. We have to keep in mind that this operation has to be performed for every particle and every time step, leading to a high computational effort. For the computation of angular velocities, also derivatives of eq. (14.30) would have to be considered. Specifically, we would have to compute

$$\begin{aligned} \dot{\phi} &= -\tilde{\omega}_x \frac{\sin\phi\cos\theta}{\sin\theta} + \tilde{\omega}_y \frac{\cos\phi\cos\theta}{\sin\theta} + \tilde{\omega}_z \,, \\ \dot{\theta} &= \tilde{\omega}_x \cos\theta + \tilde{\omega}_y \sin\phi \,, \\ \dot{\psi} &= \tilde{\omega}_x \frac{\sin\phi}{\sin\theta} - \tilde{\omega}_y \frac{\cos\phi}{\sin\theta} \,, \end{aligned} \tag{14.31}$$

which, for some angles, θ (e. g., $\theta = 0$) gives diverging terms.

Figure 14.4 Denis J. Evans is a professor at the Australian National University in Canberra.

14.2.4 Quaternions

Denis J. Evans (see Figure 14.4) came up with a trick to optimize the computation of rotational velocities [251, 252].

This trick is based on quaternions, which are a generalization of complex numbers with four basis vectors that span a four-dimensional space. The sum of the squares of these basis vectors is equal to unity. We use the quaternion parameters

$$q_0 = \cos\left(\frac{\theta}{2}\right)\cos\left(\frac{\phi+\psi}{2}\right), \tag{14.32}$$

$$q_1 = \sin\left(\frac{\theta}{2}\right)\cos\left(\frac{\phi-\psi}{2}\right), \tag{14.33}$$

$$q_2 = \sin\left(\frac{\theta}{2}\right)\sin\left(\frac{\phi-\psi}{2}\right), \tag{14.34}$$

$$q_3 = \cos\left(\frac{\theta}{2}\right)\sin\left(\frac{\phi+\psi}{2}\right), \tag{14.35}$$

with $0 < q_i < 1$ and $\sum_i q_i^2 = 1$ for $i \in \{1,\ldots,4\}$, to represent the Euler angles (ϕ, θ, ψ). Since the Euclidean norm of $\mathbf{q} = (q_1, q_2, q_3, q_4)^T$ equals unity, there exist only three independent parameters. The rotation matrix, as defined in eq. (14.30), has a quaternion representation

$$R = \begin{pmatrix} q_0^2+q_1^2-q_2^2-q_3^2 & 2(q_1q_2+q_0q_3) & 2(q_1q_3-q_0q_2) \\ 2(q_1q_2-q_0q_3) & q_0^2-q_1^2+q_2^2-q_3^2 & 2(q_2q_3+q_0q_1) \\ 2(q_1q_3+q_0q_2) & 2(q_2q_3+q_0q_1) & q_0^2-q_1^2-q_2^2+q_3^2 \end{pmatrix}. \tag{14.36}$$

This offers a more efficient way to compute rotations without the necessity of computing products of trigonometric functions. The angular velocities can be computed according to

$$\begin{pmatrix} \dot{q}_0 \\ \dot{q}_1 \\ \dot{q}_2 \\ \dot{q}_3 \end{pmatrix} = \frac{1}{2}\begin{pmatrix} q_0 & -q_1 & -q_2 & -q_3 \\ q_1 & q_0 & -q_3 & q_2 \\ q_2 & q_3 & q_0 & -q_1 \\ q_3 & -q_2 & q_1 & q_0 \end{pmatrix}\begin{pmatrix} 0 \\ \omega_x \\ \omega_y \\ \omega_z \end{pmatrix}. \tag{14.37}$$

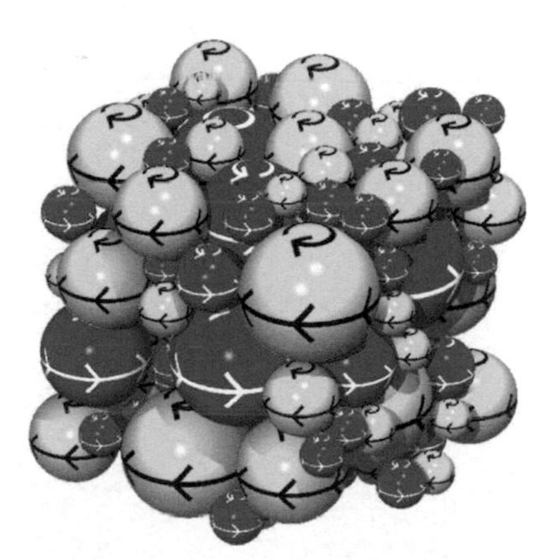

Figure 14.5 Rotating spheres in a sphere assembly as an example of rigid body dynamics [253].

Since quaternion and Eulerian angle representations are connected by a diffeomorphism, there is always the possibility of calculating the values of the Euler angles if needed:

$$\phi = \arctan\left[\frac{2(q_0q_1+q_2q_3)}{1-2\left(q_1^2+q_2^2\right)}\right], \tag{14.38}$$

$$\theta = \arcsin\left[2(q_0q_2-q_1q_3)\right], \tag{14.39}$$

$$\psi = \arctan\left[\frac{2(q_0q_3+q_1q_2)}{1-2\left(q_2^2+q_3^2\right)}\right]. \tag{14.40}$$

There is no need of calculating the Euler angles except if one wants to render the rotating object in space graphically. With the help of quaternions, we can now simulate rigid body dynamics in the following way:

Rigid Body Simulation with Quaternions

- Determine the torque in the lab frame $\widetilde{\mathbf{T}}$ and the rotation matrix R (see eq. (14.36)).
- Compute the torque in the body frame, $\mathbf{T} = R^T\widetilde{\mathbf{T}}$,
- Obtain $\omega(t + \Delta t)$ according to eq. (14.28),
- Update the rotation matrix as defined in eq. (14.36) by computing $\mathbf{q}(t+\Delta t)$ according to eq. (14.37).

In Figure 14.5, we show an example of a system of rotating particles whose dynamics has been simulated using quaternions.

15 Long-Range Potentials

In Chapter 13, we discussed different optimization techniques to speed up our MD simulations. If the potential decays sufficiently fast, it is possible to define a cutoff to neglect vanishingly small force contributions at large distances. However, if the potential of a system in d dimensions decays as $V(r) \propto r^{-d}$ or even slower, it is not possible to define a cutoff anymore. The reason is that there is a nonnegligible energy contribution at any distance. Examples of such potentials occur in electrostatic, gravitational, and dipolar interactions. To simulate such systems, we discuss the following methods:

- Ewald summation,
- particle-mesh methods (PM, PPPM & APPPM),
- reaction field method.

15.1 Ewald Summation

Paul Peter Ewald (see Figure 15.1) developed a method to compute long-range interactions in periodic systems. It has been originally developed to study ionic crystals and our goal is to apply this method to MD simulations.

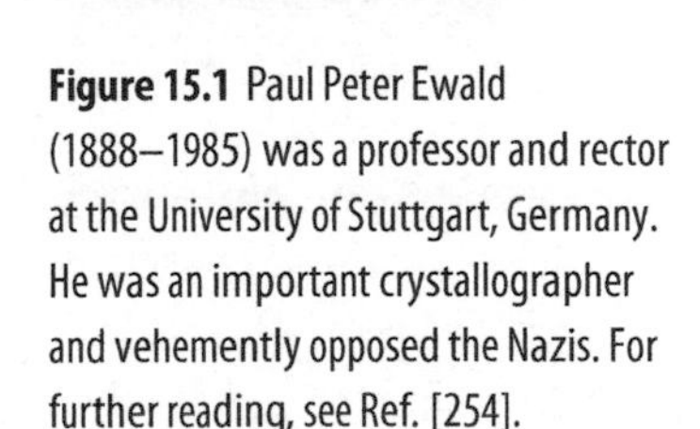

Figure 15.1 Paul Peter Ewald (1888–1985) was a professor and rector at the University of Stuttgart, Germany. He was an important crystallographer and vehemently opposed the Nazis. For further reading, see Ref. [254].

Until now, we used periodic boundary conditions for systems of finite size. This was only possible, because the interaction correlations were decaying with increasing distance. For long-range potentials, the particles are no longer uncorrelated since they are able to interact even at large distances. What we can do is periodically repeating our system by attaching its own image at its borders. This leads to a larger system size and allows us to model farther-reaching interactions. We illustrate this procedure in Figure 15.2. We treat the aforementioned repeated systems as really existing, and compute interactions between particles in our "box" of interest with all remaining particles in the system. Every copy of the system is characterized by a vector that ranges from the origin of the central system of interest to the origin of the copy. The resulting potential is thus the sum over particle interactions that repeat over several boxes. For gravitational or electrostatic $V(r) \propto r^{-1}$ potentials, we obtain

$$V = \frac{1}{2} \sum_{\mathbf{n}}{}' \sum_{i,j} z_i z_j \left| \mathbf{r}_{ij} + \mathbf{n} \right|^{-1}, \tag{15.1}$$

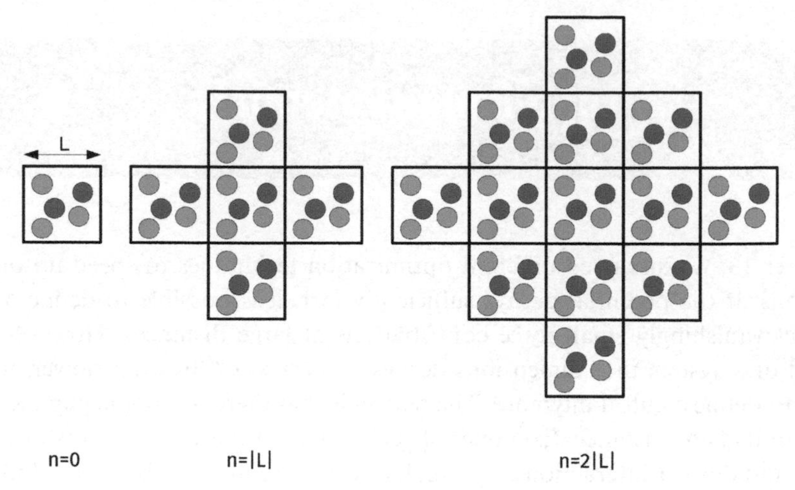

Figure 15.2 An example of an Ewald summation procedure. From left to right, images of the system are periodically attached at the boundaries.

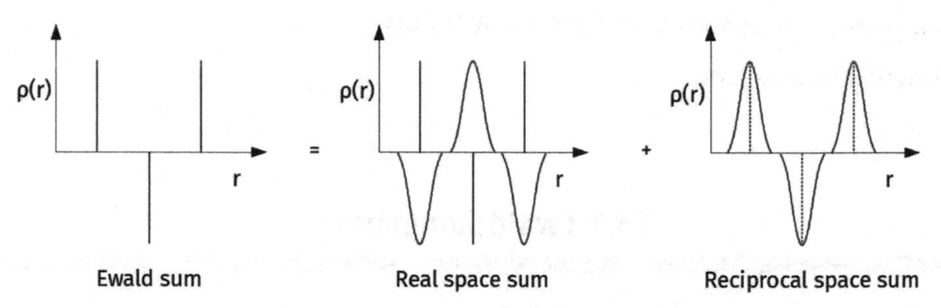

Figure 15.3 The Ewald sum is decomposed into one sum in real and another sum in reciprocal space. The charges are screened with Gaussian charge distributions. Other distributions are also possible.

where z_i and z_j represent the corresponding charges or masses. For the sake of brevity, we are using units, in which $4\pi\epsilon_0$ or $4\pi G$ are unity in eq. (15.1). The sum over

$$\mathbf{n} = \left(n_x L, n_y L, n_z L\right) \tag{15.2}$$

denotes the summation over all copies of the system of linear size L and integers n_x, n_y, and n_z. The prime indicates that we are excluding the potential-energy contribution for $\mathbf{n} = 0$ when $i = j$. The first sum comprises the central box at $\mathbf{n} = 0$ and the following terms are those for which $|\mathbf{n}| = L$. That is, $\mathbf{n} = (\pm L, 0, 0)$, $\mathbf{n} = (0, \pm L, 0)$, and $\mathbf{n} = (0, 0, \pm L)$. The Ewald sum is only conditionally convergent (i.e., the convergence depends on the order of the summation) and the convergence is very slow. Since Ewald never used a computer, he intended to sum over an infinite number of cells. In practice, we have to truncate the sum at some point and try to estimate the error. Because of the numerical difficulties of the algorithm, it is only used for systems consisting of a few particles. Several approaches are possible to improve the convergence [255, 256].

Additional Information: Screening with Gaussian Charge Distributions

One can reduce the long-range effect of point-like charges by screening them with Gaussian charge distributions of opposite sign

$$\rho_i(r) = \frac{z_i \kappa^3}{\pi^{\frac{3}{2}}} \exp\left(-\kappa^2 r^2\right) \tag{15.3}$$

with some appropriate inverse width κ. After having introduced the screening charges, we have to cancel them out again using a charge distribution like the one of eq. (15.3) but of opposite sign. This is done in Fourier space. The basic idea behind this summation is illustrated in Figure 15.3. The result is one sum in real and another sum in reciprocal space which converges faster [257]

$$\begin{aligned} V = \frac{1}{2} \sum_{i,j} \Bigg[& \sum_{\mathbf{n}}{}' z_i z_j \frac{\operatorname{erfc}\left(\kappa \left|\mathbf{r}_{ij} + \mathbf{n}\right|\right)}{\left|\mathbf{r}_{ij} + \mathbf{n}\right|} \\ & + \frac{1}{\pi L^3} \sum_{\mathbf{k} \neq 0} z_i z_j \frac{4\pi^2}{k^2} \exp\left(-\frac{k^2}{4\kappa^2}\right) \cos\left(\mathbf{k} \mathbf{r}_{ij}\right) \Bigg] \\ & - \frac{\kappa}{\sqrt{\pi}} \sum_i z_i^2 + \frac{2\pi}{3L^3} \left| \sum_i z_i \mathbf{r}_i \right|^2 \end{aligned} \tag{15.4}$$

with the *complementary error function*

$$\operatorname{erfc}(x) = \frac{2}{\sqrt{\pi}} \int_x^\infty \exp\left(-t^2\right) dt.$$

These formulas are rather clumsy and only applicable to small systems and we therefore focus on other approaches in the next sections.

15.2 Particle-Mesh Method

Another possibility of simulating long-range particle interactions is the so-called particle-mesh (PM) method. This method was invented by Eastwood and Hockney, and is very well suited for reasonably homogeneous distributions of particles [249, 258]. It is based on the following steps.

The mesh concept is similar to the linked-cell method (see Figure 13.2). Criteria for a particle-mesh scheme are as follows:

- Errors should vanish for large distances between particles.
- Momentum should always be conserved ($\mathbf{F}_{ij} = -\mathbf{F}_{ji}$).
- Charges or masses on the mesh and the interpolated forces should vary smoothly.

This method is not very efficient and time consuming for

- inhomogeneous distributions of charges or masses,
- strong correlations, like bound states in molecules,

Particle-Mesh Method

- Put a fine mesh on top of the system so that only a few particles are in each cell.
- Distribute the charges onto the mesh points.
- Calculate the electrostatic potential by solving the Poisson equation on the mesh using FFT.
- Calculate the force on each particle by numerically differentiating the potential and interpolating back from the mesh to the particle position.

- deviation from the point-like object approximation,
- complex geometries of the system.

In those cases, we may use the *particle-particle-particle-mesh* (P^3M) or *adaptive particle-particle-particle-mesh* (AP^3M) algorithms. These methods are presented later in the subsequent sections. The PM method is well suited for charged gases, because such systems are characterized by overall homogeneous particle densities.

15.2.1 Implementation

Once we implemented the grid for our d-dimensional system, there exists more than one possibility to assign charge (or mass) values to vertices. Two common options are as follows:

- nearest grid point: assign the charge to nearest grid point.
- cloud-in-cell: assign charge proportionally to the 2^d nearest grid points.

In two dimensions, a cloud in cell implementation of a charge q located at (x, y) is given by the charge distribution

$$\rho_{i\,j} = \frac{q}{2\Delta l^2}(x_{i+1} - x)(y_{i+1} - y)\,, \tag{15.5}$$

$$\rho_{i+1\,j} = \frac{q}{2\Delta l^2}(x - x_i)(y_{i+1} - y)\,, \tag{15.6}$$

$$\rho_{i\,j+1} = \frac{q}{2\Delta l^2}(x_{i+1} - x)(y - y_i)\,, \tag{15.7}$$

$$\rho_{i+1\,j+1} = \frac{q}{2\Delta l^2}(x - x_i)(y - y_i)\,, \tag{15.8}$$

where Δl is the grid spacing. The potential at a certain point $\mathbf{r}$ can be computed by the convolution of $\rho(\mathbf{r})$ with the Green's function:

$$\phi(\mathbf{r}) = \int \rho(\mathbf{r}')\, G(\mathbf{r}, \mathbf{r}')\, \mathrm{d}\mathbf{r}'\,. \tag{15.9}$$

For electrostatic and gravitational interactions, one has to solve the Poisson equation for which the corresponding Green's function is $G(\mathbf{r}, \mathbf{r}') \propto |\mathbf{r} - \mathbf{r}'|^{-1}$. In Fourier space, eq. (15.9) corresponds to a multiplication with the Fourier transform $\widehat{G}(\mathbf{k}) \propto \mathbf{k}^{-2}$ such that $\hat{\phi}(\mathbf{k}) = \hat{\rho}(\mathbf{k})\,\widehat{G}(\mathbf{k})$ (convolution theorem). After applying the inverse Fourier

Additional Information: Fourier Transform

Let $f(t)$ be a square-integrable function (i. e., $\int_{-\infty}^{\infty} |f(t)|\, \mathrm{d}t < \infty$), then the Fourier transform is

$$\hat{f}(\omega) = \int_{-\infty}^{\infty} f(t) e^{-i\omega t}\, \mathrm{d}t, \tag{15.10}$$

and the inverse Fourier transform is

$$f(t) = \frac{1}{2\pi} \int_{-\infty}^{\infty} \hat{f}(\omega) e^{i\omega t}\, \mathrm{d}\omega. \tag{15.11}$$

transform, one obtains ϕ. If one performs the Fourier transform using fast Fourier transform (FFT) [259, 260], the method has computational complexity $O(N \log N)$, where N is the number of mesh points. See the gray box on page 174 and Ref. [261] for more details on the Fourier transform.

Knowing the potential, we compute the force

$$\mathbf{F}\left(\mathbf{r}_{ij}\right) = -\nabla\phi\left(\mathbf{r}_{ij}\right) \tag{15.12}$$

at every vertex of the mesh and then interpolate the forces on the corners of the mesh to find the force that is exerted by the field on a particle inside the cell. Note that the PM algorithm is not very well suited for inhomogeneous distributions of particles, strong correlations like bound states, and complex geometries. In these cases, the P³M and AP³M methods, tree codes, and multiple expansions constitute possible extensions of the method. One important application is the very heterogeneous distribution of stars and galaxies in astrophysics. For further reading, we recommend Ref. [262].

15.2.2 P³M (Particle-Particle-Particle-Mesh)

For particles with close neighbors (e. g., binary stars), we decompose the force acting on it into two terms:

$$\mathbf{F} = \mathbf{F}_s + \mathbf{F}_l\,, \tag{15.17}$$

where $\mathbf{F}_s$ accounts for short-range interactions and $\mathbf{F}_l$ describes interactions over large distances. The short-range interactions are simulated by solving Newton's equation of motion whereas the PM method is applied to the long-range forces. This approach has the drawback that the long-range potential field has to be computed for each close particle and we still have not solved the problem of strong heterogeneity.

15.2.3 AP³M (Adaptive Particle-Particle-Particle-Mesh)

For simulations of the mass distribution in the Universe, most of the grid points will be empty, making the previous algorithms very inefficient. A possibility to

Additional Information: Discrete and Fast Fourier Transform

Computing the Fourier transform of discrete data points on a computer is done in terms of the *discrete Fourier transform* (DFT). To define the DFT of a sequence of N data points v_k, we first construct an appropriate basis using the Nth roots of unity

$$\omega_N := \exp\left(-\frac{2\pi i}{N}\right). \tag{15.13}$$

Then, the DFT $\hat{v}$ of a vector $v \in \mathbb{C}^N$ is defined as

$$\hat{v}_j = \frac{1}{\sqrt{N}} \sum_{k=0}^{N-1} \omega_N^{kj} v_k . \tag{15.14}$$

The computation of $\hat{v}_j$ requires $\mathcal{O}(N^2)$ operations. Fast Fourier transform methods such as the algorithm of Cooley and Tukey [259] achieve performances of $\mathcal{O}(N \log N)$ by computing the DFT in terms of smaller even and odd-indexed parts of v_k:

$$\hat{v}_j = \frac{1}{\sqrt{N}} \left(\sum_{k=0}^{N/2-1} \omega_N^{2kj} v_{2k} + \sum_{k=0}^{N/2-1} \omega_N^{(2k+1)j} v_{2k+1} \right). \tag{15.15}$$

Defining an even-indexed term

$$e_j = \frac{1}{\sqrt{N}} \sum_{k=0}^{N/2-1} \omega_N^{2kj} v_{2k}$$

and an odd-indexed term

$$o_j = \frac{1}{\sqrt{N}} \omega_N^j \sum_{k=0}^{N/2-1} \omega_N^{2kj} v_{2k+1} ,$$

we can rewrite eq. (15.15) in terms of

$$\hat{v}_j = e_j + \omega_N^j o_j . \tag{15.16}$$

Similarly, we can write $\hat{v}_{j+N/2} = e_j - \omega_N^j o_j$. We can further decompose e_j and o_j in even- and odd-indexed parts and so forth. This recursive way of computing $\hat{v}_j$ is the basis of the Cooley and Tukey FFT algorithm.

deal with strong heterogeneities is to adapt the mesh to the particle density (i.e., using an adaptive mesh with higher resolution where the particle density is higher). One implementation is to refine a cell as long as the number of particles in the cell exceeds a certain threshold. Tree codes can be used to locate the cells that emerge in the refinement process [262]. The advances in the simulation of many-body systems often depend on the development of such bookkeeping techniques. Current simulation frameworks are able to simulate the interaction of up to 10^{12} particles [263]. For further reading, we recommend the textbook *Computer Simulations Using Particles* [249].

Exercise: Simulating Gravitational Interactions with the Particle-Mesh Method

Start from a set of massive point-like particles randomly distributed in space. Next, generate a grid with $N = L \times L$ grid cells. You can, for instance, start with $N = 128 \times 128$ grid cells.

For each grid cell, aggregate the total mass density according to eq. (15.8), solve the Poisson equation for Newtonian gravity

$$\nabla^2 \phi = 4\pi G \rho \,, \tag{15.18}$$

where ϕ is the gravitational potential, G is the gravitational constant, and ρ is the local mass density. Use the gravitational potential ϕ to compute the resulting gravitational forces according to eq. (15.12). Note that one possible discretization of eq. (15.12) is ($\Delta x = \Delta y = 1$)

$$\mathbf{F}\left(\mathbf{r}_{ij}\right) = -\begin{pmatrix} \phi(\mathbf{r}_{i+1j}) - \phi(\mathbf{r}_{ij}) \\ \phi(\mathbf{r}_{ij+1}) - \phi(\mathbf{r}_{ij}) \end{pmatrix}. \tag{15.19}$$

After determining the forces, advance the particle positions in time and determine the new particle densities in each grid cell.

Figure 15.4 Lars Onsager (1903–1976) was a professor for theoretical chemistry at Yale University. He received the Nobel Prize in Chemistry in 1968.

15.3 Reaction Field Method

For more complex interactions such as composed molecules or non-point-like particles, a good solution is to ignore the complexity of distant particles and only take into account their mean effect while explicitly calculating the interaction with close particles. The concept has its root in the work of Onsager (see Figure 15.4) on the dielectric constant but it was introduced as an explicit computational technique in the 1970s [264–266]. This method is mostly used for the simulation of dipole–dipole interactions.

To outline this method, we consider a sphere N_i of radius r_c around particle i (see Figure 15.5). The region outside the sphere is described by an effective dielectric medium and the dipole moments $\boldsymbol{\mu}_j$ of the N_i particles j within the sphere induce an electric field (reaction field) in the external dielectric medium

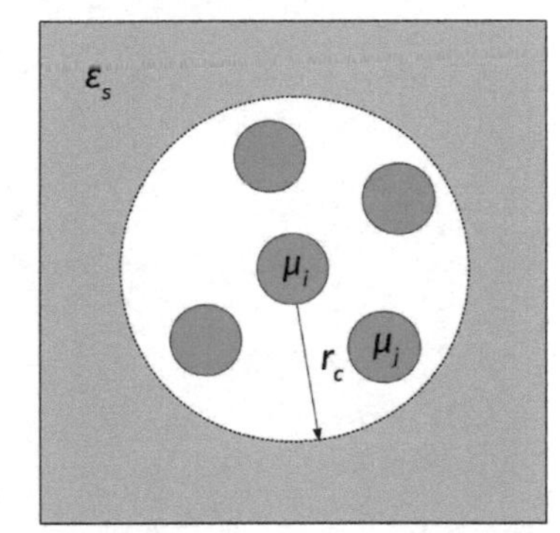

Figure 15.5 In the reaction-field method, particle i is surrounded by a spherical region of radius r_c. The interactions between all particles within the sphere N_i (blue disks) are treated as dipole interactions with dipole moments $\boldsymbol{\mu}_j$ ($j \in N_i$). The region outside the sphere is described by an effective dielectric constant ϵ_s.

$$\mathbf{E}_i = \frac{2(\epsilon_s - 1)}{2\epsilon_s + 1} \frac{1}{r_c^3} \sum_{j \in N_i} \boldsymbol{\mu}_j \,, \tag{15.20}$$

where ϵ_s is the effective permittivity of the dielectric continuum outside the sphere.

The resulting total force exerted on particle i is then

$$\mathbf{F}_i = \sum_{j \in N_i} \mathbf{F}_{ij} + \boldsymbol{\mu}_i \wedge \mathbf{E}_i \,. \tag{15.21}$$

As the particles are moving, the number of particles inside the cavity changes. This causes force discontinuities because of the sudden addition or removal of a particle.

To avoid this effect, one can weight interactions between particles i and j in eq. (15.21) according to

$$g(r_j) = \begin{cases} 1, & \text{for } r_j < r_t, \\ \frac{r_c - r_j}{r_c - r_t}, & \text{for } r_t \le r_j \le r_c, \\ 0, & \text{for } r_c < r_j, \end{cases} \tag{15.22}$$

where $r_t = 0.9 r_c$.

16 Canonical Ensemble

After having discussed various techniques to simulate the interaction of particles, we now focus again on their statistical description. Often we are not interested in the properties of single particles but rather in macroscopic quantities such as temperature, pressure, and density. Up until now, we have not introduced any method to control such macroscopic properties in our MD simulations. We only imposed energy conservation and were thus simulating the microcanonical ensemble. Experiments are, however, usually conducted at constant temperature and not at constant energy because systems are usually able to exchange energy with their environment. In this chapter, we will therefore discuss some important methods such as the Nosé–Hoover thermostat and the Parrinello–Rahman barostat to control temperature and pressure in our MD simulations.

To be able to control temperature, we must first couple our system to a heat bath. There are various options to do this [267]:

- rescaling of velocities,
- introducing constraints (Hoover),
- Nosé–Hoover thermostat,
- stochastic method (Anderson).

However, before focusing on a more detailed discussion of these methods, we will precisely define the concept of temperature used in the subsequent sections. We start from the equipartition theorem [268]

$$\left\langle q_\mu \frac{\partial \mathcal{H}}{\partial q_\nu} \right\rangle = \left\langle p_\mu \frac{\partial \mathcal{H}}{\partial p_\nu} \right\rangle = \delta_{\mu\nu} k_B T, \tag{16.1}$$

for a Hamiltonian $\mathcal{H}$ with the generalized coordinates $\mathbf{p}$ and $\mathbf{q}$. We consider a classical system whose Hamiltonian is given by

$$\mathcal{H} = \sum_{i=1}^{N} \frac{\mathbf{p}_i^2}{2m_i} + V(\mathbf{x}_1, \ldots, \mathbf{x}_N), \tag{16.2}$$

where $\mathbf{p}_i = m_i \dot{\mathbf{x}}_i$. We define the instantaneous temperature [257]

$$\mathcal{T} = \frac{2}{3k_B (N-1)} \sum_{i=1}^{N} \frac{\mathbf{p}_i^2}{2m_i}, \tag{16.3}$$

which, after inserting eq. (16.2) into eq. (16.1), is, when averaged, equal to T.

16.1 Velocity Rescaling

We now discuss a straightforward yet physically incorrect method to simulate a canonical ensemble of particles. Equation (16.3) shows how the temperature depends on momentum (i.e., on velocity). Intuitively, we should be able to adjust the instantaneous temperature of the system by rescaling the particle velocity. The algorithm is as follows:

Velocity Rescaling

Rescale the velocities of all particles according to

$$\mathbf{v}_i \rightarrow \alpha \mathbf{v}_i. \tag{16.4}$$

The measured temperature is proportional to the squared velocities, so

$$\mathcal{T} \rightarrow \alpha^2 \mathcal{T}. \tag{16.5}$$

Therefore, we have to set

$$\alpha = \sqrt{\frac{T}{\mathcal{T}}}, \tag{16.6}$$

to simulate at a certain desired temperature T.

This method is very easy to implement. However, the problem is that the resulting velocity distribution deviates from the Maxwell–Boltzmann distribution that one expects in the canonical ensemble. A modification of this method makes use of an additional parameter τ_T, which describes the coupling to a heat bath.

Berendsen Thermostat

For the so-called Berendsen thermostat, the velocity-scaling factor is [269]

$$\alpha = \sqrt{1 + \frac{\Delta t}{\tau_T}\left(\frac{T}{\mathcal{T}} - 1\right)}. \tag{16.7}$$

Still, we do not recover the canonical velocity distribution. Velocity rescaling should only be applied to initialize an MD simulation at a given temperature.

16.2 Constraint Method

Another possibility to keep the temperature of the system constant is to add a friction term to the equations of motion:

$$\dot{\mathbf{p}}_i = \mathbf{f}_i - \xi \mathbf{p}_i\,. \tag{16.8}$$

Figure 16.1 William G. Hoover is a computational physicist who worked at the University of California's Livermore Laboratory. The photograph is taken from Ref. [270].

Various definitions of the friction coefficient ξ are possible. The original proposal of Hoover (see Figure 16.1) is based on the following constant temperature condition [271, 272]:

$$\dot{\mathcal{T}} \propto \frac{\mathrm{d}}{\mathrm{d}t}\left(\sum_{i=1}^{N} \mathbf{p}_i^2\right) \propto \sum_{i=1}^{N} \dot{\mathbf{p}}_i \mathbf{p}_i = 0\,. \tag{16.9}$$

Combining Eqs. (16.8) and (16.9) yields

$$\xi = \frac{\sum_{i=1}^{N} \mathbf{f}_i \mathbf{p}_i}{\sum_{i=1}^{N} |\mathbf{p}_i|^2}\,. \tag{16.10}$$

This method makes it necessary to already start at the desired temperature. Instead of eq. (16.10), another possibility is to determine the friction coefficient according to ([269, 273], see Figure 16.2)

$$\xi = \gamma\left(1 - \frac{T}{\mathcal{T}}\right), \tag{16.11}$$

or [273]

$$\dot{\xi} = \frac{f k_B}{Q}\left(\mathcal{T} - T\right)\,. \tag{16.12}$$

Figure 16.2 Herman Berendsen (1934–2019) was a professor of physical chemistry at the University of Groningen. He was one of the pioneers of MD simulations of molecules [250]. Photograph courtesy the University of Groningen.

The parameters γ and Q determine how fast the temperature attains the desired one, and f is the number of degrees of freedom. Still, all these methods have the drawback that they either do not recover the Maxwell–Boltzmann velocity distribution or not satisfy the H-theorem. In addition, the approaches of Eqs. (16.11) and (16.12) are not time reversible.

16.3 Nosé–Hoover Thermostat

In order to overcome the problem of the wrong velocity distribution, we are now going to discuss the Nosé–Hoover thermostat as the correct method to simulate heat bath particle dynamics. Shuichi Nosé (see Figure 16.3) introduced a new degree of freedom s that describes the heat bath [273–275]. The potential and kinetic energies associated with s are

$$\begin{aligned} \mathcal{V}(s) &= (3N+1)\,k_B T \ln s\,,\\ K(s) &= \frac{1}{2} Q \dot{s}^2\,, \end{aligned} \tag{16.13}$$

Figure 16.3 Shuichi Nosé (1951–2005) was a professor of theoretical physics at the Keio University, Tokyo.

where the parameter Q is called thermal inertia. The new degree of freedom s is used to apply a Sundman time-transformation and a change of variables [276]:

$$\mathrm{d}t' = s\,\mathrm{d}t \quad \text{and} \quad \mathbf{p}_i' = s\,\mathbf{p}_i\,. \tag{16.14}$$

The resulting rescaled velocities are

$$\mathbf{v}_i' = \frac{\mathrm{d}\mathbf{x}_i}{\mathrm{d}t'} = \frac{\mathrm{d}\mathbf{x}_i}{\mathrm{d}t}\frac{\mathrm{d}t}{\mathrm{d}t'} = \frac{\mathbf{v}_i}{s}\,. \tag{16.15}$$

The Hamiltonian is thus

$$\mathcal{H} = \sum_{i=1}^{N} \frac{\mathbf{p}_i'^2}{2m_i s^2} + \frac{p_s{}^2}{2Q} + V(\mathbf{x}_1, \ldots, \mathbf{x}_N) + \mathcal{V}(s)\,, \tag{16.16}$$

where we have set $p_s = Q\dot{s}$. Applying the Hamilton equations (12.9), the velocities are

$$\begin{aligned} \frac{\mathrm{d}\mathbf{x}_i}{\mathrm{d}t'} &= \nabla_{\mathbf{p}_i'}\mathcal{H} = \frac{\mathbf{p}_i'}{m_i s^2}, \\ \frac{\mathrm{d}s}{\mathrm{d}t'} &= \frac{\partial \mathcal{H}}{\partial p_s} = \frac{p_s}{Q}\,. \end{aligned} \tag{16.17}$$

With $\mathbf{p}_i' = m_i s^2 \dot{\mathbf{x}}_i$ (in virtual time t') we find

$$\mathbf{f}_i = \frac{\mathrm{d}\mathbf{p}_i'}{\mathrm{d}t'} = -\nabla_{\mathbf{x}_i}\mathcal{H} = -\nabla_{\mathbf{x}_i}V(\mathbf{x}_1, \ldots, \mathbf{x}_N) = 2m_i s\dot{s}\dot{\mathbf{x}}_i + m_i s^2 \ddot{\mathbf{x}}_i, \tag{16.18}$$

and

$$\frac{\mathrm{d}\mathbf{p}_s}{\mathrm{d}t'} = -\frac{\partial \mathcal{H}}{\partial s} = \frac{1}{s}\left[\sum_{i=1}^{N} \frac{\mathbf{p}_i'^2}{m_i s^2} - (3N+1)\,k_B T\right]. \tag{16.19}$$

Based on these equations, we find for the equations of motion in rescaled time t' (see eq. (16.14))

$$m_i s^2 \ddot{\mathbf{x}}_i = \mathbf{f}_i - 2m_i \dot{s} s \dot{\mathbf{x}}_i \quad \text{with} \quad i \in \{1, \ldots, N\}, \tag{16.20}$$

and

$$Q\ddot{s} = \sum_{i=1}^{N} m_i s \dot{\mathbf{x}}_i^2 - \frac{1}{s}(3N+1)\,k_B T\,. \tag{16.21}$$

Note that Eqs. (16.20) and (16.21) are coupled. This reflects the fact that the two systems are not isolated and exchange energy in form of heat. In order to obtain the equations of motion in real time, we have to remind ourselves that $\mathrm{d}t = \mathrm{d}t'/s$ and $\mathbf{p}_i' = s\mathbf{p}_i$. Thus, we find for the velocities

$$\begin{aligned} \frac{\mathrm{d}\mathbf{x}_i}{\mathrm{d}t} &= s\frac{\mathrm{d}\mathbf{x}_i}{\mathrm{d}t'} = \frac{\mathbf{p}_i'}{m_i s} = \frac{\mathbf{p}_i}{m_i}, \\ \frac{\mathrm{d}s}{\mathrm{d}t} &= s\frac{\mathrm{d}s}{\mathrm{d}t'} = s\frac{\mathbf{p}_s}{Q}, \end{aligned} \tag{16.22}$$

and for the forces,

$$\frac{\mathrm{d}\mathbf{p}_i}{\mathrm{d}t} = s\frac{\mathrm{d}}{\mathrm{d}t'}\left(\frac{\mathbf{p}'_i}{s}\right) = \frac{\mathrm{d}\mathbf{p}'_i}{\mathrm{d}t'} - \frac{1}{s}\frac{\mathrm{d}s}{\mathrm{d}t'}\mathbf{p}'_i = \mathbf{f}_i - \frac{1}{s}\frac{\mathrm{d}s}{\mathrm{d}t}\mathbf{p}_i\,,$$

$$\frac{\mathrm{d}\mathbf{p}_s}{\mathrm{d}t} = s\frac{\mathrm{d}p_s}{\mathrm{d}t'} = \sum_{i=1}^{N}\frac{\mathbf{p}_i^2}{m_i} - (3N+1)\,k_B T\,. \tag{16.23}$$

Nosé–Hoover Equations of Motion

We use $\xi = \frac{\mathrm{d}\ln(s)}{\mathrm{d}t} = \frac{\dot{s}}{s}$ to denote a friction term and rewrite the equations of motion (16.20) and (16.21) in real time:

$$\ddot{\mathbf{x}}_i = \frac{\mathbf{f}_i}{m_i} - \xi\dot{\mathbf{x}}_i\,, \tag{16.24}$$

and

$$\frac{1}{2}Q\dot{\xi} = \frac{1}{2}\sum_{i=1}^{N} m_i\dot{\mathbf{x}}_i^2 - \frac{1}{2}(3N+1)\,k_B T\,. \tag{16.25}$$

The first term in eq. (16.25) represents the measured kinetic energy and the second term corresponds to the desired kinetic energy. The thermal inertia Q describes the coupling to the heat bath. The smaller its value, the stronger the system reacts to temperature fluctuations. For $Q \to \infty$, we recover microcanonical MD, and for too small values of Q we find spurious temperature oscillations. We show such oscillations in Figure 16.4. A reasonable value of Q is characterized by temperature fluctuations that are normally distributed:

$$\overline{\Delta T} = \sqrt{\frac{2}{Nd}}\overline{T}\,, \tag{16.26}$$

where d is the system's dimension and N is the number of particles.

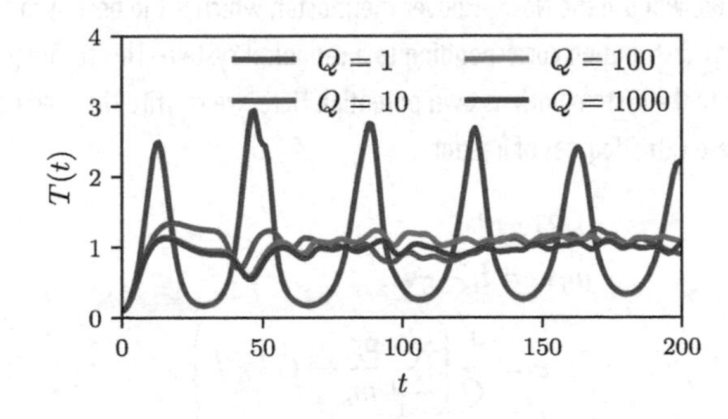

Figure 16.4 Temperature fluctuations of the Nosé–Hoover thermostat set at $T = 1$ for 64 particles and different values of the thermal inertia Q. Particle interactions are described by a Lennard–Jones potential.

Additional Information: Nosé–Hoover Particles Follow a Canonical Velocity Distribution

We now show that the Nosé–Hoover thermostat recovers the canonical partition function. Therefore, we start from microcanonical MD of the closed system of particles and heat bath, and the corresponding partition function

$$Z = \int \delta(\mathcal{H} - E)\, \mathrm{d}s\, \mathrm{d}p_s\, \mathrm{d}^{3N}x'\, \mathrm{d}^{3N}p', \tag{16.27}$$

where integration with respect to x and p has to be taken over a three-dimensional space for N particles. With $\mathcal{H} = \mathcal{H}_1 + (3N+1)\, k_B T \ln(s)$ in real time, we find

$$\begin{aligned} Z &= \int \delta\left[(\mathcal{H}_1 - E) + (3N+1)\, k_B T \ln(s)\right] s^{3N} \mathrm{d}s\, \mathrm{d}p_s\, \mathrm{d}^{3N}x\, \mathrm{d}^{3N}p \\ &= \int \delta\left[s - e^{-\frac{\mathcal{H}_1 - E}{(3N+1)k_B T}}\right] \frac{s^{3N+1}}{(3N+1)\, k_B T} \mathrm{d}s\, \mathrm{d}p_s\, \mathrm{d}^{3N}x\, \mathrm{d}^{3N}p, \end{aligned} \tag{16.28}$$

where we used the identity $\delta\left[f(s)\right] = \delta(s - s_0) / f'(s)$ with $f(s_0) = 0$ in the second step. Integrating eq. (16.28) over s yields

$$\begin{aligned} Z &= \int \frac{1}{(3N+1)\, k_B T} e^{-\frac{\mathcal{H}_1 - E}{k_B T}} \mathrm{d}p_s\, \mathrm{d}^{3N}x\, \mathrm{d}^{3N}p \\ &= \int e^{-\frac{\mathcal{H}_0}{k_B T}} \mathrm{d}^{3N}x\, \mathrm{d}^{3N}p \int \frac{1}{(3N+1)\, k_B T} e^{\frac{E - \frac{p_s^2}{2Q}}{k_B T}} \mathrm{d}p_s, \end{aligned} \tag{16.29}$$

with $\mathcal{H}_1 = \mathcal{H}_0 + \frac{p_s^2}{2Q}$. The first part of the last expression is the canonical partition function and the last part is a constant factor. In 1985, Hoover proved that the Nosé–Hoover thermostat is the only method with a single friction parameter that gives the correct canonical distribution. Hoover also showed that this thermostat satisfies the Liouville equation. That is, the density of states is conserved in phase space.

Exercise: Canonical MD Simulations

We will now simulate a system in contact with a thermal reservoir at fixed temperature T. There are several techniques to do this. We will use the Nosé–Hoover thermostat, which is the best in the sense that it reproduces the correct energy distribution corresponding to a canonical system. This technique introduces a new degree of freedom s into the system with its own potential. Here, we rewrite the final equations of motions for a system having these extra degrees of freedom:

$$\begin{aligned} \dot{\mathbf{x}}_i &= \mathbf{v}_i, \\ m_i \dot{\mathbf{v}}_i &= \mathbf{f}_i - \xi \mathbf{v}_i, \\ \dot{\xi} &= \frac{1}{Q}\left(\sum_{i=1}^{N} \frac{\mathbf{p}_i^{\,2}}{m_i} - G\, k_B\, T\right), \end{aligned}$$

Exercise: Canonical MD Simulations (*cont.*)

where $G = 3N + 1$ is the number of degrees of freedom, T is the (desired) temperature of the system, k_B is the Boltzmann factor, and Q is the inertia corresponding to the extra degree of freedom s. The friction coefficient ξ is defined as

$$\xi = \frac{\dot{s}}{s} .$$

Note that the coordinate s of the extra degree of freedom is not needed in the simulation. One can discretize the above equations easily with the Verlet method.
Extend your MD code to simulate a canonical system with a given temperature.

1. Start with an initial configuration with instant temperature $\mathcal{T}_0$, and set the desired temperature T to a different value and observe the behavior of T in time.
2. When the system reaches the equilibrium state, compute the total energy of the system and compute its distribution. Do you get the correct distribution?
3. Compute the heat capacity of the system as a function of temperature using the fluctuation-dissipation theorem (see eq. (3.27)).

Please consider the following points:

- Try to adjust the value of Q so that the system reaches the equilibrium fast and still has reasonably small fluctuations (in T and/or the total energy). (For large Q, the convergence to equilibrium is slow and the fluctuations are small while for small Q it is the other way around.)

16.4 Stochastic Method

Another method to perform canonical MD simulations has been proposed by Andersen [277] (see Figure 16.5). The so-called *stochastic method* is a combination of MD and a Monte Carlo algorithm.

Stochastic Method

At every nth time step, randomly select one particle and give it a new momentum p, which is distributed according to the Maxwell–Boltzmann distribution

$$P(\mathbf{p}) = \frac{1}{(\pi k_B T)^{\frac{3}{2}}} e^{-\frac{\mathbf{p}^2}{k_B T}}, \tag{16.30}$$

where T is the desired temperature.

At every nth time step, a particle is selected uniformly at random and given a new momentum according to eq. (16.30). If n is too small, we recover pure Monte Carlo and loose the ability to simulate particle interactions in real physical time. If n is too large, the coupling to the heat bath is too weak, equilibration is slow, and we will essentially simulate a microcanonical system.

16.5 Constant Pressure

Figure 16.5 Hans C. Andersen is a professor emeritus of chemistry at Stanford University.

In addition to the simulation of MD systems with constant temperature, another important situation is that of constant pressure. To formulate constant pressure MD, we again consider the equipartition theorem and a Hamiltonian as defined by Eqs. (16.1) and (16.2), and find

$$\frac{1}{3}\left\langle \sum_{i=1}^{N} \mathbf{x}_i \cdot [\nabla_{\mathbf{x}_i} V(\mathbf{x})] \right\rangle = Nk_B T \,. \tag{16.31}$$

We now distinguish between particle–particle and particle–wall forces $\mathbf{f}^{\mathrm{part}}$ and $\mathbf{f}^{\mathrm{ext}}$, respectively. This leads to

$$\begin{aligned} \frac{1}{3}\left\langle \sum_{i=1}^{N} \mathbf{x}_i \cdot [\nabla_{\mathbf{x}_i} V(\mathbf{x})] \right\rangle &= -\frac{1}{3}\left\langle \sum_{i=1}^{N} \mathbf{x}_i \cdot \left(\mathbf{f}_i^{\mathrm{part}} + \mathbf{f}_i^{\mathrm{ext}}\right) \right\rangle \\ &= -\frac{1}{3}\left\langle \sum_{i=1}^{N} \mathbf{x}_i \cdot \mathbf{f}_i^{\mathrm{part}} \right\rangle - \frac{1}{3}\left\langle \sum_{i=1}^{N} \mathbf{x}_i \cdot \mathbf{f}_i^{\mathrm{ext}} \right\rangle . \end{aligned} \tag{16.32}$$

We define

$$w = \frac{1}{3}\sum_{i=1}^{N} \mathbf{x}_i \cdot \mathbf{f}_i^{\mathrm{part}} \tag{16.33}$$

as the *virial*. Using

$$\frac{1}{3}\left\langle \sum_{i=1}^{N} \mathbf{x}_i \cdot \mathbf{f}_i^{\mathrm{ext}} \right\rangle = -\frac{1}{3}\int_{\Gamma} p\mathbf{x}\mathrm{d}\mathbf{A} = -\frac{1}{3}p\int_{V} (\nabla \cdot \mathbf{x})\,\mathrm{d}\mathbf{V} = -pV \,, \tag{16.34}$$

we define the instantaneous pressure $\mathcal{P}$ via

$$\mathcal{P}V = Nk_B T + \langle w \rangle \,. \tag{16.35}$$

Similar to the thermal inertia Q of the Nosé–Hoover thermostat, we introduce a parameter W which adjusts the pressure of the system. The ratio $\mathcal{V}$ between the new and old volumes now becomes the new degree of freedom and instead of time, we now rescale the spatial coordinate x according to

$$\mathbf{x} \to \mathcal{V}^{\frac{1}{3}}\mathbf{x} \,. \tag{16.36}$$

Similar to rescaling the time derivative in Section 16.3, here we have to rescale the spatial derivative. The outlined rescaling is only valid in isotropic systems where a change in length is the same in every direction. The rescaled Hamiltonian is then given by

$$\mathcal{H} = \sum_{i=1}^{N} \frac{1}{2} m_i \dot{\mathbf{x}}_i^2 + \frac{1}{2} W \dot{\mathcal{V}}^2 + V(\mathbf{x}_1, \ldots, \mathbf{x}_N) + p\mathcal{V}, \tag{16.37}$$

where the new variable $\mathcal{V}$ is the relative volume change controlled by a piston of mass W (see Figure 16.6) that also defines the canonical momentum

$$p_{\mathcal{V}} = W\dot{\mathcal{V}}. \tag{16.38}$$

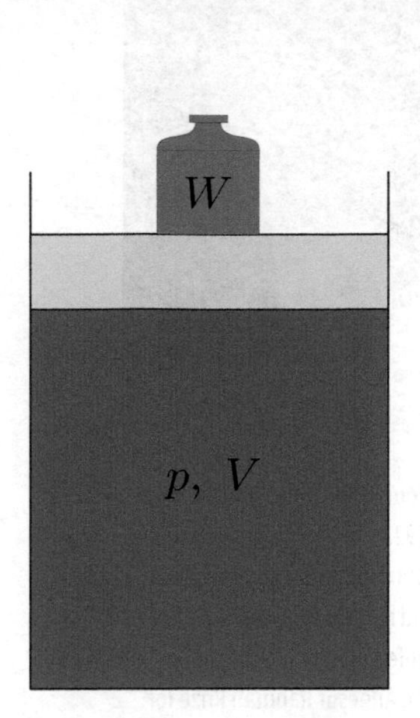

Figure 16.6 A weight of mass W exerts a pressure p on the system with volume V.

We again derive the equations of motion from the Hamiltonian. The velocities are

$$\begin{aligned} \frac{d\mathbf{x}_i}{dt} &= \nabla_{\mathbf{p}_i} \mathcal{H} = \frac{\mathbf{p}_i}{m_i \mathcal{V}^{\frac{1}{3}}}, \\ \frac{d\mathcal{V}}{dt} &= \frac{\partial \mathcal{H}}{\partial p_{\mathcal{V}}} = \frac{p_{\mathcal{V}}}{W}, \end{aligned} \tag{16.39}$$

and the corresponding forces are

$$\begin{aligned} \frac{d\mathbf{p}_i}{dt} &= -\nabla_{\mathbf{x}_i} \mathcal{H} = -\nabla_{\mathbf{x}_i} V\left(\mathcal{V}^{\frac{1}{3}} \mathbf{x}_i\right) = \mathbf{f}_i, \\ \frac{dp_{\mathcal{V}}}{dt} &= -\frac{\partial \mathcal{H}}{\partial \mathcal{V}} = -\frac{1}{3\mathcal{V}} \sum_{i=1}^{N} \left[\mathbf{x}_i \cdot \nabla_{\mathbf{x}_i} V\left(\mathcal{V}^{\frac{1}{3}} \mathbf{x}_i\right)\right] - p. \end{aligned} \tag{16.40}$$

Finally, the equations of motion of the so-called Berendsen barostat are

$$\begin{aligned} \ddot{\mathbf{x}}_i &= \frac{\mathbf{f}_i}{m_i} - \frac{\dot{\mathcal{V}}}{3\mathcal{V}} \dot{\mathbf{x}}_i, \\ W\ddot{\mathcal{V}} &= \frac{1}{3\mathcal{V}} \sum_{i=1}^{N} m_i \dot{\mathbf{x}}_i^2 + \frac{1}{3\mathcal{V}} \sum_{i=1}^{N} \mathbf{f}_i \mathbf{x}_i - p. \end{aligned} \tag{16.41}$$

16.6 Parrinello–Rahman Barostat

For the Berendsen barostat, an isotropic rescaling of space and an isotropic medium are assumed. However, this assumption may not be satisfied for all systems since, for instance, the volume of certain crystals cannot be rescaled equally in every direction. The corresponding generalization of the Berendsen barostat is an anisotropic scaling of a box, spanned by three vectors, **a**, **b**, and **c**, of volume

$$\mathcal{V} = \mathbf{a} \cdot (\mathbf{b} \wedge \mathbf{c}) = \det(H) \quad \text{with} \quad H = \{\mathbf{a}, \mathbf{b}, \mathbf{c}\}. \tag{16.42}$$

We describe the position of a particle i in the box by

$$\mathbf{r}_i = H\mathbf{s}_i = x_i\mathbf{a} + y_i\mathbf{b} + z_i\mathbf{c} \quad \text{with} \quad 0 < x_i, y_i, z_i < 1. \tag{16.43}$$

Figure 16.7 Aneesur Rahman (1927–1987) is known for his contributions to computational physics and molecular dynamics. He was a professor at the University of Minnesota. The Aneesur Rahman Prize for Computational Physics awarded by the American Physical Society is the most prestigious prize for computational physics.

Therefore, the distance between two particles i and j is

$$\mathbf{r}_{ij}^2 = \mathbf{s}_{ij}^T G \, \mathbf{s}_{ij} \quad \text{with } G = H^T H. \tag{16.44}$$

According to Parrinello (see Figure 22.3) and Rahman (see Figure 16.7) [331], the Hamiltonian that we defined previously in eq. (16.37) can be rewritten in terms of the rescaled coordinates (16.44):

$$\mathcal{H} = \frac{1}{2}\sum_i m_i \dot{\mathbf{s}}_i^T G \dot{\mathbf{s}}_i + \sum_{ij} V\left(\mathbf{r}_{ij}\right) + \frac{1}{2} W \mathrm{Tr}\left(\dot{H}^T \dot{H}\right) + p\mathcal{V}. \tag{16.45}$$

The corresponding equations of motion are

$$\begin{aligned} m_i \ddot{\mathbf{s}}_i &= H^{-1}\mathbf{f}_i - m_i G^{-1}\left(\dot{G}\dot{\mathbf{s}}_i\right), \\ W\ddot{H} &= p\mathcal{V}\left(H^{-1}\right)^T. \end{aligned} \tag{16.46}$$

The Parrinello–Rahman barostat is important for simulating crystals with applications in solid-state physics and material science. Unlike the Hoover thermostat, here the new degree of freedom is not a simple scalar but a matrix H that accounts for the anisotropy of the system.

Inelastic Collisions in Molecular Dynamics 17

In many cases, loss of energy is not just a minor unwanted effect of some otherwise well-behaved and converging mathematical method. From landslides to turbulent flow in aircraft engines, dissipation is often necessary for the simulations to be stable and important in practice. Often macroscopic systems are not closed and energy leaks out in the form of wear, shape deformation, acoustic emission, and other degrees of freedom that one cannot or does not want to include explicitly in the calculation. To mathematically describe inelastic collisions, we outline some concepts such as damped oscillators and plastic deformations.

17.1 Restitution Coefficient

The kinetic energy of colliding billiard balls is not constant because of friction, plastic deformation, and sound emission. As an example, we could think of a basketball dropped from a certain height (see Figure 17.1). Due to energy dissipation, the ball will not reach the same height as before after bouncing back from the ground. In our MD simulations, we account for energy dissipation effects in an effective manner by introducing the *restitution coefficient*

$$r = \frac{E^{\text{after}}}{E^{\text{before}}} = \left(\frac{v^{\text{after}}}{v^{\text{before}}}\right)^2 , \tag{17.1}$$

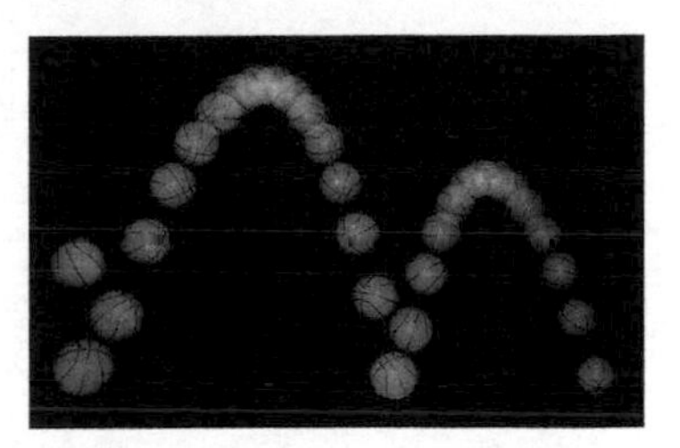

Figure 17.1 A basketball dropped from a certain height will not reach its initial height again because of dissipation effects.

where E^{after} and E^{before} are the energies before and after the collision event, and v^{after} and v^{before} are the corresponding velocities. Elastic collisions correspond to $r = 1$, whereas perfect plasticity is described by $r = 0$. We can also distinguish between normal and tangential velocity components v_n and v_t, and define the corresponding coefficients

$$e_n = \frac{v_n^{\text{after}}}{v_n^{\text{before}}} , \tag{17.2}$$

$$e_t = \frac{v_t^{\text{after}}}{v_t^{\text{before}}} . \tag{17.3}$$

For a bouncing ball, the restitution coefficient accounts for effects such as wear, plastic deformations, and sound emission. These coefficients strongly depend on the material, angle of impact, and weakly on other factors like particle shape and relative impact velocity, and can be determined experimentally. In MD simulations, we incorporate inelasticity with the help of a damped oscillator. For the corresponding equations of motion, we use $r = R_i + R_j - |\mathbf{x}_i - \mathbf{x}_j|$ and $m_{\text{eff}} = \frac{m_1 m_2}{m_1 + m_2}$ to denote the radial distance between two particles and the effective mass, respectively. The equation of motion is then

$$m_{\text{eff}}\ddot{r} = -kr - \gamma\dot{r}, \tag{17.4}$$

where k is the spring constant and γ is the viscous damping coefficient. The solution of this differential equation is given by

$$r(t) = \frac{v^{\text{before}}}{\omega}\sin(\omega t)\exp(-\delta t), \tag{17.5}$$

where $\omega = \sqrt{\omega_0^2 - \delta^2}$, $\omega_0 = \sqrt{\frac{k}{m_{\text{eff}}}}$, and $\delta = \frac{\gamma}{2m_{\text{eff}}}$ are the oscillator frequency, the ground frequency of the undamped oscillator, and the damping coefficient, respectively. Damping the oscillator reduces its frequency and generally increases the period T.

We regard the collision between two particles as a half-cycle of a damped oscillator, for which some energy is dissipated due to damping. The collision lasts half the period of an oscillation. The kinetic energy of the two particles decreases when they come together, and when the particles bounce back their kinetic energy increases. We therefore compute the restitution coefficient according to

$$e_n = \frac{\dot{r}(t+T)}{\dot{r}(t)} = \exp(-\delta T) = \exp\left(-\frac{\gamma\pi}{\sqrt{4m_{\text{eff}}k - \gamma^2}}\right). \tag{17.6}$$

This relation implies that a restitution coefficient uniquely determines a viscous damping coefficient according to

$$\gamma = 2\ln(e_n)\sqrt{\frac{m_{\text{eff}}k}{\ln(e_n)^2 + \pi^2}}. \tag{17.7}$$

An important assumption that we made is that the restitution coefficient is constant throughout the complete interaction. This assumption is violated for a Hertzian contact between spheres in three dimensions for which the contact area depends on their distance. The dependence between energy loss and interaction overlap becomes nonlinear and our approach has then to be corrected since the restitution coefficient is not constant anymore [278].

17.2 Plastic Deformation

In many engineering and industrial applications, collisions between two deformable objects are characterized by an irreversible change of their shapes, a so-called *plastic deformation*. We show an example of a plastic deformation in Figure 17.2. One approach to model plasticity has been introduced by Walton and Braun [279, 280]. The stiffnesses before and after the collision are modeled by springs with two different spring constants, k_1 and k_2, so that after the collision the particle does not come back to its original position, leaving a remnant deformation as we illustrate in Figure 17.3. The interaction is dissipative and the force is

$$f = \begin{cases} k_1 \delta \,, & \text{loading}\,, \\ k_2 (\delta - \delta_0)\,, & \text{unloading}\,. \end{cases} \tag{17.8}$$

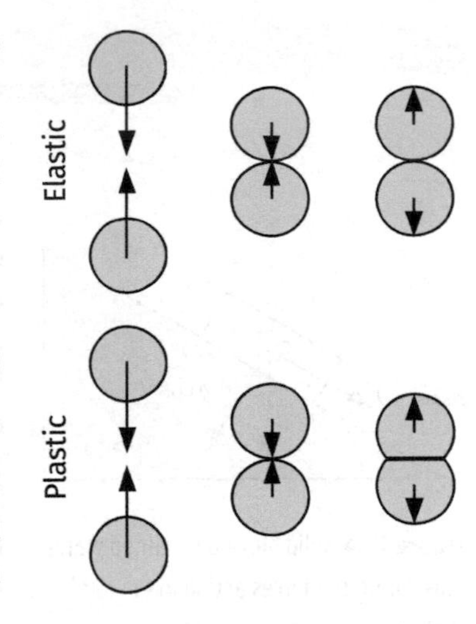

Figure 17.2 Two particles undergoing elastic and plastic interactions.

First, the objects approach each other according to Hooke's law and overlap up to $\delta_{\max}$ so that $k_1 = f_{\max}/\delta_{\max}$. During this initial contact, the objects are deformed and therefore repulsion occurs with a different spring constant $k_2 = f_{\max}/(\delta_{\max} - \delta_0)$ such that the repulsive forces vanish before the overlap does.

The collision time is not symmetric since loading lasts longer than unloading. We therefore describe the contact time by

$$t_c = \frac{\pi}{2}\left(\sqrt{\frac{m_{ij}}{k_1}} + \sqrt{\frac{m_{ij}}{k_2}}\right). \tag{17.9}$$

The dissipated energy is equal to the area enclosed by the two curves (i.e., the blue area in Figure 17.3). From this observation, we obtain the restitution coefficient

$$r = \frac{E^{\text{after}}}{E^{\text{before}}} = \frac{k_1}{k_2}. \tag{17.10}$$

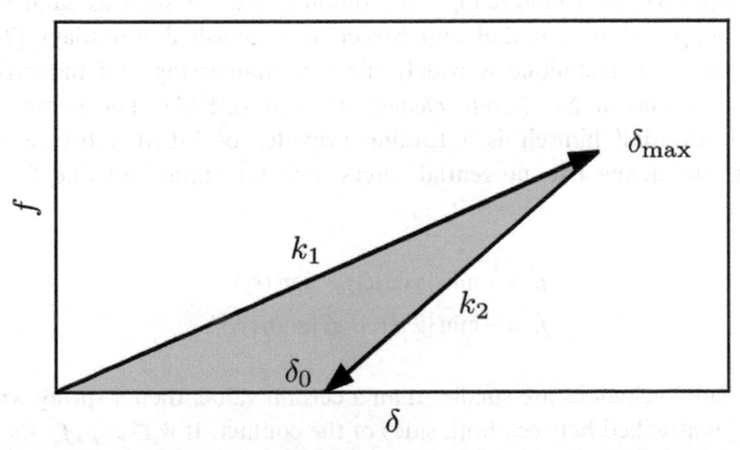

Figure 17.3 Force f as a function of deformation δ for a plastic collision. The diagram shows a plastic deformation according to Walton and Braun [279].

17.3 Coulomb Friction and Discrete Element Method

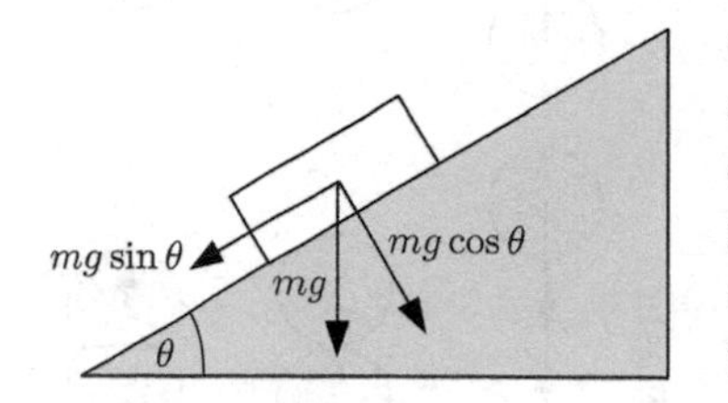

Figure 17.4 Solid block on inclined plane illustrating the forces acting in Coulomb friction.

Coulomb friction occurs when two solids are sheared against each other, that is, when there is a difference in tangential velocity at their contact point. The energy dissipation due to Coulomb friction is proportional to the normal force. Two friction coefficients are used to take such dissipation effects into account. The static friction coefficient describes friction effects of nonmoving objects and the dynamic coefficient accounts for friction of moving objects. The classic example illustrating Coulomb friction is an object located on an inclined plane (see Figure 17.4). At a certain inclination angle, the static *friction angle*, the object begins to slide down the plane. The tangent of this angle gives the static friction coefficient. The object only stops moving when the angle is decreased to the dynamic friction angle, the tangent of which defines the dynamic friction coefficient [281]. Note that the friction force is independent of the contact area.

Numerically, the problem is very difficult to handle because the friction coefficient μ is described by a discontinuous function of the relative tangential velocity v_t. For nonzero tangential velocities $v_t > 0$, the dynamic friction coefficient μ_d is different from the static one μ_s when $v_t = 0$.

One approach of implementing Coulomb friction is to distinguish between a small shear velocity v_t below which static friction is implemented and above which dynamic friction is used [282]. For two colliding particles, the tangential momentum transfer is then given by

$$\begin{aligned} \Delta p_t &= \mu_s m_{\text{eff}} (e_t + 1) \left(\mathbf{v}_i^{\text{before}} - \mathbf{v}_j^{\text{before}} \right) \mathbf{t} \quad \text{for} \quad v_t \ll 1 \,, \\ \Delta p_t &= \mu_d m_{\text{eff}} (e_t + 1) \left(\mathbf{v}_i^{\text{before}} - \mathbf{v}_j^{\text{before}} \right) \mathbf{t} \quad \text{for} \quad v_t \gg 1 \,, \end{aligned} \tag{17.11}$$

where $\mathbf{t}$ is the tangential unit vector.

For multiple particles interacting via Coulomb friction such as sand or stones, a method proposed by Cundall and Strack is established nowadays [283]. This forward integration technique is widely used in engineering and industrial applications and is known as *discrete element method* (DEM). The company Itasca founded by Cundall himself is a leading provider of DEM software. In DEM simulations, we define two tangential forces, one for static and one for dynamic friction:

$$f_t = -\min\left[\gamma v_t, \mu_d f_n\right] \operatorname{sign}(v_t) \,, \tag{17.12}$$

$$f_t = -\min\left[|k_s \xi|, \mu_s f_n\right] \operatorname{sign}(v_t) \,. \tag{17.13}$$

If the tangential velocities are smaller than a certain value, then a spring with spring constant k_s is attached between both sides of the contact. If $|k_s \xi| > \mu_s f_n$, the spring is removed and the particles are free to move with dynamic friction coefficient. Equations (17.12) and (17.13) only describe the tangential force component, and for a complete

DEM program, one has to add the normal forces using eq. (17.7). Discrete-element-method simulations are used very much in mining and geotechnical and chemical engineering.

Exercise: Sliding on an Inclined Chute

In this exercise, we study the train model on an inclined chute with inclination angle θ. The train model consists of a chain of blocks of mass m. All blocks are connected through linear springs with spring constant k. The first block is pulled with constant velocity v. We account for Coulomb friction between each block and the chute by using static and dynamic friction coefficients μ_s and μ_d. Use eq. (17.11) to add Coulomb friction in your simulations.

Tasks

- Simulate the motion of the train model for different inclination angles θ.
- Vary the spring constant k, and compare the motion of the train for large and small spring constants. The observed stop and go motion is called stick-slip.

18 Event-Driven Molecular Dynamics

Standard molecular dynamics (MD) methods, such as Verlet or leapfrog methods, are based on explicit forward integration schemes with a finite time step Δt to numerically solve certain MD problems (see Sections 12.4 and 12.5). These approaches are, however, not suited to handle hard-core potentials because of infinite repelling forces that occur if particles collide. Instead, event-driven methods are the right choice to simulate rigid particle collisions. As opposed to *batch* programming, the flow of an event-driven program is not determined by loops but by events. Therefore, event-driven MD simulations are characterized by branching points and conditional logic and are intrinsically sequential. In the next sections, we discuss different examples of event-driven dynamics and their application to simulate rigid particles.

18.1 Event-Driven Procedure

As a first example of the application of event-driven methods to MD, we consider a work by Alder [243] (see Figure 12.1). In this work, Alder studied rigid bodies of finite volume such as billiard balls. A hard-core potential is assumed to model elastic collisions between rigid particles not taking friction into account. Our standard MD integration schemes fail to describe such systems and we therefore focus on an event-driven formulation of particle interactions. We regard particle collisions as instantaneous events between which particles are not interacting. In this method, only the exchange of the particles' momenta is taken into account and no forces are computed. Furthermore, only binary collisions are considered and simultaneous interactions between three or more particles are neglected. Between two collision events, particles follow ballistic trajectories. To perform an event-driven MD simulation, we need to determine the time t_c at which the next collision will occur to then obtain the velocities of the two particles after the collision based on the particle velocities before the collision. A look-up table can be used to perform these computations efficiently. To determine t_c, we have to identify the next collision event.

Let us consider a two-dimensional collision problem involving two disks i and j with radii R_i and R_j, respectively. We show such a collision in Figure 18.1. The *collision angle* θ is the angle between the connecting vector $\mathbf{r}_{ij} = \mathbf{r}_i - \mathbf{r}_j$ and the relative velocity $\mathbf{v}_{ij} = \mathbf{v}_i - \mathbf{v}_j$. For the moment, we are not taking into account the influence of friction and thus neglect the exchange of angular momentum. We compute the times t_{ij} at

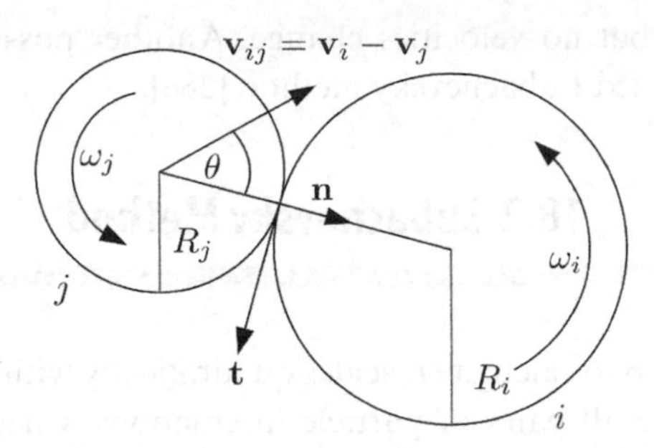

Figure 18.1 Two particles i and j collide. Particle i has radius R_i and angular velocity ω_i, and particle j has radius R_j and angular velocity ω_j. The relative velocity of both particles is $\mathbf{v}_{ij} = \mathbf{v}_i - \mathbf{v}_j$, and the collision angle θ is the angle between $\mathbf{r}_{ij} = \mathbf{r}_i - \mathbf{r}_j$ and $\mathbf{v}_{ij}$. We use $\mathbf{n} = \mathbf{r_{ij}}/|\mathbf{r}_{ij}|$ to denote the normal vector that points from the center of particle i to the center of particle j. The corresponding tangential unit vector is $\mathbf{t}$.

which the next collision between particles i and j will occur. At time t_{ij}, the distance between the two particles is

$$\left|\mathbf{r}_{ij}\left(t_{ij}\right)\right| = \left|R_i + R_j\right| . \tag{18.1}$$

Given a relative velocity $\mathbf{v}_{ij}$ at time t_0, we obtain $\mathbf{r}_{ij}(t_{ij}) = \mathbf{r}_{ij}(t_0) + \mathbf{v}_{ij}t_{ij}$. The corresponding contact time t_{ij} for particles i and j can be found via

$$v_{ij}^2 t_{ij}^2 + 2\left[\mathbf{r}_{ij}(t_0)\,\mathbf{v}_{ij}\right] t_{ij} + \left[r_{ij}(t_0)\right]^2 - \left(R_i + R_j\right)^2 = 0 . \tag{18.2}$$

If solutions t_{ij} of eq. (18.2) are not real (i.e., $t_{ij} \notin \mathbb{R}$), the corresponding trajectories of particles i and j never cross. The next collision occurs at time

$$t_c = \min_{ij}\left(t_{ij}\right) , \tag{18.3}$$

where the minimum is taken over all real-valued t_{ij}. Thus, in the time interval $[t_0, t_c]$, positions and angular orientations of all particles evolve according to

$$\mathbf{r}_i\left(t_0 + t_c\right) = \mathbf{r}_i\left(t_0\right) + \mathbf{v}_i\left(t_0\right) t_c \quad \text{and} \quad \phi_i\left(t_0 + t_c\right) = \phi_i\left(t_0\right) + \omega_i\left(t_0\right) t_c . \tag{18.4}$$

It is also possible to add gravitational and electrostatic force terms. The main bottleneck is the computation of contact times t_{ij} because this operation scales quadratically with the number of particles. For each particle i, we have to identify the particle j that first collides with particle i to then compute t_{ij}. After performing this procedure for all particles, we have to find the minimum in the set of all collision times t_{ij} according to eq. (18.3). For high-density particle systems, very distant particles will not collide and it is thus not necessary to determine corresponding collision times t_{ij}. Still, the distances between all particles need to be determined and thus the algorithm may still be very inefficient. Instead of looking at distances between particles, it is also possible to divide the system into sectors and treat those separately. The crossing of particles through the boundaries of a sector is then considered as a "collision" event at which lists are recalculated but no velocities change. Another possibility to speed up this algorithm is the so-called Lubachevsky method [284].

18.2 Lubachevsky Method

Unfortunately, the loop to calculate t_c scales quadratically with the number of particles when simply checking all pairs of particle interactions. Checking and updating all particles is, however, very inefficient because (i) most collisions would only occur after $O(N)$ steps and (ii) most particles are not participating in a finally chosen collision but their positions are always updated. This can be improved by only considering and updating those particles that are part of a collision event.

An algorithm that has been developed by Boris Lubachevsky [284] reduces the computational complexity for this improved algorithm is based on lists of events and binary stacks. In addition to the particle position and velocity, Lubachevsky's method stores the last and next events for each particle. In this way, we are keeping track of the time of the event and the partner particle involved in the event. In practice, this can be implemented by updating at each collision six arrays (*event time*, *new partner*, and for each particle *position* and *velocity*) of dimension N (number of particles in the system). Alternatively, we can create a list of pointers pointing to a data structure for each particle consisting of six variables. Storing the last event is needed as particles are only updated after being involved in an event. For each particle i, the time $t^{(i)}$ is the minimal time of all possible collisions involving this particle:

$$t^{(i)} = \min_j \left(t_{ij}\right) . \tag{18.5}$$

Considering the collision of particle i with $N - 1$, others can be improved by dividing the computational domain in sectors such that only neighboring sectors have to be considered in this step. Sector boundaries have to be treated similar to obstacles such that a collision event happens during which no velocities change when particles cross sector boundaries. For each particle i, this step would then be of order $O(1)$ instead of $O(N)$. The next collision occurs at time

$$t_c = \min_i \left(t^{(i)}\right) . \tag{18.6}$$

To summarize, the Lubachevsky method is as follows:

Lubachevsky Trick

We store $t^{(i)}$ in increasing order in a stack:

- The vector $\mathrm{PART}[m]$ points to particle i which is at position m in the stack. (Sometimes also a vector $\mathrm{POS}[i]$ is used to store position m of particle i in the stack.)

Lubachevsky Trick (*cont.*)

- Establish an implicit ordering of the collision times $t^{(i)}$, where $m = 1$ points to the shortest time.
- PART[1] is the particle with minimal collision time: $t_c = t^{(\mathrm{PART}[1])}$.
- After the collision events, all six entries (event times, new partners, positions, and velocities) have to be updated for both particles. Additionally, the vector PART[m] has to be reordered.

Reordering the times $t^{(i)}$ after each event is of order $O(\log N)$ when using, for example, binary tournament trees for sorting [284]. The advantages of this method are that it is neither necessary to calculate the collision times for all particle pairs at every step nor to update the positions of particles that do not collide. Only position and velocity of the particles involved in the collision event are updated.

18.3 Collision with Perfect Slip

After having identified the next collision event, we have to update the positions, velocities, and angular orientations of the colliding particles. With eq. (18.4), we only described the time evolution until the collision occurs. In this section, we now focus on the particle dynamics after a collision event. In a first approximation, we are assuming perfect slip and neglect any tangential exchange of momentum. Only linear momentum and no angular momentum is exchanged. Conservation of momentum implies that the velocities after the collision are

$$\mathbf{v}_i^{\mathrm{after}} = \mathbf{v}_i^{\mathrm{before}} + \frac{\Delta \mathbf{p}}{m_i}, \tag{18.7}$$

$$\mathbf{v}_j^{\mathrm{after}} = \mathbf{v}_j^{\mathrm{before}} - \frac{\Delta \mathbf{p}}{m_j}, \tag{18.8}$$

and energy conservation yields

$$\frac{1}{2} m_i \left(\mathbf{v}_i^{\mathrm{before}}\right)^2 + \frac{1}{2} m_j \left(\mathbf{v}_j^{\mathrm{before}}\right)^2 = \frac{1}{2} m_i \left(\mathbf{v}_i^{\mathrm{after}}\right)^2 + \frac{1}{2} m_j \left(\mathbf{v}_j^{\mathrm{after}}\right)^2 . \tag{18.9}$$

Thus, the exchanged momentum is

$$\Delta \mathbf{p} = -2 m_{\mathrm{eff}} \left[\left(\mathbf{v}_i^{\mathrm{before}} - \mathbf{v}_j^{\mathrm{before}} \right) \cdot \mathbf{n} \right] \mathbf{n}, \tag{18.10}$$

where $m_{\mathrm{eff}} = \frac{m_i m_j}{m_i + m_j}$ is the *effective mass* and $\mathbf{n} = \mathbf{r}_{ij}/|\mathbf{r}_{ij}|$ is the collision normal vector. If $m_i = m_j$, the velocity updates are

$$\mathbf{v}_i^{\mathrm{after}} = \mathbf{v}_i^{\mathrm{before}} - v_{ij}^n \cdot \mathbf{n}, \tag{18.11}$$

$$\mathbf{v}_j^{\mathrm{after}} = \mathbf{v}_j^{\mathrm{before}} + v_{ij}^n \cdot \mathbf{n}, \tag{18.12}$$

with $v_{ij}^n = \left(\mathbf{v}_i^{\mathrm{before}} - \mathbf{v}_j^{\mathrm{before}}\right) \cdot \mathbf{n}$. The values can be stored once in a look-up table such that there is no need of calculating the new velocities at every collision.

18.4 Collision with Rotation

We now consider two spherical particles i and j of the same radius R and mass m. Due to friction, the tangential component of angular momentum is exchanged if particles collide with nonzero tangential velocity. To account for such an angular momentum exchange in our MD simulations, we begin with the equation of motion for rotation [280]

$$I\frac{\mathrm{d}\omega_i}{\mathrm{d}t} = \mathbf{r} \wedge \mathbf{f}_i\,, \tag{18.13}$$

where $\wedge$ denotes the cross product and ω_i, I, and $\mathbf{f}_i$ are angular velocity, moment of inertia, and the forces exerted on particle i, respectively. For more details on rigid body dynamics, see Section 14.2. For the two colliding spherical particles of radius R, moment of inertia I, and mass m, angular momentum conservation yields

$$\begin{aligned} I\left(\omega_i' - \omega_i\right) &= -mR\mathbf{n} \wedge \left(\mathbf{v}_i' - \mathbf{v}_i\right)\,,\\ I\left(\omega_j' - \omega_j\right) &= mR\mathbf{n} \wedge \left(\mathbf{v}_j' - \mathbf{v}_j\right)\,, \end{aligned} \tag{18.14}$$

with the primed velocities representing velocities after the collision. Using conservation of momentum

$$\mathbf{v}_i' + \mathbf{v}_j' = \mathbf{v}_i + \mathbf{v}_j\,, \tag{18.15}$$

the angular velocities after the collision are

$$\omega_i' - \omega_i = \omega_j' - \omega_j = -\frac{Rm}{I}\mathbf{n} \wedge \left(\mathbf{v}_i' - \mathbf{v}_i\right)\,. \tag{18.16}$$

The relative velocity between particles i and j is

$$\mathbf{u}_{ij} = \mathbf{v}_i - \mathbf{v}_j - R\left(\omega_i + \omega_j\right) \wedge \mathbf{n}\,, \tag{18.17}$$

where $\mathbf{n}$ is the unit vector pointing from particle i to particle j. We decompose the relative velocity into its normal and tangential components $\mathbf{u}_{ij}^n$ and $\mathbf{u}_{ij}^t$, respectively. For the angular momentum exchange, we have to consider the relative velocity of the particle surfaces at the contact point. The normal and tangential velocities are given by

$$\begin{aligned} \mathbf{u}_{ij}^n &= \left(\mathbf{u}_{ij}\mathbf{n}\right)\mathbf{n}\,,\\ \mathbf{u}_{ij}^t &= \mathbf{u}_{ij} - \mathbf{u}_{ij}^n = -\mathbf{n} \wedge \left(\mathbf{n} \wedge \mathbf{u}_{ij}\right)\\ &= \mathbf{v}_i^t - \mathbf{v}_j^t - R\left(\omega_i + \omega_j\right) \wedge \mathbf{n}\,, \end{aligned} \tag{18.18}$$

and general slips are described by

$$\mathbf{u}_{ij}^{t\,\prime} = e_t \mathbf{u}_{ij}^t\,, \tag{18.19}$$

where the *tangential restitution coefficient* e_t accounts for different slip types. The perfect slip collision is recovered for $e_t = 1$ which implies that no rotation energy is transferred from one particle to the other. No slip corresponds to $e_t = 0$. Energy conservation only holds if $e_t = 1$ and energy is dissipated if $e_t < 1$.

If we compute the difference between the relative tangential velocities before and after the slip event, we find

$$\begin{aligned}(1-e_t)\,\mathbf{u}^t_{ij} &= \mathbf{u}^t_{ij}-\mathbf{u}^{t\,\prime}_{ij}\\ &= -\left[\left(\mathbf{v}^{t\prime}_i-\mathbf{v}^t_i-\mathbf{v}^{t\prime}_j+\mathbf{v}^t_j\right)-R\left(\omega'_i-\omega_i+\omega'_j-\omega_j\right)\wedge\mathbf{n}\right].\end{aligned} \tag{18.20}$$

Combining this equation with Eqs. (18.15) and (18.16), we obtain an expression without angular velocities

$$\begin{aligned}(1-e_t)\,\mathbf{u}^t_{ij} &= \mathbf{u}^t_{ij}-\mathbf{u}^{t\,\prime}_{ij}\\ &= -\left[2\left(\mathbf{v}^{t\prime}_i-\mathbf{v}^t_i\right)+2q\left(\mathbf{v}^{t\prime}_i-\mathbf{v}^t_i\right)\right]\end{aligned} \tag{18.21}$$

and thus

$$\mathbf{v}^{t\prime}_i=\mathbf{v}^t_i-\frac{(1-e_t)}{2\,(1+q)}\mathbf{u}^t_{ij}\quad\text{with}\quad q=\frac{mR^2}{I}. \tag{18.22}$$

Analogously, we find for the remaining quantities

$$\begin{aligned}\mathbf{v}^{t\prime}_j &= \mathbf{v}^t_j+\frac{(1-e_t)}{2\,(1+q)}\mathbf{u}^t_{ij},\\ \omega'_i &= \omega_i-\frac{(1-e_t)}{2R\,(1+q^{-1})}\mathbf{u}^t_{ij}\wedge\mathbf{n},\\ \omega'_j &= \omega_j-\frac{(1-e_t)}{2R\,(1+q^{-1})}\mathbf{u}^t_{ij}\wedge\mathbf{n}.\end{aligned} \tag{18.23}$$

Finally, the updated velocities are

$$\begin{aligned}\mathbf{v}_i{}' &= \mathbf{v}_i-\mathbf{u}^n_{ij}-\frac{(1-e_t)}{2\,(1+q)}\mathbf{u}^t_{ij},\\ \mathbf{v}_j{}' &= \mathbf{v}_j+\mathbf{u}^n_{ij}+\frac{(1-e_t)}{2\,(1+q)}\mathbf{u}^t_{ij}.\end{aligned} \tag{18.24}$$

18.5 Inelastic Collisions

The relative velocity of two colliding particles at their contact point is

$$\mathbf{u}^n_{ij}=\left(\mathbf{u}_{ij}\mathbf{n}\right)\mathbf{n}=\left[\left(\mathbf{v}_i-\mathbf{v}_j\right)\mathbf{n}\right]\mathbf{n}. \tag{18.25}$$

For an inelastic collision, dissipation effects due to various physical mechanisms lead to reduced normal velocities

$$\mathbf{u}^{n\,\prime}_{ij}=-e_n\mathbf{u}^n_{ij}, \tag{18.26}$$

where e_n is the *normal restitution coefficient* defined in eq. (17.2). For $e_n=1$, there is no dissipation, whereas dissipation occurs for $e_n<1$. The velocities after the collision are given by

$$\begin{aligned}\mathbf{v}_i{}' &= \mathbf{v}_i-\frac{(1+e_n)}{2}\mathbf{u}^n_{ij},\\ \mathbf{v}_j{}' &= \mathbf{v}_j+\frac{(1+e_n)}{2}\mathbf{u}^n_{ij}.\end{aligned} \tag{18.27}$$

In the normal direction, we include the normal restitution coefficient in eq. (18.10) and describe the momentum exchange according to

$$\Delta \mathbf{p}_n = -m_{\text{eff}}(1+e_n)\left[\left(\mathbf{v}_i - \mathbf{v}_j\right)\mathbf{n}\right]\mathbf{n}\,. \tag{18.28}$$

With $q = \frac{mR^2}{I}$, the generalized set of Eqs. (18.22) and (18.23) is given by

$$\begin{aligned}
\mathbf{v}_i' &= \mathbf{v}_i - \frac{(1+e_n)}{2}\mathbf{u}_{ij}^n - \frac{(1-e_t)}{2\,(1+q)}\mathbf{u}_{ij}^t\,,\\
\mathbf{v}_j' &= \mathbf{v}_j + \frac{(1+e_n)}{2}\mathbf{u}_{ij}^n + \frac{(1-e_t)}{2\,(1+q)}\mathbf{u}_{ij}^t\,,\\
\omega_i' &= \omega_i - \frac{(1-e_t)}{2R\,(1+q^{-1})}\mathbf{u}_{ij}^t \wedge \mathbf{n}\,,\\
\omega_j' &= \omega_j - \frac{(1-e_t)}{2R\,(1+q^{-1})}\mathbf{u}_{ij}^t \wedge \mathbf{n}\,,
\end{aligned} \tag{18.29}$$

describing inelastic collisions of rotating particles [278].

Because dissipation occurs in macroscopic particle systems, it is important to incorporate energy dissipation effects in simulations by accounting for normal and tangential restitution coefficients [280].

18.6 Inelastic Collapse

A prominent consequence of dissipation in hard-sphere event-driven MD simulations is the so-called *inelastic collapse*, which is characterized by particles that collide infinitely often in finite time. This effect is often confused with the clustering instability [285, 286] that can occur in regions of high particle densities, where the large number of inelastic collisions leads to large dissipation effects and allows particles to form clusters of locally even higher densities. Such clustering effects are common in the formation of mass concentrations out of intergalactic dust (for more details, see Figure 18.5). Without clustering, stars, galaxies, and other celestial objects cannot be formed. Note, however, that clustering is a hydrodynamic instability that occurs due to dissipation if both the density and the system are large enough [287], notably without long-range forces like gravity [288].

Figure 18.2 Sean McNamara is a professor at the University of Rennes 1.

When simulating inelastic collisions of hard, rigid particles, we may encounter so-called finite-time singularities as described by McNamara (see Figure 18.2) and Young [287, 289]. Such singularities are likely to occur in event-driven simulations of high-density particle systems because of the hard-sphere interaction potential and the associated vanishing interaction duration. Instantaneous interactions in event-driven simulations are neither resembling particle collisions that we can observe in nature nor those that we simulated using soft-sphere, MD/DEM-type frameworks.

Unlike the clustering instability, the inelastic collapse can occur in systems that only involve a small number of particles. One could, for instance, think of a particle bouncing between two other particles that approach each other along a straight line (see Figure 18.3). To understand the emergence of finite-time singularities with

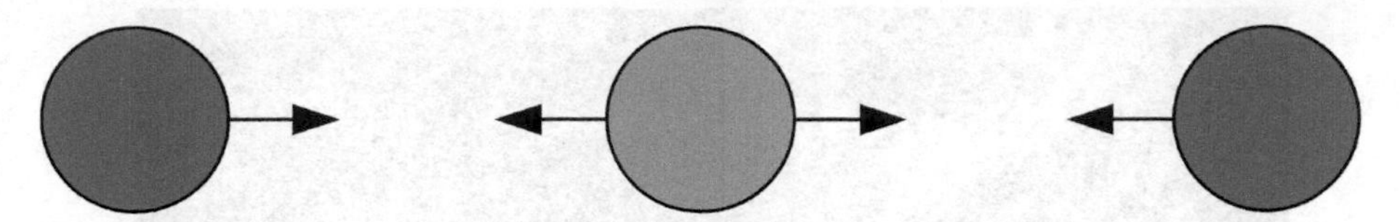

Figure 18.3 The orange particle at the center is bouncing between the blue particles, which approach each other.

an even simpler model, we consider a ball bouncing vertically on a hard surface (see Figure 17.1) [280]. Every time the ball hits the surface, its kinetic energy is lowered according to eq. (17.1). As a consequence, the ball will not reach the initial height anymore and the time between two contacts with the surface decreases after each collision. After a finite time, the ball comes to a rest, but the simulation takes an infinite time to run. In an event-driven simulation, the ball only stops its motion after infinitely many collisions, which means that the simulation never ends. A related problem is the famous *Zenon paradox*.[1] The total time needed for the bouncing ball to come to rest is the sum over an infinite number of times between two successive surface contacts t_i. Since the height is proportional to the potential energy, it scales with the restitution coefficient at every surface contact. Consequently, after the ith surface contact, the damping of the height is proportional to r^i. The total time is thus

$$\begin{aligned} t_{\text{tot}} &= \sum_{i=1}^{\infty} t_i \\ &= 2\sqrt{\frac{2h^{\text{initial}}}{g}} \sum_{i=1}^{\infty} \sqrt{r^i} \\ &= 2\sqrt{\frac{2h^{\text{initial}}}{g}} \left(\frac{1}{1-\sqrt{r}} - 1 \right), \end{aligned} \tag{18.30}$$

where h^{initial} is the initial height and g the gravitational acceleration. Mathematically, the bouncing ball jumps infinitely often during the finite time given by eq. (18.30).

In event-driven particle simulations, interactions between particles are modeled as instantaneous events. This is the origin of the inelastic collapse, a wrong assumption that lies at the heart of event-driven rigid-particle models. However, real collisions have a certain duration. In order to handle the inelastic collapse numerically, Luding (see Figure 18.4) and McNamara introduced a restitution coefficient that depends on the time elapsed since the last event [286]. In their so-called TC model, a collision takes a finite time t_{contact}. If the time since the last collision of both interacting particles $t^{(i)}$ is less than t_{contact}, the restitution coefficient is set to unity and otherwise to r. We thus obtain

$$r^{(i,j)} = \begin{cases} r, & \text{for} \quad t^{(i)} > t_{\text{contact}} \quad \text{or} \quad t^{(j)} > t_{\text{contact}}, \\ 1, & \text{otherwise}. \end{cases} \tag{18.31}$$

Figure 18.4 Stefan Luding is a professor at the University of Twente.

With this new definition of the restitution coefficient, the collision type changes from inelastic to elastic if too many collisions occur during t_{contact}. Depending on the

[1] The Zenon paradox is also known as the paradox of Achilles and the turtle.

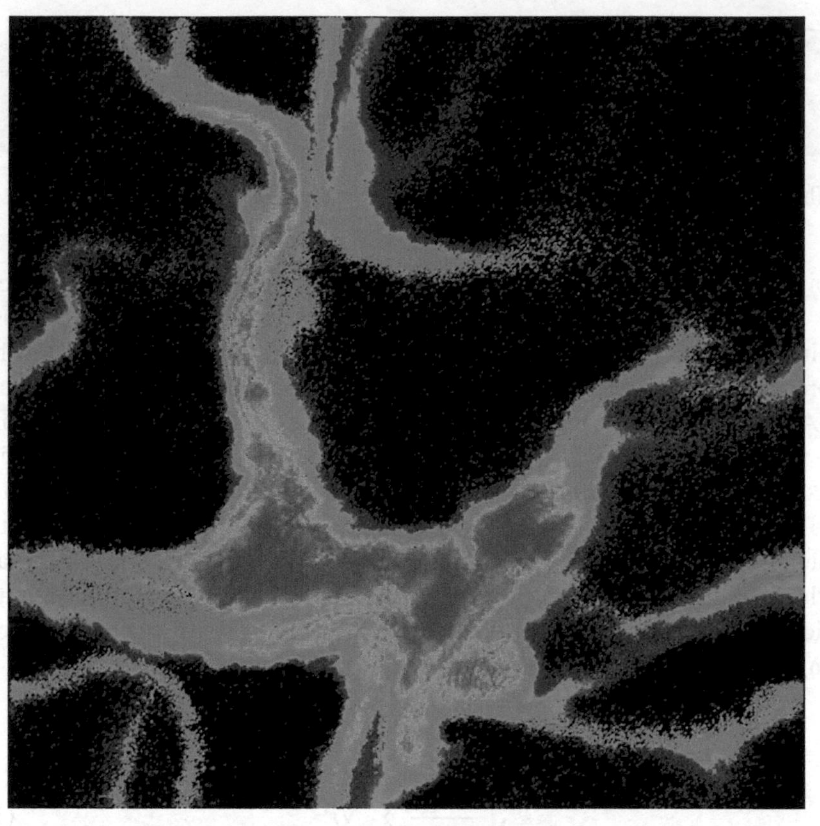

Figure 18.5 Snapshot of an event-driven simulation with $N = 99{,}856$ particles in a 2D system with periodic boundary conditions. The gas cools due to dissipation so that 10 percent of the relative velocity is lost per collision, i. e., the restitution coefficient is $r = 0.9$. Red, green, and blue correspond to large, medium, and small particle densities, respectively. Figure courtesy Stefan Luding [290].

material properties of the colliding particles, it is also possible to use more complex functional dependencies of the restitution coefficient. For very dense and viscous materials, setting the restitution coefficient to zero instead of unity may be a better choice. In this case, the particles form clusters and stick together.

In Figure 18.5, we show an example of a simulation involving 99,856 particles that collide inelastically. We observe the formation of clusters in certain regions of the simulation snapshot shown. Such simulations are important to model the instability of intergalactic dust before gravity becomes relevant. For certain systems, it may be important to account for additional long-range repulsive interactions between particles (see Chapter 15), which counteract clustering, while attractive interactions do enhance clustering [288].

We now understand the main differences between hard-sphere event-driven simulations and soft-potential approaches. Binary and instantaneous collisions determine the dynamics in an event-driven simulation, whereas multiple interactions might occur at the same time in the other case.

Exercise: Event-Driven MD

So far, in our MD simulations, we were using particles interacting with a smooth potential (Lenard-Jones potential). Event-driven simulations are able to model particles interacting by a hard-core potential considering binary collisions (i.e., instantaneous contacts). Between binary collisions, particle trajectories can be calculated analytically. In systems without external forces (e.g., gravity) the particles follow straight paths between collisions. The collision time between pairs can be calculated by solving the quadratic equation

$$\left|\mathbf{r}_{ij}(t_0) + \mathbf{v}_{ij}\, t_{ij}\right| = R_i + R_j \,,$$

where $\mathbf{r}_{ij}(t_0)$ is the distance vector between particles at time t_0, $\mathbf{v}_{ij}$ the velocity difference, t_{ij} the possible collision time, and R_{ij} are the particle radii. In this method one has to calculate the collision time between all particle pairs getting the "global" collision time $t_c = \min_{ij}(t_{ij})$.

1. Implement the event-driven dynamics either in one, two, and three dimensions. You can use your old code as basis or start implementing a completely new code (e.g., for the one-dimensional case).
2. Several simplifications can be applied:
   - For a one-dimensional system, the calculation of the collision time is easier (there is no need to solve a quadratic equation and only neighboring particles interact). You can also set the radii to zero in this case.
   - In two- and three-dimensional simulations, consider collisions with perfect slip. Then, you do not need angular velocities.
   - Consider particles with the same mass m and radius R.
3. The aim of this exercise is to solve a few simple one-dimensional problems (you can also try higher-dimensional setups, if you like):
   - Chain of N beads with restitution coefficient $e_n = 1$ ($N = 5$ resembles a Newton's cradle). How do you implement the synchronous motion of multiple particles? You can also vary the restitution coefficient in this system, if you like.
   - Chain of N beads with restitution coefficient $e_n < 1$ hitting a resting wall ($e_n = 1$): Measure the effective restitution coefficient $e_{\text{eff}} = \sqrt{E_f/E_i}$ of multiple synchronously moving particles, where E_i and E_f are the initial and final kinetic energies of the chain, respectively. Vary the number N of the beads (for fixed e_n). Above which value of N does the effective restitution coefficient practically vanish (i.e., synchronously moving particles do not rebounce any more)?
4. Please consider the following points:
   - With all the above simplifications, in the one-dimensional case, the velocities are simply exchanged for perfect restitution ($e_n = 1$). For $e_n < 1$, the relation is also quite simple.
   - To speed up your code, you can store the events for each particle in a priority queue. While checking, all particle pairs require $O(N^2)$ time, generating events only for the colliding particles and inserting them in the queue has a runtime of order $O(N \log N)$ instead. Methods with insertion time of $O(1)$ (and $O(N)$ in total) have been proposed in Ref. [291]. Only interacting particles need to be advanced, but bear in mind that the collision can invalidate previously predicted events. You can either remove them manually from the queue in time $O(N)$ or, more efficiently, keep a counter of each particle's collisions and identify an event as invalid only when resolving it.

19 Nonspherical Particles

Up until now, we focused on the simulation of spherical particles. However, it would be also desirable to simulate interactions between arbitrarily shaped particles. Such methods are relevant to describe realistic granular systems consisting of particles of various shapes. For some cases such as ellipsoidal particles, it is possible to obtain analytical solutions of the interaction dynamics. For the more general treatment of arbitrarily shaped particles, only numerical approximations are feasible.

19.1 Ellipsoidal Particles

To simulate the interaction between ellipsoidal particles, we consider two ellipses whose parameterization is given by

$$\frac{[(x-x_a)\cos(\theta_a)+(y-y_a)\sin(\theta_a)]^2}{a_1^2}+\frac{[(x-x_a)\sin(\theta_a)-(y-y_a)\cos(\theta_a)]^2}{a_2^2}=1\,, \quad (19.1)$$

$$\frac{[(x-x_b)\cos(\theta_b)+(y-y_b)\sin(\theta_b)]^2}{b_1^2}+\frac{[(x-x_b)\sin(\theta_b)-(y-y_b)\cos(\theta_b)]^2}{b_2^2}=1\,, \quad (19.2)$$

where θ_a and θ_b are the rotation angles, a_1 and b_1 are the major axis, and a_2 and b_2 are the minor axis of ellipses A and B, respectively. We show the two ellipses in Figure 19.1. As proposed by Perram and Wertheim [292], we take the overlap of the two ellipses as being proportional to the interaction force. To calculate the overlap, we transform the ellipses into circles using an appropriate metric. Let $\mathbf{u}_1$, $\mathbf{u}_2$ and $\mathbf{v}_1$, $\mathbf{v}_2$ be orthonormal vectors along the principal axes of the ellipses A and B, respectively. The rotated ellipses are defined by the matrices

$$A=\sum_k a_k^{-2}\mathbf{u}_k\otimes\mathbf{u}_k^T \quad \text{and} \quad B=\sum_k b_k^{-2}\mathbf{v}_k\otimes\mathbf{v}_k^T\,. \quad (19.3)$$

Ellipse A can be described by the functional

$$G_A(\mathbf{r})=(\mathbf{r}-\mathbf{r}_a)^T A\,(\mathbf{r}-\mathbf{r}_a)\,, \quad (19.4)$$

which is smaller than 1 inside the ellipse, larger then 1 outside and exactly 1 on the ellipse. That is,

$$G_A(\mathbf{r})=\begin{cases}<1\,, & \text{if } \mathbf{r} \text{ is inside the ellipse}\,,\\ 1\,, & \text{if } \mathbf{r} \text{ is on the ellipse}\,,\\ >1\,, & \text{if } \mathbf{r} \text{ is outside the ellipse}\,.\end{cases} \quad (19.5)$$

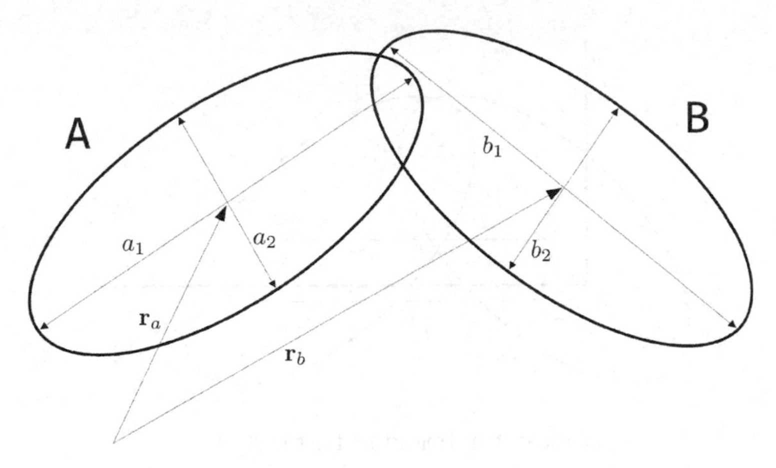

Figure 19.1 Two overlapping ellipses.

Analogously, we use $G_B(\mathbf{r})$ to denote the corresponding functional for ellipse B. Next, we define another functional

$$G(\mathbf{r}, \lambda) = \lambda G_A(\mathbf{r}) + (1 - \lambda) G_B(\mathbf{r}) , \tag{19.6}$$

which combines the two ellipses with a parameter $\lambda \in [0, 1]$ that interpolates between the two centers of the ellipses. With the help of $G(\mathbf{r}, \lambda)$, we compute the contact point. To do this, we have to minimize the functional according to

$$\nabla_{\mathbf{r}} G(\mathbf{r}, \lambda)|_{\mathbf{r}=\mathbf{r}_m} = 0 . \tag{19.7}$$

This yields

$$\mathbf{r}_m(\lambda) = [\lambda A + (1 - \lambda) B]^{-1} [\lambda A \mathbf{r}_a + (1 - \lambda) B \mathbf{r}_b] . \tag{19.8}$$

If we start from $\lambda = 0$ and arrive at $\lambda = 1$, the quantity $\mathbf{r}_m(\lambda)$ describes a path from the center of the first ellipse to the center of the second one. We rewrite the path in terms of $\mathbf{r}_{ab} = \mathbf{r}_b - \mathbf{r}_a$ and find

$$\begin{aligned} \mathbf{r}_m(\lambda) &= \mathbf{r}_a + (1 - \lambda) A^{-1} \left[(1 - \lambda) A^{-1} + \lambda B^{-1}\right]^{-1} \mathbf{r}_{ab}, \\ \mathbf{r}_m(\lambda) &= \mathbf{r}_b - \lambda B^{-1} \left[(1 - \lambda) A^{-1} + \lambda B^{-1}\right]^{-1} \mathbf{r}_{ab} . \end{aligned} \tag{19.9}$$

If the value of the functional along this path between the centers is always smaller than unity, we know that we never left the interior of the ellipses, and hence they overlap. We define the *overlap function*

$$S(\lambda) = G(\mathbf{r}_m(\lambda), \lambda), \tag{19.10}$$

and by inserting eq. (19.9) we obtain

$$S(\lambda) = \lambda (1 - \lambda) \mathbf{r}_{ab}^T \left[(1 - \lambda) A^{-1} + \lambda B^{-1}\right]^{-1} \mathbf{r}_{ab} . \tag{19.11}$$

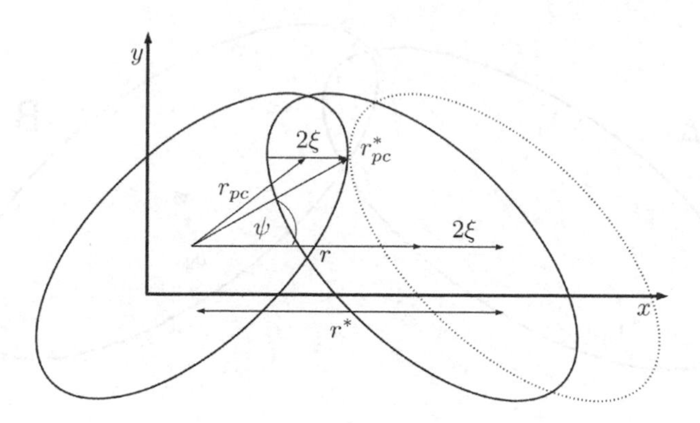

Figure 19.2 Overlap vector and contact point for two ellipses.

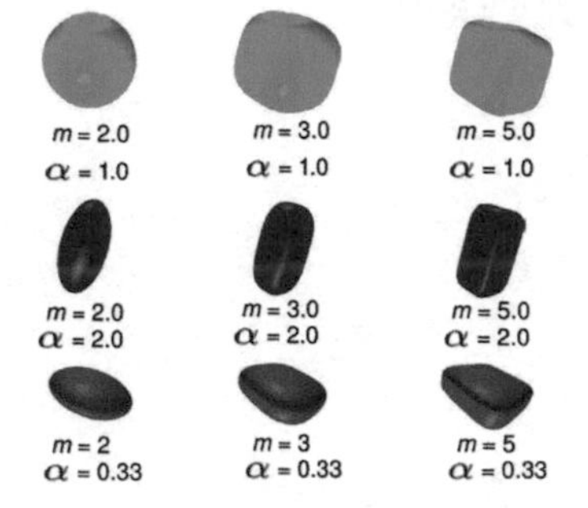

Figure 19.3 Examples of superellipsoids with parameters as defined by eq. (19.16). The figure is taken from Ref. [293].

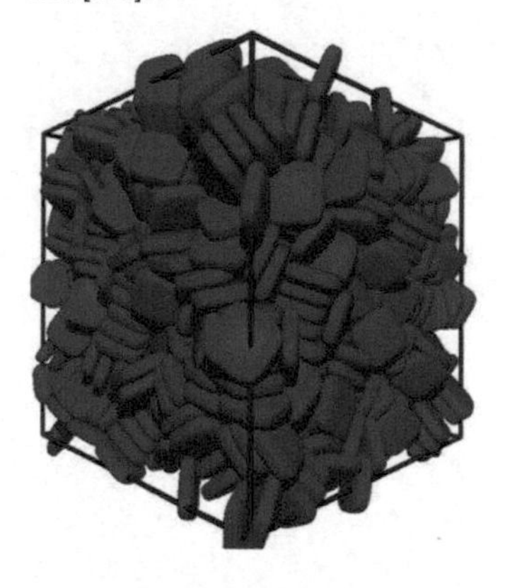

Figure 19.4 An example of a superellipsoid packing with 1,125 particles. The shape of the superellipsoids is determined by eq. (19.16) with $m = 4$ and $\alpha = 0.3$. The figure is taken from Ref. [293].

This corresponds to the value of the functional (19.6) on the path that connects the two centers. We are interested to know if the path ever exhibits values larger than unity. If $\lambda_{\max}$ is the value of $\lambda \in [0, 1]$ at which S has its maximum, we will be able to tell if the ellipses are overlapping or not:

$$S\left(\lambda_{\max}\right) = \begin{cases} < 1\,, & \text{if the ellipses overlap}\,, \\ 1\,, & \text{if the ellipses touch}\,, \\ > 1\,, & \text{if the ellipses are separated}\,. \end{cases} \tag{19.12}$$

Based on this result, we compute the contact point for stiff ellipses for which overlaps in the simulations are not allowed. We set $S\left(\lambda_{\max}\right)$ to unity and find the distance between the centers of the ellipses for which they just touch:

$$r^* = \left\{\lambda_{\max}\left(1 - \lambda_{\max}\right) \mathbf{e}_r^T \left[\left(1 - \lambda_{\max}\right) A^{-1} + \lambda_{\max} B^{-1}\right]^{-1} \mathbf{e}_r\right\}^{-\frac{1}{2}}\,, \tag{19.13}$$

where $\mathbf{e}_r = \mathbf{r}_{ab}/|\mathbf{r}_{ab}|$. We define the overlap vector as

$$\xi = \frac{r^* \mathbf{e}_r - \mathbf{r}_{ab}}{2}\,. \tag{19.14}$$

The contact points of the overlapping and the rigid case are related via (see Figure 19.2)

$$\mathbf{r}_{pc} = \mathbf{r}^*_{pc} + \xi\,. \tag{19.15}$$

The algorithm described above can be easily expanded to three-dimensional ellipsoids. Moreover, there exist generalized versions of ellipsoids, so-called *superellipsoids*, which are mathematically defined by [293]

$$x^m + y^m + \left(\frac{z}{\alpha}\right)^m = 1\,, \tag{19.16}$$

where the parameters m and α control the shape (see Figure 19.3). The macroscopic properties of superellipsoids exhibit interesting features such as high packing densities [293, 294]. We show an example of a superellipsoid packing in Figure 19.4. The

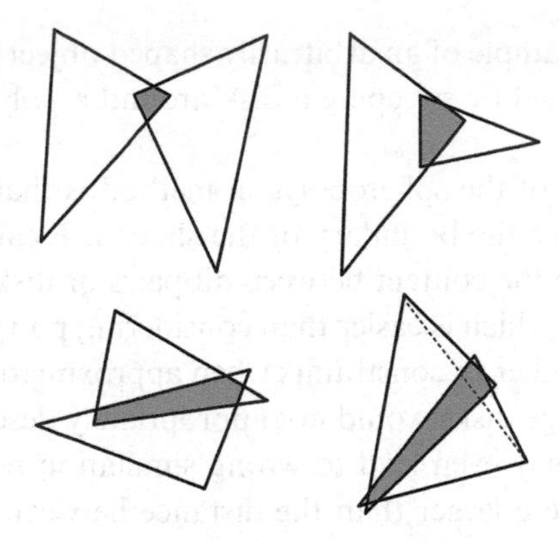

Figure 19.5 Possible overlaps of two triangles.

contact of superellipsoids can be treated by generalizing the outlined algorithm for ellipsoids.

We saw that even for the simplest generalization of spheres, it was already necessary to develop complex analytical methods. This substantially increases the computational complexity of corresponding simulation algorithms. Ellipsoids and superellipsoids can describe only a limited number of the shapes encountered in nature and, therefore, it is necessary to develop approaches to handle objects of even more complex shapes.

19.2 Polygons

A particular class of macroscopic particles are those described by polygons in two dimensions or polyhedra in three dimensions. Examples of such objects include dice and crystalline grains like sugar. Again, a measure for the repulsive forces between "particles" is the overlap area. The advantage of simple polygons is that it is possible to compute the overlap area dividing them into triangles. However, between triangles, there can be various different types of contacts and their identification is cumbersome as we show in Figure 19.5.

Additional complexities arise when the overlap area (blue) does not include all parts that were deformed (see lower panels of Figure 19.5). Furthermore, rotations of particles require an additional treatment, as the torque strongly depends on the shape of the particles. Due to the mentioned reasons, simulating objects that are composed of polygons is rather complex.

Figure 19.6 Fernando Alonso-Marroquin is currently a senior lecturer at the University of Sydney.

19.3 Spheropolygons

Another class of methods that is more efficient than the mere division into polygons relies on *spheropolygons* [295]. This technique was invented by Fernando Alonso-Marroquin (see Figure 19.6), who uses the two-dimensional *Minkowski cow*

Figure 19.7 The Minkowski cow is an example of an irregular shape composed of spheropolygons. The figure is taken from Ref. [295].

(see Figure 19.7) as an example of an arbitrarily shaped object in his simulations. This particular shape is obtained by sweeping a disk around a polygon that resembles the shape of a cow.

The important feature of the spheropolygon method is that it is possible to simulate arbitrary shapes. Once the boundary of the shape is locally described by a disk, we only have to compute the contact between all pairs of disks that define the edges and vertices of the shape, which is easier than considering polygons (see Section 19.2). There are of course a number of constraints when approximating the shapes in such a way. For example, too large disks would not appropriately describe the original shape of a certain object and may even lead to wrong simulation results. Imagine that the diameters of the disks were larger than the distance between the hooves or the distance between the tail and the rest of the cow of Figure 19.7. Substantial features of the shape would be lost. A generalization of this technique to three dimensions is straightforward.

Many other techniques have been developed to describe arbitrarily shaped objects. Effective implementations have been developed by engineers, physicists, and mathematicians. An important example is the field of *mathematical morphology* developed by Jean Serra in the 1960s [296]. The techniques of Marroquin are related to the so-called *dilation techniques*, but there are also other methods like the *erosion technique*.

Exercise: Triangle Jumping on a Plane

We simulate the interaction of a pair of particles under the influence of gravitational acceleration g using spheropolygons.

The spheropolygons fall under the influence of gravity from an initial height with zero initial velocity and hit a "sphero-segment" at the bottom. We use the following shapes: (i) a circular particle, (ii) a convex spheropolygon square, and (iii) a nonconvex spheropolygon boomerang.

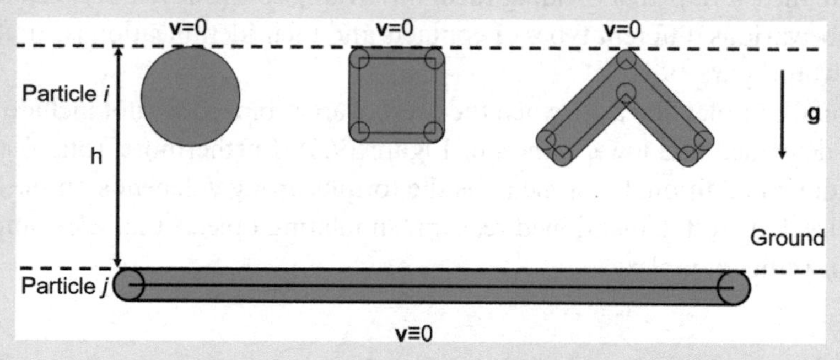

The contact force $\mathbf{F}$ between one pair of edge-vertex contact can be written as:

$$\mathbf{F} = k_n \delta_n \mathbf{e}_n \,. \tag{19.17}$$

Exercise: Triangle Jumping on a Plane (*cont.*)

where k_n is the normal stiffness, δ_n is the length of the overlap of the contact layer, and $\mathbf{e}_n$ is the normal unit vector pointing in the direction of the contact force. The corresponding equations of motion of particle i are

$$m\mathbf{a} = \sum_c \mathbf{F}_c + m\mathbf{g}\,. \tag{19.18}$$

where m is the mass of particle i and $\mathbf{a}$ is its acceleration. The index c accounts for potentially multiple contact pairs between particles i and j.

Task 1: How many contact points define the interaction between particles i and j for each case? Where is the location of the contact, and how is $\mathbf{F}_c$ defined (edge-to-vertex or vertex-to-edge) after a contact?

Task 2: Implement the numerical simulation of the falling spheropolygon. Investigate the influence of the radius of particle i. How does decreasing the radius affect the numerical stability of the simulation?

Task 3: Measure the potential and kinetic energies of particle i and check whether the energy is conserved.

20 Contact Dynamics

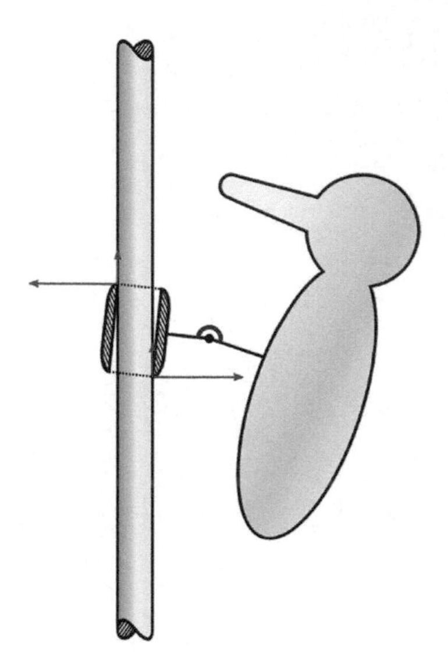

Figure 20.1 A woodpecker toy is a benchmark example of contact dynamics simulations [297]. An animation can be found at https://github.com/gabyx/Woodpecker.

We started our discussion of MD simulations by introducing integration methods to directly solve Newton's equations of motion for a certain number of interacting particles. Such direct integration methods are able to simulate the dynamics of particles which interact via soft potentials. To overcome the problem of infinite forces in the case of hard-sphere potentials, we discussed event-driven MD techniques in Chapter 18. The advantage of event-driven MD is that it is applicable to rigid bodies, but it is not capable to compute forces since we are only considering binary collision events and accounting for the corresponding momentum and angular momentum transfer. Event-driven methods also have difficulties to describe long-lasting contacts. However, very often one actually wants to compute forces and model contact phenomena that may occur in granulates, rocks, and machine parts. An example of a system that can be modeled with contact dynamics is a woodpecker toy (see Figure 20.1).

Simulating such and related systems requires one to resolve the observed contact interactions. This is done in the context of *contact dynamics*. The basic idea is that particle interactions are fully determined by constraints. To avoid overlapping particles, we impose constraint forces at their contacts. Per Lötstedt first described the idea of the method and Jean Jacques Moreau (see Figure 20.2) provided a concrete formulation and wrote the first working program. Contact dynamics is an application of nonsmooth mechanics where the time evolution of the particles' positions and momenta are not assumed to be described by smooth functions anymore. An important problem in this field is the *ambiguous boundary condition problem* or *Signorini problem* which has been originally formulated by Alberto Signorini. It consists of determining the shape of a rigid body. Perfectly rigid bodies with perfect volume exclusion are described by the *Signorini graph*. If particles are not in contact, there are also no forces between them. Forces only occur if the distance between both particles is zero. As we show in Figure 20.3, this causes force discontinuities which are difficult to handle numerically. Such force discontinuities may also occur due to friction between particles. Let us consider two particles that are in contact. If their relative velocity is zero, many values of the friction forces are possible. We show the resulting discontinuous behavior in the *Coulomb graph* in Figure 20.3.

20.1 One-Dimensional Contact

To familiarize ourselves with the concepts of contact dynamics, we first focus on the example of a one-dimensional contact as shown in Figure 20.4. In this example, we neglect friction and only consider normal forces at the contact point of the two particles. Both particles have mass m and velocities v_1 and v_2, respectively. We have to make sure that these two particles do not overlap. Therefore, we impose constraint forces $R_i(t)$ in such a way that they compensate all other forces which would lead to overlaps. These constraint forces should be defined in such a way that they have no influence on the particle dynamics before and after the contact. The time evolution of the particles' positions and velocities is described by an implicit Euler scheme which is given by

Figure 20.2 Jean Jacques Moreau (1923–2014) was working in Montpelier and mainly focused on elliptic equations. After his retirement, he developed contact dynamics [298]. Photograph courtesy Franck (Farhang) Radjai.

$$\begin{aligned} \mathbf{v}_i(t+\Delta t) &= \mathbf{v}_i(t) + \Delta t \frac{1}{m_i} \mathbf{F}_i(t+\Delta t), \\ \mathbf{r}_i(t+\Delta t) &= \mathbf{r}_i(t) + \Delta t \mathbf{v}_i(t+\Delta t). \end{aligned} \tag{20.1}$$

The forces $\mathbf{F}_i(t) = \mathbf{F}_i^{\text{ext}}(t) + \mathbf{R}_i(t)$ consist of an external force $\mathbf{F}_i^{\text{ext}}(t)$ and a contact force $\mathbf{R}_i(t)$.

So far, we only considered forces that act on the center of mass. However, contact forces act locally on the contact point and not on the center of mass. We therefore introduce a matrix H which transforms local contact forces into particle forces, and the corresponding transpose H^T transforms particle velocities into relative velocities. We proceed with example of Figure 20.4 and find

$$v_n^{\text{loc}} \equiv v_2 - v_1 = \begin{bmatrix} -1 & 1 \end{bmatrix} \begin{bmatrix} v_1 \\ v_2 \end{bmatrix} = H^T \begin{bmatrix} v_1 \\ v_2 \end{bmatrix} \tag{20.2}$$

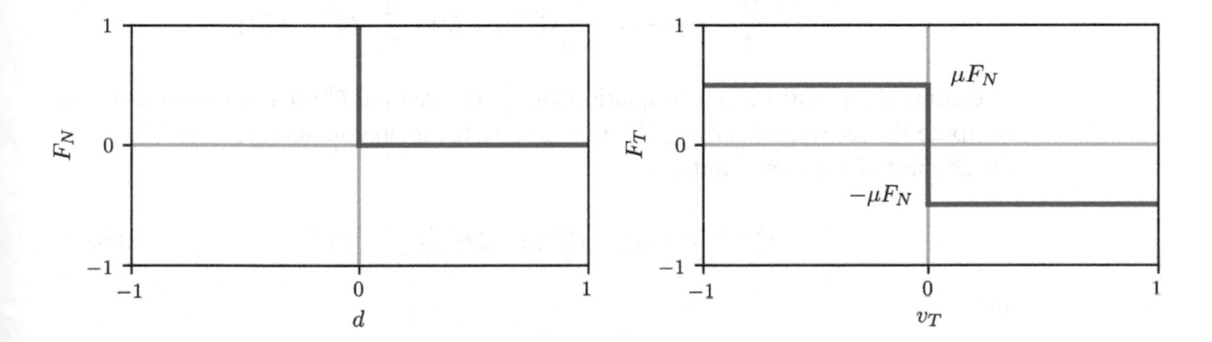

Signorini (left) and Coulomb (right) graphs.

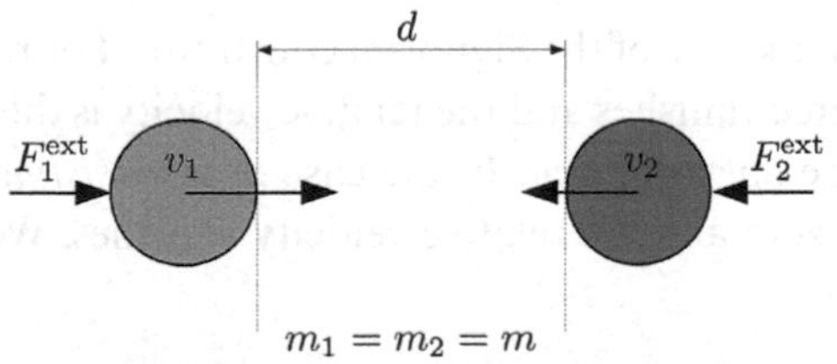

Figure 20.3

A one-dimensional contact of two particles that are a distance d apart. Both particles have mass m and velocities v_1 and v_2, respectively. Additional external forces are F_1^{ext} and F_2^{ext}.

Figure 20.4

and

$$\begin{bmatrix} R_1 \\ R_2 \end{bmatrix} = \begin{bmatrix} -R_n^{\text{loc}} \\ R_n^{\text{loc}} \end{bmatrix} = \begin{bmatrix} -1 \\ 1 \end{bmatrix} R_n^{\text{loc}} = H R_n^{\text{loc}}, \tag{20.3}$$

where the superscript "loc" refers to variables in the space of contacts. Note that we used momentum conservation in the first step of eq. (20.3). The equations of motion for both particles are

$$\frac{\mathrm{d}}{\mathrm{d}t} \begin{bmatrix} v_1 \\ v_2 \end{bmatrix} = \frac{1}{m} \left[\begin{bmatrix} R_1 \\ R_2 \end{bmatrix} + \begin{bmatrix} F_1^{\text{ext}} \\ F_2^{\text{ext}} \end{bmatrix} \right]. \tag{20.4}$$

We combine this equation with the transformation of eq. (20.3) to obtain the equation of motion of the contact

$$\begin{aligned} \frac{\mathrm{d}v_n^{\text{loc}}}{\mathrm{d}t} &= \begin{bmatrix} -1 & 1 \end{bmatrix} \frac{1}{m} \left[\begin{bmatrix} -1 \\ 1 \end{bmatrix} R_n^{\text{loc}} + \begin{bmatrix} F_1^{\text{ext}} \\ F_2^{\text{ext}} \end{bmatrix} \right] \\ &= \frac{1}{m_{\text{eff}}} R_n^{\text{loc}} + \frac{1}{m} \left(F_2^{\text{ext}} - F_1^{\text{ext}} \right), \end{aligned} \tag{20.5}$$

where $m_{\text{eff}} = m/2$ is the effective mass and $\frac{1}{m}\left(F_2^{\text{ext}} - F_1^{\text{ext}}\right)$ is the acceleration without contact forces. We integrate the prior equation with an implicit Euler method (see eq. (20.1)) and find

$$\frac{v_n^{\text{loc}}(t + \Delta t) - v_n^{\text{loc}}(t)}{\Delta t} = \frac{1}{m_{\text{eff}}} R_n^{\text{loc}}(t + \Delta t) + \frac{1}{m} \left(F_2^{\text{ext}} - F_1^{\text{ext}} \right). \tag{20.6}$$

The unknown quantities in this equation are $v_n^{\text{loc}}(t + \Delta t)$ and $R_n^{\text{loc}}(t + \Delta t)$. First, one can compute the relative velocity $v_n^{\text{loc, free}}(t + \Delta t)$ and the local contact force $R_n^{\text{loc}}(t + \Delta t)$ in the absence of any constraints:

$$v_n^{\text{loc, free}}(t + \Delta t) = v_n^{\text{loc}}(t) + \Delta t \frac{1}{m} \left(F_2^{\text{ext}} - F_1^{\text{ext}} \right) \tag{20.7}$$

and

$$R_n^{\text{loc}}(t + \Delta t) = m_{\text{eff}} \frac{v_n^{\text{loc}}(t + \Delta t) - v_n^{\text{loc, free}}(t + \Delta t)}{\Delta t}. \tag{20.8}$$

To find a solution, we make use of the Signorini constraint. For particles that are not in contact, the contact force vanishes and the relative velocity is different from zero. This case corresponds to the *open solution*. In the case of a *persisting contact*, the contact force is different from zero and the relative velocity vanishes. We distinguish between the possible cases:

- particles are not in contact,
- particles are closing contact,

- particles are in persisting contact,
- particles are opening contact.

It is not necessary to account for the influence of contact forces if the particles are not in contact. However, for particles that are approaching each other, we have to take into account the possibility of overlaps at every time step. In the case of a potential overlap, constraint forces have to be applied such that the overlap vanishes. No overlap (e. g., no contact or closing contact) at time $t + \Delta t$ corresponds to

$$d(t + \Delta t) = d(t) + v_n^{\text{loc}}(t + \Delta t)\Delta t \geq 0, \tag{20.9}$$

and $R_n^{\text{loc}}(t+\Delta t) = 0$. A vanishing or negative distance d describes persisting and forming contacts during the time step (i. e., the gap d closes). Thus, we impose

$$v_n^{\text{loc}}(t + \Delta t) = -\frac{d(t)}{\Delta t}\,. \tag{20.10}$$

The contact force is then given by

$$R_n^{\text{loc}}(t + \Delta t) = m_{\text{eff}}\frac{-d(t)/\Delta t - v_n^{\text{loc, free}}(t + \Delta t)}{\Delta t}\,. \tag{20.11}$$

The Signorini condition can be reformulated in terms of velocities to find the intersection of the Signorini graph with eq. (20.8) (see Figure 20.5). In addition to the normal contact forces, we may also consider tangential contact forces to model friction. For simplicity, we may assume that the static and dynamic friction coefficients are equal ($\mu_s = \mu_d$). Similar to the solution of eq. (20.8), the solution for the tangential contact force can be obtained with the help of the Coulomb graph of Figure 20.3.

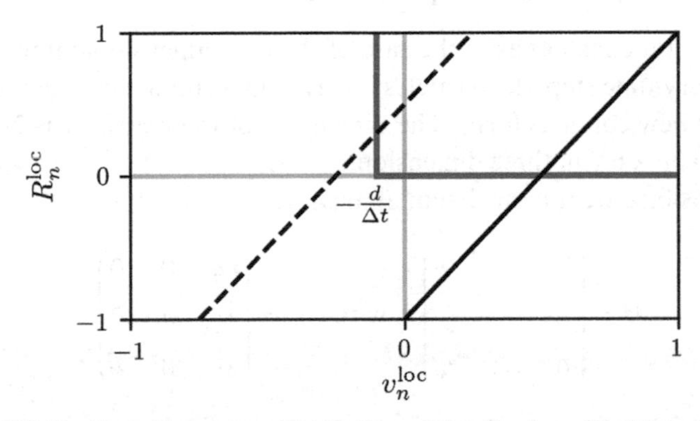

Figure 20.5 Intersection of eq. (20.8) with the Signorini graph (shifted by $-d/\Delta t$). The slope of the black solid line is $m_{\text{eff}}/\Delta t$, and the intersection point between this line and the Signorini graph occurs at $v_n^{\text{loc}} = v_n^{\text{loc, free}}$ and $R_n^{\text{loc}} = 0$ (see eq. (20.8)). This solution (i. e., no constraint forces) corresponds to no contact or an opening contact. For a persisting or forming contact (black dashed line), the local velocity is $v_n^{\text{loc}} = -d/\Delta t$ (see eq. (20.10)), but R_n^{loc} is finite.

20.2 Generalization to N Particles

To describe the interactions of N particles utilizing contact dynamics, we use the generalized coordinates

$$\dot{q} = \begin{pmatrix} \mathbf{v}_1 \\ \omega_1 \\ \vdots \\ \mathbf{v}_N \\ \omega_N \end{pmatrix}, \quad R = \begin{pmatrix} \mathbf{R}_1 \\ \mathbf{T}_1 \\ \vdots \\ \mathbf{R}_N \\ \mathbf{T}_N \end{pmatrix}, \quad \text{and} \quad F^{\text{ext}} = \begin{pmatrix} \mathbf{F}_1^{\text{ext}} \\ 0 \\ \vdots \\ \mathbf{F}_N^{\text{ext}} \\ 0 \end{pmatrix}, \tag{20.12}$$

where ω_i are the angular velocities and $\mathbf{T}_i$ are the torques. The number of components depends on the dimension of the simulated system. In two dimensions, we have $3N$ components (two translational and one rotational per particle), whereas in three dimensions there are $6N$ components (three translational and three rotational per particle). Let c be the number of contacts. In two dimensions, there exist $2c$ contact variables (one normal and one tangential at each contact), while in three dimensions the number of contact variables is $3c$ (one normal and two tangential for each contact). The contact velocities and forces are

$$\mathbf{u} = \begin{bmatrix} \mathbf{v}_1^{\text{loc}} \\ \vdots \\ \mathbf{v}_c^{\text{loc}} \end{bmatrix} \quad \text{and} \quad \mathbf{R}^{\text{loc}} = \begin{bmatrix} \mathbf{R}_1^{\text{loc}} \\ \vdots \\ \mathbf{R}_c^{\text{loc}} \end{bmatrix}. \tag{20.13}$$

The corresponding transformation from contact into particle variables is given by

$$\mathbf{u} = H^T \dot{\mathbf{q}} \quad \text{and} \quad \mathbf{R} = H \mathbf{R}^{\text{loc}}. \tag{20.14}$$

The matrix H is not constant as in the case of the one-dimensional interaction of two particles. At every time step, the particles involved in a contact change; some contacts open and some new contacts form. The dimension of the matrix H is $2c \times 3N$ in two dimensions and $3c \times 6N$ in three dimensions.

In two dimensions, we use the diagonal mass

$$M = \begin{pmatrix} \xi_1 & \cdots & 0 \\ \vdots & \ddots & \vdots \\ 0 & \cdots & \xi_N \end{pmatrix} \quad \text{with} \quad \xi_i = \begin{pmatrix} m_i & 0 & 0 \\ 0 & m_i & 0 \\ 0 & 0 & I_i \end{pmatrix}. \tag{20.15}$$

to obtain

$$M \ddot{\mathbf{q}}(t) = \mathbf{R}(t) + \mathbf{F}^{\text{ext}}. \tag{20.16}$$

Based on the relation

$$\dot{\mathbf{u}} = H^T \ddot{\mathbf{q}}, \tag{20.17}$$

we find

$$\begin{aligned}\dot{\mathbf{u}} &= H^T\left(M^{-1}\mathbf{R}(t) + M^{-1}\mathbf{F}^{\text{ext}}\right)\\ &= H^T M^{-1}\mathbf{R}(t) + H^T M^{-1}\mathbf{F}^{\text{ext}}\,.\end{aligned} \tag{20.18}$$

Using contact variables yields

$$\dot{\mathbf{u}} = H^T M^{-1} H \mathbf{R}^{\text{loc}} + H^T M^{-1}\mathbf{F}^{\text{ext}}\,. \tag{20.19}$$

We define the effective inverse mass matrix

$$M_{\text{eff}}^{-1} = H^T M^{-1} H, \tag{20.20}$$

and solve the resulting system of equations with an implicit Euler scheme (as for the one-dimensional contact):

$$\mathbf{R}^{\text{loc}}(t + \Delta t) = M_{\text{eff}}\frac{\mathbf{u}(t+\Delta t) - \mathbf{u}^{\text{free}}(t+\Delta t)}{\Delta t}\,. \tag{20.21}$$

The dimensions of the vectors change over time because the number of contacts is not constant. To solve eq. (20.21), we perform a direct inversion of $M_{\text{eff}}^{-1} = H^T M^{-1} H$ and to determine $\mathbf{u}(t + \Delta t)$, we have to account for all constraints by checking if the contacts close. Friction, rolling friction, and cohesion can be simulated by modifying the constraint forces.

Up until now, we have only focused on contact dynamics simulations of spherical particles. As described in Section 19.3, nonspherical objects can be composed of spherical objects (e. g., by introducing certain constraints between particles). For general nonspherical rigid objects of finite volume (e. g., polyhedra), one can generalize the above procedure systematically by determining the conditions for touching and slipless motion for each contact using the technique described in Chapter 19 [299].

Since in general the number of particle variables is larger than the number of contact variables, the solution for particle positions and velocities is not unique. Obtaining the physically correct solution requires excluding unphysical cases by imposing adequate conditions, which adds some subtlety to the programming. For further reading, we recommend the textbook *The Physics of Granular Media* [300].

Exercise: Contact Dynamics

1. Implement the contact dynamics for spherical particles either in 2D or 3D. For more information, see Ref. [300].
2. Consider the following simplifications:
   - Use a frictionless contact because then the transformation matrices are very simple even for the 3D case. Then you do not need angular velocities.
   - For the frictionless contact, there is only a normal force component and one constraint (Signorini). That is, the force calculation is the same as described in Section 20.1 for a 1D contact.

Exercise: Contact Dynamics (*cont.*)

- Consider particles with the same mass m and radius R (possible extension: vary R).
- For dealing with contact networks (more than one contact per particle), when calculating the force at one contact, treat the other contact forces acting on the two particles involved as external forces.
- Start with a verification of your force calculation within a system including only two particles, or one particle (under the influence of gravitation) and a wall, or similar systems.
- Then apply the method to several particles, where the forces for each time step have to be calculated iteratively. The simplest implementation is to use a fixed number of iterations for each time step.
- A simple 3D example would be settling of particles under gravity, supported by a wall in vertical direction and periodic boundaries in horizontal directions (alternatively you can also use walls as horizontal boundaries instead).

Please consider the following points:

- You could first use a 2D setup within your 3D code to test it. Note that for settling of frictionless disks with the same radius one gets strong crystallization effects, which are avoided in 3D (or using disks with slightly different radii in 2D).
- Using this iterative scheme for the force calculation, the exact solution will usually not be reached, leading to small particle overlaps. The average and maximal overlap in the system can be used as an simulation error measure.
- To avoid an artificially large repulsive force by these overlaps, one typically treats overlapping contacts as being directly in contact (in practice: set distance d to zero if $d < 0$), so that the contact force only acts to avoid further overlap (not compensating already existing overlap).
- When using a fixed number of iterations N_I the system macroscopically behaves like a system of elastic particles. For a given time step Δt, the effective elasticity scales like $\kappa \propto N_I/(\Delta t)^2$. That is, you can either use a large number of iterations or a small time step to reduce "quasi-elasticity." As within the force iteration, the information is spreading diffusively, equalizing the diffusion length of this process. The maximal system extension L_{max} (measured in number of particles with diameter d_{particle}) leads to a lower bound for the number of iterations needed to really model perfectly rigid particles: $N_I > (L_{max}/d_{\text{particle}})^2$ [301].
- It is useful to store the total force $\mathbf{R}_i$ acting on a particle i due to all its contacts in a contact network. This is particularly helpful when calculating the contact force at a contact considering all other contact forces like external forces. Before updating the force at one contact (involving particles i and j), one simply subtracts the current contact's contribution from $\mathbf{R}_i$ and $\mathbf{R}_j$ (for both particles i and j) involved in the contact, defining $\mathbf{R}'_i$ and $\mathbf{R}'_j$. Then calculate the force at the contact by setting $\mathbf{F}_i^{\text{ext}} + \mathbf{R}'_i$ and $\mathbf{F}_j^{\text{ext}} + \mathbf{R}'_j$ as external forces, and add the new contact's contribution. Then you can continue in the same way with the next contact.

21 Discrete Fluid Models

There are two possibilities to model fluid flow. First, in a *continuum approach* the fluid is described by its equations of motion, namely, the Navier–Stokes equations. Second, it is possible to use *particle-based* models of fluids. These two approaches are equivalent, since they are both based on local momentum conservation. In fact, there exist many different discrete particle methods today, including

- lattice gas automata (LGA),
- lattice Boltzmann method (LBM),
- stochastic rotation dynamics (SRD),
- direct simulation Monte Carlo (DSMC),
- dissipative particle dynamics (DPD),
- smooth particle hydrodynamics (SPH).

21.1 Lattice Gas Automata

Lattice gas automata (LGA) use cellular automata methods to simulate fluid flows. When particles meet, they follow the collision rules shown in Figure 21.1. With a Chapman–Enskog expansion, we can retrieve the Navier–Stokes equations in the continuum limit of LGA [302].

Unfortunately, the anisotropy of a certain lattice has strong influence on LGA. In two dimensions, the only triangular lattice is Galilean invariant and there exists no regular three-dimensional Galilean-invariant lattice. Only by making a projection from a four-dimensional lattice, one can achieve Galilean-invariant three-dimensional simulations. For these reasons, LGA has been of little practical use.

21.2 Lattice Boltzmann Method

Based on the Chapman–Enskog theory, it is possible to derive the Navier–Stokes equations from the Boltzmann equation [303]. This connection between fluid dynamics and Boltzmann transport theory allows us to simulate the motion of fluids by solving the corresponding Boltzmann equation on a lattice with a discrete set of velocities $\mathbf{v}_i$, where i labels a certain bond. In the lattice Boltzmann method (LBM), we use a

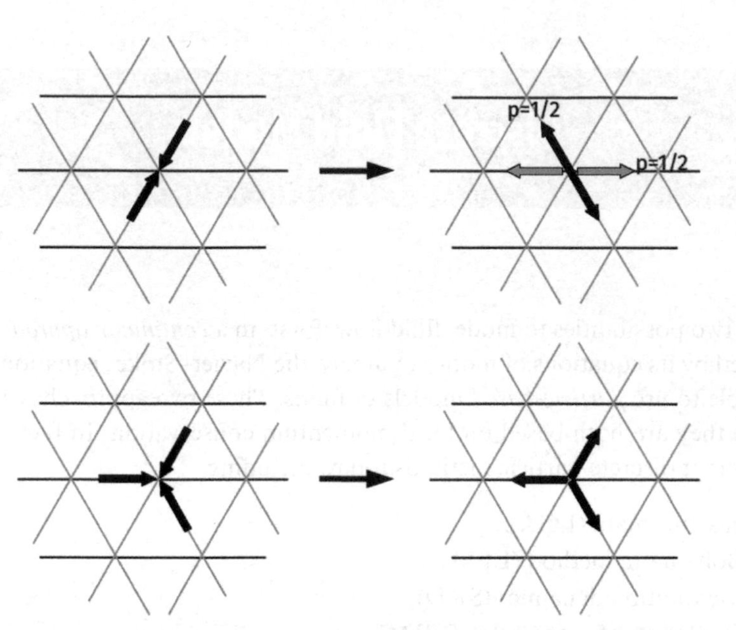

Figure 21.1 Rules for collisions in a lattice gas automaton. The rules are analogously applied in the other lattice directions. Only in the shown cases, momentum is exchanged – in all other cases, particles do not interact. The upper rule is probabilistic in the sense that the black and the gray outcomes occur each with probability 1/2.

velocity distribution function $f(\mathbf{x}, \mathbf{v}_i, t)$ to describe the density of particles with velocity $\mathbf{v}_i$ at position $\mathbf{x}$ and time t. The mass and momentum densities are

$$\rho(\mathbf{x}, t) = \sum_i f(\mathbf{x}, \mathbf{v}_i, t) \quad \text{and} \quad \rho\mathbf{u}(\mathbf{x}, t) = \sum_i \mathbf{v}_i f(\mathbf{x}, \mathbf{v}_i, t), \tag{21.1}$$

respectively. The velocity distribution can be updated following Bhatnagar, Gross, and Krook (BGK) [304] as

$$f(\mathbf{x} + \Delta t\mathbf{v}_i, \mathbf{v}_i, t + \Delta t) = f(\mathbf{x}, \mathbf{v}_i, t) + \frac{1}{\tau}\left[f_i^{\text{eq}} - f(\mathbf{x}, \mathbf{v}_i, t)\right]\Delta t, \tag{21.2}$$

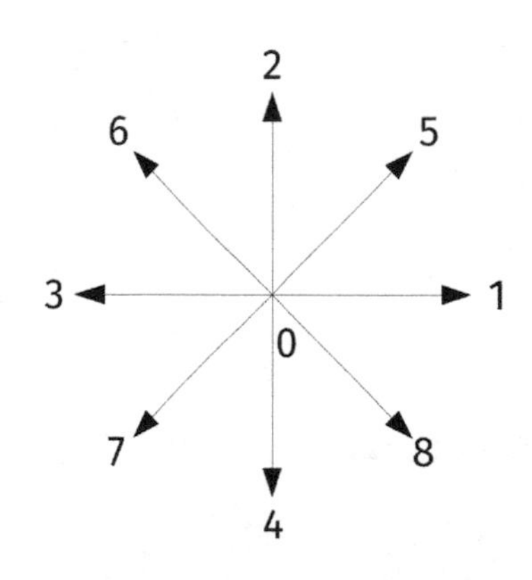

Figure 21.2 Directions of the lattice Boltzmann weights on a square lattice with next-nearest neighbors (D2Q9).

where f_i^{eq} is the equilibrium distribution and τ a characteristic relaxation time scale. An expansion of the Maxwell–Boltzmann distribution in Hermite polynomials is used as the equilibrium distribution

$$f_i^{\text{eq}} = w_i\rho\left[1 + \frac{3}{c_s^2}\mathbf{u}\mathbf{v}_i + \frac{9}{2c_s^4}(\mathbf{u}\mathbf{v}_i)^2 - \frac{3}{2c_s^2}\mathbf{u}\mathbf{u}\right], \tag{21.3}$$

where $c_s^2 = (1/3)\Delta x^2/\Delta t^2$ is the lattice speed and w_i are weights that depend on the lattice and on the direction of the bond. The weights are obtained through Gaussian quadrature (see the gray box on page 217). One possible choice of the weights in two dimensions is (D2Q9, see Figure 21.2)

$$w_i = \begin{cases} 4/9 & i = 0\,, \\ 1/9 & i = 1,2,3,4\,, \\ 1/36 & i = 5,6,7,8\,. \end{cases} \tag{21.4}$$

The first step in a lattice Boltzmann simulation is the so-called *collision* (or relaxation) step:

$$f^*(\mathbf{x}, \mathbf{v}_i, t) = f(\mathbf{x}, \mathbf{v}_i, t) + \frac{1}{\tau}\left[f_i^{\text{eq}} - f(\mathbf{x}, \mathbf{v}_i, t)\right]\Delta t\,. \tag{21.5}$$

The second step is *propagation* (or streaming):

$$f(\mathbf{x} + \Delta t \mathbf{v}_i, \mathbf{v}_i, t + \Delta t) = f^*(\mathbf{x}, \mathbf{v}_i, t)\,. \tag{21.6}$$

Implementing the LBM on a computer is straightforward.

Lattice Boltzmann Method

1. Compute $\rho(\mathbf{x}, t)$ and $\mathbf{u}(\mathbf{x}, t)$ according to eq. (21.1).
2. Determine the equilibrium distribution f_i^{eq} (see eq. (21.3)).
3. Perform collision (or relaxation) step (see eq. (21.5)).
4. Perform propagation (or relaxation) step (see eq. (21.6)).
5. Advance time from t to $t + \Delta t$ and go back to step 1.

In the D2Q9 model, the characteristic relaxation time scale τ and kinematic viscosity ν are connected via

$$\nu = \frac{2\tau - 1}{6} c_s^2 \Delta t\,. \tag{21.7}$$

Figure 21.3 Sauro Succi is working at the Italian Institute of Technology.

The LBM is very well suited to simulate fluids in complex geometries like porous media and due to the underlying regular lattice it is particularly well suited for parallelization. For a detailed treatment of lattice Boltzmann approaches, we recommend *The Lattice Boltzmann Equation: For Fluid Dynamics and Beyond* [303] (see Figure 21.3).

Additional Information: Derivation of Lattice Boltzmann Weights

For the derivation of the weights w_i, we start with the Gaussian quadrature theorem. For a polynomial $g(x)$ of at most degree $2n + 1$ and weight function $w(x)$, we have

$$\int_a^b g(x) w(x)\,\mathrm{d}x = \sum_{i=0}^{n} w_i g(x_i), \tag{21.8}$$

with

$$w_i = \int_a^b w(x) \prod_{k \neq i}^{n} \frac{x - x_k}{x_i - x_k}\,\mathrm{d}x\,,\ i = 0, \ldots, n, \tag{21.9}$$

Additional Information: Derivation of Lattice Boltzmann Weights (*cont.*)

if there exists a polynomial $p(x)$ of degree $n+1$ such that

$$\int_a^b x^k p(x) w(x)\,\mathrm{d}x = 0\,,\ \forall k = 0,\ldots,n, \tag{21.10}$$

and x_i are the zeros of $p(x)$. The equilibrium distribution $\mathbf{f}^{\mathrm{eq}}(\mathbf{v})$ is given by the Maxwell–Boltzmann velocity distribution

$$\mathbf{f}^{\mathrm{eq}}(\mathbf{v}) = \frac{\rho}{m(2\pi c_s^2)^{d/2}} e^{-\frac{(\mathbf{v}-\mathbf{u})^2}{2c_s^2}}\,. \tag{21.11}$$

A Taylor expansion for small $||\mathbf{u}||/c_s$ yields

$$\mathbf{f}^{\mathrm{eq}}(\mathbf{v}) \approx \frac{\rho}{m(2\pi c_s^2)^{d/2}} e^{-\frac{\mathbf{v}^2}{2c_s^2}} \left[1 + \frac{\mathbf{vu}}{c_s^2} + \frac{(\mathbf{vu})^2}{2c_s^4} - \frac{\mathbf{u}^2}{2c_s^2}\right]. \tag{21.12}$$

Identifying the expansion in Hermite polynomials in the square brackets of eq. (21.12) with $p(x)$ and the prefactor with $w(x)$, one can use the above theorem of Gaussian quadrature to determine the weights w_i and thus approximate eq. (21.11).

Exercise: Flow around an Obstacle

Figure 21.4 Example of a two dimensional vortex street simulation in which a fluid flows around a circular obstacle (white) [305].

Use the lattice Boltzmann method to simulate the fluid flow through a two-dimensional channel around a rectangular obstacle in a two-dimensional box. Discretize the computational domain using a square lattice and apply no-slip boundary conditions at the obstacle and walls. The bottom and top parts of the box are solid walls. To create a fluid flow with a certain entry velocity, set the fluid velocity at the left boundary of the system to a constant value. The resulting flow pattern is known under the name "von Kárman vortex street" (see Figure 21.4).

Implement the D2Q9 method and simulate vortex streets for different Reynolds numbers

$$\mathrm{Re} = \frac{uD}{\nu}\,,$$

where D is the width of the channel and u the flow velocity. You can use eq. (21.7) to determine a characteristic relaxation time τ that corresponds to a certain viscosity ν and thus fix the Reynolds number.

21.3 Stochastic Rotation Dynamics

Stochastic rotation dynamics (SRD) is a particle-based fluid modeling approach [306]. This method has been developed by Malevanets and Kapral (see Figure 21.5). It is also known as multiparticle collision dynamics (MPC). In this method, space is being

Stochastic Rotation Dynamics

The particle positions and velocities are updated according to

$$\mathbf{x}_i' = \mathbf{x}_i + \Delta t \mathbf{v}_i , \tag{21.13}$$

$$\mathbf{v}_i' = \mathbf{u} + \Omega (\mathbf{v}_i - \mathbf{u}) , \tag{21.14}$$

where $\mathbf{u} = \langle \mathbf{v} \rangle$ is the mean velocity of particles in the respective cell and Ω is the rotation matrix. It is given by

$$\Omega = \begin{pmatrix} \cos(\alpha) & \pm\sin(\alpha) & 0 \\ \mp\sin(\alpha) & \cos(\alpha) & 0 \\ 0 & 0 & 1 \end{pmatrix} , \tag{21.15}$$

where α is a randomly chosen angle.

Figure 21.5 Raymond Kapral is a professor of chemistry at the University of Toronto.

discretized into cells and fluids are composed of N point-like particles with mass m, positions $\mathbf{x}_i$, and velocities $\mathbf{v}_i$.

The momentum exchange between particles is modeled by rotations of local particle velocities. In this model, a certain stochastic motion is intrinsic, so that one can effectively model small Péclet numbers. Since eq. (21.14) conserves the momentum after coarse graining, SRD recovers hydrodynamics correctly.

21.4 Direct Simulation Monte Carlo

Direct simulation Monte Carlo (DSMC), as introduced by Graeme Bird, is a particle-based simulation technique, which is appropriate to model rarified gases (i.e., particle systems at large Knudsen numbers) [307, 308]

$$\mathrm{Kn} = \frac{\lambda}{L} , \tag{21.16}$$

where λ is the mean free path and L is a characteristic system length scale. This method is very popular in aerospace modeling because the atmosphere has very low density at high altitudes and the corresponding Knudsen numbers are large.

In this method, we divide the system into cells and generate particles in each cell according to a desired density. To impose fluid velocities and temperatures, we assign a velocity to each particle which is distributed according to the Maxwell–Boltzmann distribution. In DSMC the number of point-like simulation particles (*simulators*) is much smaller than the number of physical molecules. Each simulator represents a certain number of physical molecules. For traditional DSMC, about 20 particles per collision cell is the rule of thumb.

Direct Simulation Monte Carlo

Collisions are modeled by sorting particles into cells. We iterate over all cells and

1. randomly select collision partners within a cell,
2. implement a collision that conserves momentum.

We note that collision pairs with a large relative velocity are more likely to collide. The temperature T_W may be implemented via collisions with a thermal wall. This wall resets the momentum of a particle by choosing it from a Maxwell–Boltzmann distribution such as

$$\begin{aligned} P_{v_x}(v_x) &= \pm \frac{m}{k_B T_W} v_x e^{-\frac{m v_x^2}{2k_B T_W}}, \\ P_{v_y}\left(v_y\right) &= \sqrt{\frac{m}{2\pi k_B T_W}} e^{-\frac{m(v_y - u_W)^2}{2k_B T_W}}, \\ P_{v_z}(v_z) &= \sqrt{\frac{m}{2\pi k_B T_W}} e^{-\frac{m v_z^2}{2k_B T_W}}. \end{aligned} \tag{21.17}$$

Here, u_W is the velocity of the thermal wall in y-direction. For more information on DSMC, we refer the reader to Bird's textbook [308].

21.5 Dissipative Particle Dynamics

Figure 21.6 Julia M. Yeomans is a professor of physics at the University of Oxford and works on the physics of soft matter and biological systems.

Another particle-based fluid simulation technique is *dissipative particle dynamics* (DPD) [309]. In this method, particle interactions are described by

$$\mathbf{F}_i = \sum_{i \neq j} \left(\mathbf{f}_{ij}^C + \mathbf{f}_{ij}^R + \mathbf{f}_{ij}^D \right), \tag{21.18}$$

where $\mathbf{f}_{ij}^C$ denotes the conservative forces, $\mathbf{f}_{ij}^R$ a random force, and $\mathbf{f}_{ij}^D$ the dissipative forces. In DPD, momentum is conserved. The dissipative forces are proportional to the particle velocities and the weights of the random and dissipative forces must be chosen such that thermal equilibrium is reached. More details on DPD and associated remaining challenges can be found in Ref. [310]. This method is mostly used in soft condensed matter physics to study polymers and colloids. Important contributions in this field were made by Julia Yeomans (see Figure 21.6) [311], who also contributed to the development of other discrete fluid models, including LBM and SRD [312].

21.6 Smoothed Particle Hydrodynamics

Another important technique in the field of computational fluid dynamics is *smoothed particle hydrodynamics* (SPH) [313]. This method uses *smooth kernel functions* W to represent properties of particles in a weighted sense [314, 315]. Instead of localized positions and velocities, the particle characteristics are smoothened over a smoothing length h. The value of an arbitrary quantity A is then given by

$$A(r) = \int_{\Omega} W\left(\left|r - r'\right|, h\right) A(r')\,\mathrm{d}r' \approx \sum_j \frac{m_j}{\rho_j} W\left(\left|r - r_j\right|, h\right) A_j \,. \tag{21.19}$$

In this method, no spatial discretization is necessary and even free surfaces and fluid drops can be simulated easily with SPH. This makes this method broadly applicable in many different fields where fluids exhibit interfaces.

Examples of kernel functions include Gaussians and spline functions like [316]

$$W(r, h) = \frac{\sigma}{h^d} \begin{cases} 1 - \frac{3}{2}q^2 + \frac{3}{4}q^3 \,, & \text{if } 0 \leq q \leq 1 \,, \\ \frac{1}{4}(2 - q)^3 \,, & \text{if } 1 < q \leq 2 \,, \\ 0 \,, & \text{otherwise} \,, \end{cases} \tag{21.20}$$

where d is the system dimension, $\sigma = [\frac{2}{3}, \frac{10}{7\pi}, \frac{1}{\pi}]$, $q = \frac{r}{h}$, and $r = |r_a - r_b|$. Kernels may be changed without much effort for different purposes [317].

22 Ab Initio Simulations

Simulating interactions of (classical) objects such as billiard balls, sand grains, and stars is based on classical mechanics (i.e., solving Newton's equations of motion). To a certain extent, it is also possible to consider classical approaches for the simulation of atoms and molecules if one can describe interactions well by a known potential or if the bond length can be assumed to be constant given certain environmental conditions (see Section 14.1). However, for an ab initio (i.e., from first principles) simulation of molecular and atomic interactions, we have to take quantum-mechanical effects into account. In the subsequent sections, we therefore briefly discuss the main quantum-mechanical methods which are necessary for ab initio MD simulations. In particular, we introduce some basic concepts of quantum mechanics including the Born–Oppenheimer and Kohn–Sham approximations. We conclude this chapter by discussing the Car–Parrinello method.

22.1 Introduction

In quantum mechanics, the wave function $\psi(\mathbf{r}, t)$ at position $\mathbf{r}$ is represented by a vector $|\psi\rangle$ that belongs to the *Hilbert space* $L^2\left(\mathbb{R}^3\right)$, with scalar product

$$\langle\psi|\phi\rangle = \int_{\mathbb{R}^3} \psi^*(\mathbf{r})\,\phi(\mathbf{r})\,\mathrm{d}\mathbf{r}\,. \tag{22.1}$$

In general, a quantum state is a vector $|\psi\rangle$ of a Hilbert space with scalar product $\langle\psi|\phi\rangle$. All states satisfy the normalization condition

$$\langle\psi|\psi\rangle = \int_{\mathbb{R}^3} \psi^*(\mathbf{r})\,\psi(\mathbf{r})\,\mathrm{d}\mathbf{r} = 1. \tag{22.2}$$

Additional Information: Square-Integrable Functions

The Hilbert space $L^2\left(\mathbb{R}^3\right)$ is the space of square-integrable functions. That is, elements $|\psi\rangle \in L^2\left(\mathbb{R}^3\right)$ satisfy

$$\langle\psi|\psi\rangle = \int_{\mathbb{R}^3} \psi^*(\mathbf{r})\,\psi(\mathbf{r})\,\mathrm{d}\mathbf{r} < \infty\,.$$

and are uniquely defined up to phase $e^{i\alpha}$ with $\alpha \in [0, 2\pi)$. That is,

$$|\psi\rangle \propto e^{i\alpha} |\psi\rangle . \tag{22.3}$$

According to the statistical interpretation of the wave function (*Born rule*), the probability density $|\psi(\mathbf{r})|^2$ denotes the probability of finding a particle described by ψ at position $\mathbf{r}$. Observables are represented by self-adjoint operators A with the expectation value $\langle\psi|A|\psi\rangle$.

Given a *Hamilton operator*

$$H = \frac{p^2}{2m} + V(\mathbf{r}) \quad \text{with} \quad p = -i\hbar\nabla , \tag{22.4}$$

the time evolution of the function $|\psi_t\rangle$ is described by the time-dependent Schrödinger equation

$$i\hbar \frac{\mathrm{d}\,|\psi_t\rangle}{\mathrm{d}t} = H\,|\psi_t\rangle . \tag{22.5}$$

To determine the time evolution of $\psi_t(\mathbf{r})$, we use separation of variables with $\psi_t(\mathbf{r}) = \psi(\mathbf{r})\,\phi(t)$ and find

$$\phi(t) = e^{-iEt/\hbar} , \tag{22.6}$$

where E is the expectation value of the Hamilton operator. The time-independent Schrödinger equation of a single particle is thus

$$-\frac{\hbar^2}{2m}\nabla^2 |\psi\rangle + V(\mathbf{r})\,|\psi\rangle = E\,|\psi\rangle . \tag{22.7}$$

For simulating quantum-mechanical particle interactions, we have to consider the many-body wave function of N particles

$$\psi(\mathbf{r}_1, s_1, \mathbf{r}_2, s_2, \ldots, \mathbf{r}_N, s_N) , \tag{22.8}$$

where s_i is the spin of particle i. Many-body operators A are defined by their action on all individual particles. For the following discussion of many-body wave functions, we use $\psi(1, 2, \ldots, N)$ to abbreviate the wave function of eq. (22.8) and $A(1, 2, \ldots, N)$ as an abbreviation of a corresponding many-body operator. We now define the exchange operator P_{ij} $(1 \leq i, j \leq N)$ according to

$$\begin{aligned} P_{ij}\psi(1, 2, \ldots, i, \ldots, j, \ldots, N) &= \psi(1, 2, \ldots, j, \ldots, i, \ldots, N) , \\ P_{ij}A(1, 2, \ldots, i, \ldots, j, \ldots, N) &= A(1, 2, \ldots, j, \ldots, i, \ldots, N) . \end{aligned} \tag{22.9}$$

The Hamiltonian commutes with the exchange operation because it is invariant under particle exchange. That is,

$$\left[H, P_{ij}\right] = 0 . \tag{22.10}$$

We note that $\left(P_{ij}\right)^{-1} = P_{ij}$ and $\left(P_{ij}\right)^2 = 1$. The two possible eigenvalues are thus $+1$ and -1, which are both realized in nature with the corresponding wave functions

$$\psi(1, 2, \ldots, i, \ldots, j, \ldots, N) = \begin{cases} +\psi(1, 2, \ldots, j, \ldots, i, \ldots, N) , & \text{(bosons)} , \\ -\psi(1, 2, \ldots, j, \ldots, i, \ldots, N) , & \text{(fermions)} . \end{cases} \tag{22.11}$$

Additional Information: Commutator

The commutator of two operators A and B is defined as $[A, B] = AB - BA$. If $[A, B] = 0$, the two operators A and B commute since $AB = BA$.

The wave functions of *bosons* are completely symmetric under particle exchange, whereas *fermions* exhibit completely antisymmetric wave functions. For fermions, this implies that two particles cannot be in the same state since

$$\psi(1, 2, \ldots, i, \ldots, i, \ldots, N) = -\psi(1, 2, \ldots, i, \ldots, i, \ldots, N) = 0 . \tag{22.12}$$

This is the so-called *Pauli exclusion principle*. Note that a quantum-mechanical description of N particles requires one to construct many-body wave functions by considering all possible particle permutations since all particles (unlike classical ones) are indistinguishable.

For N noninteracting particles, the corresponding Hamilton operator is

$$H = \sum_{i=1}^{N} H_i \quad \text{with} \quad H_i = \frac{p_i^2}{2m} + V(\mathbf{r}_i) . \tag{22.13}$$

To construct many-body wave functions, we use ψ_ν to denote the eigenstates of the single-particle Hamiltonian such that

$$H_i \psi_\nu(\mathbf{r}_i, s_i) = \epsilon_\nu \psi_\nu(\mathbf{r}_i, s_i) , \tag{22.14}$$

where ν is some quantum number to label the states. For bosons, the many-body wave function is

$$\begin{aligned} \langle 1, 2, \ldots, N | \psi_B \rangle &= \psi_B(1, 2, \ldots, N) \\ &= \sum_{P \in S_N} P \psi_{\nu_1}(\mathbf{r}_1, s_1) \psi_{\nu_2}(\mathbf{r}_2, s_2) \cdots \psi_{\nu_N}(\mathbf{r}_N, s_N) , \end{aligned} \tag{22.15}$$

and for fermions we find

$$\begin{aligned} \langle 1, 2, \ldots, N | \psi_F \rangle &= \psi_F(1, 2, \ldots, N) \\ &= \sum_{P \in S_N} \operatorname{sign}(P) P \psi_{\nu_1}(\mathbf{r}_1, s_1) \psi_{\nu_2}(\mathbf{r}_2, s_2) \cdots \psi_{\nu_N}(\mathbf{r}_N, s_N) . \end{aligned} \tag{22.16}$$

Here, P is the exchange operator for the state ν_i and $\operatorname{sign}(P)$ is the sign of the permutation P which is $+1$ and -1, if P is composed of an even and odd number of transpositions, respectively. For N particles, the set of possible exchange operators is described by the *symmetric group* S_N. For fermions, it is possible to rewrite eq. (22.16) in terms of the so-called *Slater determinant*

$$\psi_F(1, \ldots, N) = \left\| \begin{pmatrix} \psi_{\nu_1}(1) & \cdots & \psi_{\nu_N}(1) \\ \vdots & \ddots & \vdots \\ \psi_{\nu_1}(N) & \cdots & \psi_{\nu_N}(N) \end{pmatrix} \right\| . \tag{22.17}$$

We still have to normalize the many-body wave functions by dividing them by

$$\langle\psi_B|\psi_B\rangle = N!n_{\nu_1}!n_{\nu_2}!\cdots n_{\nu_N}! \quad \text{(bosons)} \tag{22.18}$$

and

$$\langle\psi_F|\psi_F\rangle = N! \quad \text{(fermions)}\,. \tag{22.19}$$

For bosons, a certain index ν_i may appear multiple times in eq. (22.15). Therefore, we have to account for the additional normalization factors n_{ν_i} which denote the number of particles in the stationary single particle state ν_i. For fermions, the occupation numbers n_{ν_i} are either zero or one. The calculation of eq. (22.17) is computationally very time consuming.

22.2 Implementation of Wave Functions

Single particle wave functions are often expanded in an orthonormal basis

$$\psi_j = \sum_k c_{jk}\chi_k\,, \tag{22.20}$$

where $\{\chi_k\}$ is a set of orthonormal basis functions and c_{jk} are the corresponding expansion coefficients for ψ_j in the considered basis. In solid state physics, a plane wave basis $\chi_k = \exp(ikx)$ is often used (plane wave functions are not localized and can therefore describe valence electrons in metals).

We typically use as many plane waves as necessary until to obtain the energy with a desired precision. For 16 water molecules, a cutoff of about 70 Ry (1 Ry = $2.179\ldots \times 10^{-18}$ J [318]) is needed. This corresponds typically to about 15,000 plain waves per electron.

Using localized basis sets can help to decrease the number of basis functions for covalent bondings. Wave functions are then dependent on the ion positions. One possible choice are Gaussian-type orbitals (GTOs)

$$\chi_l(r) = c_l r^l \exp\left(-\alpha r^2\right). \tag{22.21}$$

For localized basis sets, so-called *Pulay forces* appear [319]. These forces are due to numerical artifacts originating from a finite basis. They may be up to an order of magnitude bigger than the physical forces and have to be corrected for.

22.3 Born–Oppenheimer Approximation

For the simulation of atomic and molecular interactions, we have to describe the quantum-mechanical evolution equations of all involved particles. Instead of a full quantum-mechanical description for both nuclei and electrons, we make use of the fact that their masses differ by three orders of magnitude. In this way, we determine the electron distribution for fixed ion positions. This is the so-called *Born–Oppenheimer approximation*.

The Born–Oppenheimer approximation implies that the time step for the motion of ions has to be sufficiently small, such that electrons do not skip any energy level transition when ions move. Mathematically, the energies ϵ_i of energy level i have to satisfy

$$|\epsilon_i(\mathbf{r}+\Delta\mathbf{r})-\epsilon_i(\mathbf{r})| \ll |\epsilon_{i+1}(\mathbf{r})-\epsilon_i(\mathbf{r})| , \tag{22.22}$$

where $\mathbf{r}$ is the ion position before the update, $\Delta\mathbf{r} = \mathbf{u}\Delta t$, $\mathbf{u}$ is the ion velocity, and Δt is the corresponding time step.

22.4 Hohenberg–Kohn Theorems

Figure 22.1 Pierre C. Hohenberg (1934–2017) worked at different institutions including Bell Laboratories, TU München, Leiden University, and others. He spoke 12 languages. Photograph courtesy Michael Marsland, Yale University.

If the ground state is nondegenerate (i.e., there exists only one quantum state with the energy of the ground state), all information on this ground state is contained in its density distribution

$$n(\mathbf{r}) = \sum_{i=1}^{N} |\psi_i(\mathbf{r})|^2 . \tag{22.23}$$

It is possible to obtain all observables based on $n(\mathbf{r})$. The knowledge of the wave functions is not needed. To formalize this approach, we consider N electrons described by a wave function Ψ with normalization $\langle\Psi|\Psi\rangle = N$. These electrons orbit around a nucleus with potential

$$V(\mathbf{r}) = -\sum_i \frac{Ze^2}{|\mathbf{r}-\mathbf{r}_i|} , \tag{22.24}$$

where Z is the nuclear charge number and e the elementary charge. The corresponding Hamiltonian operator is given by

$$H = \underbrace{\sum_i \frac{p_i^2}{2m} + \frac{1}{2}\sum_i\sum_{j\neq i}\frac{e^2}{|\mathbf{r}_i-\mathbf{r}_j|}}_{F} + \underbrace{\sum_i V(\mathbf{r}_i)}_{V} . \tag{22.25}$$

We abbreviate the Hamiltonian by $H = F + V$, where F represents the kinetic and electron–electron interaction energy, and V is the potential energy. The ground state $|\Psi_0\rangle$ is thus completely determined by N and $V(\mathbf{r})$. We define the ground state particle density as

$$n_0(\mathbf{r}) = \langle\Psi_0|\rho(\mathbf{r})|\Psi_0\rangle = N\int |\Psi_0(\mathbf{r},\mathbf{r}_2,\ldots,\mathbf{r}_N)|^2\, d\mathbf{r}_2\, d\mathbf{r}_3 \cdots d\mathbf{r}_N , \tag{22.26}$$

where $\rho(\mathbf{r}) = \sum_i \delta(\mathbf{r}-\mathbf{r}_i)$ is the particle density operator.

The theorems of Hohenberg and Kohn [320] (see Figures 22.1 and 22.2) state that (i) the external potential V is uniquely determined by the ground-state electron density, and that (ii) the ground state energy can be obtained by minimizing an energy functional. Because of the Hohenberg–Kohn theorems, it is not necessary anymore to consider wave functions for nondegenerate ground states. Instead, we can use a density functional approach.

Additional Information: Proof of Hohenberg–Kohn Theorems

Claim (i) can be proven by contradiction. Thus, we assume that there exist two external potentials $V(\mathbf{r})$ and $V'(\mathbf{r})$ which give rise to the same particle density $n_0(\mathbf{r})$. The corresponding Hamiltonians are $H = F + V$ and $H' = F + V'$ and the corresponding ground-state energies are

$$E_0 = \langle \Psi_0 | H | \Psi_0 \rangle \quad \text{and} \quad E'_0 = \langle \Psi'_0 | H' | \Psi'_0 \rangle \,. \tag{22.27}$$

With these definitions, we obtain

$$\begin{aligned} E_0 &< \langle \Psi'_0 | H | \Psi'_0 \rangle = \langle \Psi'_0 | H' | \Psi'_0 \rangle + \langle \Psi'_0 | (H - H') | \Psi'_0 \rangle \\ &= E'_0 + \int n_0(\mathbf{r}) \left[V(\mathbf{r}) - V'(\mathbf{r}) \right] \mathrm{d}\mathbf{r} \end{aligned} \tag{22.28}$$

and

$$\begin{aligned} E'_0 &< \langle \Psi_0 | H' | \Psi_0 \rangle = \langle \Psi_0 | H | \Psi_0 \rangle + \langle \Psi_0 | (H' - H) | \Psi_0 \rangle \\ &= E_0 + \int n_0(\mathbf{r}) \left[V'(\mathbf{r}) - V(\mathbf{r}) \right] \mathrm{d}\mathbf{r} \,. \end{aligned} \tag{22.29}$$

Adding Eqs. (22.28) and (22.29) leads to the contradiction

$$E_0 + E'_0 < E_0 + E'_0 \,. \tag{22.30}$$

Thus, the potential V is (up to a constant) uniquely determined by $n_0(\mathbf{r})$. We are now able to recast the problem of solving the Schrödinger equation for our N particle system in variational form by defining the energy functional

$$E_V[n] = \int n(\mathbf{r}) V(\mathbf{r}) \,\mathrm{d}\mathbf{r} + F[n] \,, \tag{22.31}$$

where $F[n] = \langle \Psi | F | \Psi \rangle$. The particle density $n(\mathbf{r})$ determines the potential $V(\mathbf{r})$ and the ground-state Ψ. For another potential $\tilde{V}(\mathbf{r})$, we find

$$\begin{aligned} \langle \Psi | \tilde{H} | \Psi \rangle &= \langle \Psi | F | \Psi \rangle + \langle \Psi | \tilde{V} | \Psi \rangle \\ &= F[n] + \int n(\mathbf{r}) \tilde{V}(\mathbf{r}) \,\mathrm{d}\mathbf{r} = E_{\tilde{V}}[n] \geq E_0 \end{aligned} \tag{22.32}$$

according to the variational principle. The functional $E_{\tilde{V}}[n]$ equals E_0 if Ψ is the ground-state for $\tilde{V}(\mathbf{r})$.

22.5 Kohn–Sham Approximation

Figure 22.2 Walter Kohn (1923–2016) (top) and Lu Jeu Sham (bottom). Kohn was awarded the Nobel Prize in Chemistry in 1998.

Based on the Hohenberg–Kohn theorems, we are able to describe a nondegenerate ground state with density functionals. This is the basis of density functional theory (DFT). Earlier works by Thomas and Fermi were too inaccurate for most applications. Further developments by Walter Kohn and Lu Jeu Sham (see Figure 22.2) made it possible to approximate the electron–electron interactions in terms of non-interacting single particles moving in a potential which only depends on the density distribution [321, 322]. Additional correction terms can be used to account for many-particle effects. In the Kohn–Sham approximation, electrons are treated as single particles interacting with the effective Kohn–Sham potential V_{KS}. The corresponding wave functions $\psi_i(\mathbf{r})$ with $i \in \{1, \ldots, N\}$ and energies ϵ_i obey

$$\left[-\frac{\hbar^2}{2m}\nabla^2 + V_{KS}(\mathbf{r})\right]\psi_i(\mathbf{r}) = \epsilon_i\psi_i(\mathbf{r}) . \tag{22.33}$$

The Kohn–Sham potential is given by

$$V_{KS}(\mathbf{r}) = V(\mathbf{r}) + e^2 \int \frac{n(\mathbf{r}')}{|\mathbf{r}-\mathbf{r}'|}\mathrm{d}\mathbf{r} + V_{xc}(\mathbf{r}) , \tag{22.34}$$

where $V(\mathbf{r})$ is an external potential (e. g., a nucleus potential), $V_{xc}(\mathbf{r}) = \delta E_{xc}[n]/\delta n(\mathbf{r})$ is a potential term accounting for exchange and correlation effects, and $E_{xc}(\mathbf{r})$ is the exchange-correlation energy. To use the Kohn–Sham approximation in simulations, we have to start with an initial guess for $n(\mathbf{r})$ and then solve eq. (22.33) to obtain the wave functions $\psi_i(\mathbf{r})$. This allows us to determine the particle density (see eq. (22.23)) and the kinetic energy

$$T_s[n] = -\sum_{i=1}^{N} \int \psi_i^*(\mathbf{r})\left(-\frac{\hbar^2}{2m}\nabla^2\right)\psi_i(\mathbf{r})\,\mathrm{d}\mathbf{r} . \tag{22.35}$$

The orbital energies ϵ_i have to sum up to

$$\sum_{i=1}^{N} \epsilon_i = T_s[n] + \int n(\mathbf{r})\,V_{KS}(\mathbf{r})\,\mathrm{d}\mathbf{r} . \tag{22.36}$$

Unlike many-particle orbitals, the eigenvalues of the single particle orbitals have no physical meaning. The sum of them is not the total energy. The only justification of the single particle wave functions is that they yield the correct density.

Until here, density functional theory is describing a many-body quantum system in an exact way. However, for the computation of the exchange-correlation term $E_{xc}[n] = E_x[n] + E_c[n]$ approximations have to be applied. Under the assumption of a homogeneous electron gas, the exchange energy in the *local-density approximation* (LDA) is

$$E_x^{LDA}[n] = -\frac{3}{4}\left(\frac{3}{\pi}\right)^{1/3} \int n(\mathbf{r})^{4/3}\,\mathrm{d}\mathbf{r} . \tag{22.37}$$

For the correlation energy

$$E_c^{\mathrm{LDA}}[n] = \int \epsilon_c[n(\mathbf{r})]\,\mathrm{d}\mathbf{r}\,, \tag{22.38}$$

no analytical expression is known for $\epsilon_c[n(\mathbf{r})]$. LDA has been used to compute band structures and the total energy of systems in solid state physics. In quantum chemistry, it has been much less popular since the simulation of chemical bonds requires higher accuracy. It is possible to improve the LDA by adding a dependence on the gradient of the density. This approach is called *general-gradient approximation* (GGA). The physical intuition for this correction is that quantum-mechanical effects are very strong when there is a change in the slope of the wave function and, in particular, when two identical fermions come closer. The GGA approximation of the exchange energy is

$$E_x^{\mathrm{GGA}}[n] = \int \epsilon_x[n(\mathbf{r}), |\nabla n(\mathbf{r})|]\,\mathrm{d}\mathbf{r}\,. \tag{22.39}$$

One possibility of a GGA approximation of the exchange energy is [323]

$$E_x^{\mathrm{GGA}}[n] = E_x^{\mathrm{LDA}}[n] - \beta \int \frac{(\nabla n)^2}{n^{4/3}}\mathrm{d}\mathbf{r}, \tag{22.40}$$

where β is an additional free parameter. Very popular is also another empirical approach suggested by Axel Becke [324]

$$E_x^{\mathrm{Becke}}[n] = E_x^{\mathrm{LDA}}[n] - \beta \int n^{4/3} \frac{x^2}{1 + 6\beta\, x \sinh^{-1}(x)}\mathrm{d}\mathbf{r}\,, \tag{22.41}$$

where $x = |\nabla n| / n^{4/3}$ and $\beta = 0.0042$. The corresponding paper is among the most cited ones in computational physics. The general-gradient approximation gives much better results than LDA. They can be used to calculate all kinds of chemical bonds like covalent, ionic, metallic, and hydrogen bridges. However, DFT approaches that utilize GGA fail to describe van der Waals interactions properly. A possible improvement has been suggested by Grimme [325], which leads to an additional energy term

$$E_{\mathrm{disp}} = -s_6 \sum_{i=1}^{N-1} \sum_{j=i+1}^{N} \frac{C_6^{ij}}{R_{ij}} f_{\mathrm{dmp}}\left(R_{ij}\right) \quad \text{with} \quad f_{\mathrm{dmp}}\left(R_{ij}\right) = \frac{1}{1 + e^{-d(R_{ij}/R_r - 1)}}\,, \tag{22.42}$$

where s_6, C_6^{ij}, R_r, and d are model parameters and R_{ij} is the distance between two ions.

22.6 Hellmann–Feynman Theorem

With the previous methods, we are able to describe the motion of electrons in the potential of the (slower) moving nuclei. After a sufficiently small time step, we also have to update the position of the ions. Given a Hamiltonian operator H of the system, the forces acting on the ions are (*Hellmann–Feynman theorem*) [326, 327]

$$m_\alpha \frac{d^2\mathbf{R}_\alpha}{dt^2} = -\left\langle \psi \left| \frac{\partial H}{\partial \mathbf{R}_\alpha} \right| \psi \right\rangle , \tag{22.43}$$

where m_α and $\mathbf{R}_\alpha$ are the mass and position of ion α. Combining DFT approaches with the Hellmann–Feynman theorem yields an iterative method for solving the motion of molecules and other quantum-mechanical systems according to the following steps:

Ab Initio MD

- Solve the electron dynamics in the ion potential.
- Solve the motion of the ions using classical MD techniques in combination with the Hellmann–Feynman theorem.

The problem of this method is that the first step is computationally very expensive. For this reason, the contribution of Car and Parrinello has been very important.

22.7 Car–Parrinello Method

Roberto Car and Michele Parrinello (see Figure 22.3) reformulated the ab initio MD problem in terms of the Hamiltonian [328, 329]

$$\begin{aligned} H_{\mathrm{CP}}\left[\{\mathbf{R}_I\}, \{\dot{\mathbf{R}}_I\}, \{\psi_i\}, \{\dot{\psi}_i\}\right] &= \sum_I \frac{M_I}{2} \dot{\mathbf{R}}_I^2 + \sum_i \frac{\mu}{2} \langle \dot{\psi}_i | \dot{\psi}_i \rangle \\ &\quad + E_{\mathrm{KS}}\left(\{\mathbf{R}_I\}, \{\psi_i\}\right) + E_{\mathrm{ion}}\left(\{\mathbf{R}_I\}\right) , \end{aligned} \tag{22.44}$$

where M_I, R_I, and ψ_i are the masses and positions of the nuclei, and the orbital wave functions, respectively. The parameter μ is a fictitious mass which is used to control the energy transfer from nuclei to electrons. The remaining terms are the Kohn–Sham energy E_{KS} and the ion–ion interaction energy E_{ion}. The Kohn–Sham energy is the sum of the kinetic energy of eq. (22.35) and the potential energy of eq. (22.34). The corresponding equations of motion are [330]

Car–Parrinello Equations of Motion

$$\begin{aligned} M_I \ddot{\mathbf{R}}_I &= -\nabla_{\mathbf{R}_I} E_{\mathrm{KS}} + \sum_{ij} \Lambda_{ij} \nabla_{\mathbf{R}_I} \langle \psi_i | \psi_j \rangle , \\ \mu \ddot{\psi}_i(\mathbf{r}, t) &= -\frac{\delta E_{\mathrm{KS}}}{\delta \psi_i^*(\mathbf{r}, t)} - \frac{\delta E_{\mathrm{ion}}}{\delta \psi_i^*(\mathbf{r}, t)} + \sum_j \Lambda_{ij} \psi_j(\mathbf{r}, t) . \end{aligned} \tag{22.45}$$

The additional Lagrange multiplier Λ_{ij} is used to ensure the orthonormality constraint

$$\langle \psi_i | \psi_j \rangle = \int \psi_i^* (\mathbf{r}, t) \psi_j (\mathbf{r}, t) \, d\mathbf{r} = \delta_{ij}, \tag{22.46}$$

where δ_{ij} is the Kronecker delta.

Similar to the Nosé–Hoover approach, ions and electrons are treated as two different systems coupled by the Hamiltonian as defined in eq. (22.44). The coupling is realized with a fictitious mass μ which should not be confused with the electron mass. The fictitious mass is a tunable parameter of the method describing how stiff the electron motion is coupled to the nuclei. If $\mu \to 0$, the electronic response is very rapid and the electrons remain to a sufficiently high degree in the ground state. But numerically, we want to avoid too large accelerations, so electron dynamics is made artificially slower than in reality. Since the electron configuration is now following an equation of motion, the integration step has to be small enough to resolve the electronic motion. Since μ is usually smaller than the mass of the ions, time steps used in the Car–Parrinello method are usually smaller than the time steps of the direct Born–Oppenheimer approach. Often time steps of the order of a tenth of a femtosecond are used.

Solving the Car–Parrinello equations requires much less computational effort per time step than the previously discussed ab initio methods. In addition, the energy fluctuations are smaller since we always consider the same bonds. However, it is difficult to simulate light ions.

Figure 22.3 Roberto Car is a professor of chemistry at Princeton University (top) and Michele Parrinello was a professor at ETH Zurich (bottom).

References

[1] D. Landau and K. Binder, *A Guide to Monte Carlo Simulations in Statistical Physics*. Cambridge, UK: Cambridge University Press, 2009.

[2] D. H. Lehmer, "Mathematical methods in large-scale computing units," *Annu. Comput. Lab. Harvard Univ.*, vol. 26, pp. 141–146, 1951.

[3] W. Payne, J. R. Rabung, and T. Bogyo, "Coding the Lehmer pseudo-random number generator," *Communications of the ACM*, vol. 12, no. 2, pp. 85–86, 1969.

[4] R. D. Carmichael, "Note on a new number theory function," *Bulletin of the American Mathematical Society*, vol. 16, no. 5, pp. 232–238, 1910.

[5] G. Woltman *et al.*, "GIMPS, the Great Internet Mersenne Prime Search," 2007, www.mersenne.org/.

[6] S. K. Park and K. W. Miller, "Random number generators: Good ones are hard to find," *Communications of the ACM*, vol. 31, no. 10, pp. 1192–1201, 1988.

[7] G. Marsaglia, "Random number generators," *Journal of Modern Applied Statistical Methods*, vol. 2, no. 1, p. 2, 2003.

[8] G. Marsaglia "Random numbers fall mainly in the planes," *Proceedings of the National Academy of Sciences of the United States of America*, vol. 61, no. 1, p. 25, 1968.

[9] D. E. Knuth, *The Art of Computer Programming*. Vol. 3. London, UK: Pearson Education, 1997.

[10] R. C. Tausworthe, "Random numbers generated by linear recurrence modulo two," *Mathematics of Computation*, vol. 19, no. 90, pp. 201–209, 1965.

[11] S. Kirkpatrick and E. P. Stoll, "A very fast shift-register sequence random number generator," *Journal of Computational Physics*, vol. 40, no. 2, pp. 517–526, 1981.

[12] J. R. Heringa, H. Blöte, and A. Compagner, "New primitive trinomials of Mersenne-exponent degrees for random-number generation," *International Journal of Modern Physics C*, vol. 3, no. 03, pp. 561–564, 1992.

[13] R. P. Brent and P. Zimmermann, "Random number generators with period divisible by a Mersenne prime," in *International Conference on Computational Science and Its Applications*. Berlin, Germany: Springer, 2003, pp. 1–10.

[14] M. Matsumoto and T. Nishimura, "Mersenne twister: a 623-dimensionally equidistributed uniform pseudo-random number generator," *ACM Transactions on Modeling and Computer Simulation (TOMACS)*, vol. 8, no. 1, pp. 3–30, 1998.

[15] G. Marsaglia, "The Marsaglia random number CD-ROM including the Diehard Battery of Tests of Randomness." [Online]. Available: https://web.archive.org/web/20160125103112/http://stat.fsu.edu/pub/diehard/

[16] G. E. P. Box and M. E. Muller, "A note on the generation of random normal deviates," *The Annals of Mathematical Statistics*, vol. 29, pp. 610–611, 1958.
[17] S. R. Broadbent and J. M. Hammersley, "Percolation processes: I. crystals and mazes," in *Mathematical Proceedings of the Cambridge Philosophical Society*, vol. 53, no. 3. Cambridge, UK: Cambridge University Press, 1957, pp. 629–641.
[18] D. Stauffer and A. Aharony, *Introduction to Percolation Theory*. Oxfordshire, UK: Taylor & Francis, 2018.
[19] P. Grassberger, "On the critical behavior of the general epidemic process and dynamical percolation," *Mathematical Biosciences*, vol. 63, no. 2, pp. 157–172, 1983.
[20] C. Moore and M. E. Newman, "Epidemics and percolation in small-world networks," *Physical Review E*, vol. 61, no. 5, p. 5678, 2000.
[21] D. Sornette, *Why Stock Markets Crash: Critical Events in Complex Financial Systems*. Vol. 49. Princeton, NJ: Princeton University Press, 2017.
[22] S. Galam, "Sociophysics: a review of Galam models," *International Journal of Modern Physics C*, vol. 19, no. 03, pp. 409–440, 2008.
[23] A. S. Perelson, "Immune network theory," *Immunological Reviews*, vol. 110, no. 1, pp. 5–36, 1989.
[24] B. Gauthier-Manuel, R. Meyer, and P. Pieranski, "The sphere rheometer: I. Quasistatic measurements," *Journal of Physics E: Scientific Instruments*, vol. 17, no. 12, p. 1177, 1984.
[25] D. Stauffer, A. Coniglio, and M. Adam, "Gelation and critical phenomena," in *Polymer Networks*. Berlin, Germany: Springer, 1982, pp. 103–158.
[26] H. Herrmann, D. Hong, and H. Stanley, "Backbone and elastic backbone of percolation clusters obtained by the new method of 'burning'," *Journal of Physics A: Mathematical and General*, vol. 17, no. 5, p. L261, 1984.
[27] V. W. D. Stauffer, F. W. Hehl, and J. Zabolitzky, *Computer Simulation and Computer Algebra*. Berlin, Germany: Springer, 1988.
[28] G. Grimmett, "What is percolation?," in *Percolation*. Berlin, Germany: Springer, 1999, pp. 1–31.
[29] J. L. Jacobsen, "High-precision percolation thresholds and potts-model critical manifolds from graph polynomials," *Journal of Physics A: Mathematical and Theoretical*, vol. 47, no. 13, Article 135001, 2014.
[30] J. W. Essam, "Percolation theory," *Reports on Progress in Physics*, vol. 43, no. 7, p. 833, 1980.
[31] M. F. Sykes and J. W. Essam, "Exact critical percolation probabilities for site and bond problems in two dimensions," *Journal of Mathematical Physics*, vol. 5, no. 8, pp. 1117–1127, 1964.
[32] C. D. Lorenz and R. M. Ziff, "Universality of the excess number of clusters and the crossing probability function in three-dimensional percolation," *Journal of Physics A: Mathematical and General*, vol. 31, no. 40, p. 8147, 1998.
[33] C.D. Lorenz and R.M. Ziff, "Precise determination of the bond percolation thresholds and finite-size scaling corrections for the sc, fcc, and bcc lattices," *Physical Review E*, vol. 57, no. 1, p. 230, 1998.

[34] X. Xu, J. Wang, J.-P. Lv, and Y. Deng, "Simultaneous analysis of three-dimensional percolation models," *Frontiers of Physics*, vol. 9, no. 1, pp. 113–119, 2014.

[35] S. Mertens and C. Moore, "Percolation thresholds and fisher exponents in hypercubic lattices," *Physical Review E*, vol. 98, no. 2, Article 022120, 2018.

[36] J. Hoshen and R. Kopelman, "Percolation and cluster distribution. I. Cluster multiple labeling technique and critical concentration algorithm," *Physical Review B*, vol. 14, no. 8, p. 3438, 1976.

[37] J. Gracey, "Four loop renormalization of ϕ 3 theory in six dimensions," *Physical Review D*, vol. 92, no. 2, Article 025012, 2015.

[38] D. Stauffer, "Scaling theory of percolation clusters," *Physics Reports*, vol. 54, no. 1, pp. 1–74, 1979.

[39] H. Hu, H. W. Blöte, R. M. Ziff, and Y. Deng, "Short-range correlations in percolation at criticality," *Physical Review E*, vol. 90, no. 4, Article 042106, 2014.

[40] H. Herrmann, D. Landau, and D. Stauffer, "New universality class for kinetic gelation," *Physical Review Letters*, vol. 49, no. 6, p. 412, 1982.

[41] A. E. Ferdinand and M. E. Fisher, "Bounded and inhomogeneous Ising models. I. Specific-heat anomaly of a finite lattice," *Physical Review*, vol. 185, no. 2, p. 832, 1969.

[42] Y. Deng and H. W. Blöte, "Monte Carlo study of the site-percolation model in two and three dimensions," *Physical Review E*, vol. 72, no. 1, Article 016126, 2005.

[43] G. Paul, R. M. Ziff, and H. E. Stanley, "Percolation threshold, fisher exponent, and shortest path exponent for four and five dimensions," *Physical Review E*, vol. 64, no. 2, Article 026115, 2001.

[44] Z. Zhou, J. Yang, Y. Deng, and R. M. Ziff, "Shortest-path fractal dimension for percolation in two and three dimensions," *Physical Review E*, vol. 86, no. 6, Article 061101, 2012.

[45] J. Chalupa, P. L. Leath, and G. R. Reich, "Bootstrap percolation on a Bethe lattice," *Journal of Physics C: Solid State Physics*, vol. 12, no. 1, p. L31, 1979.

[46] J. Adler, "Bootstrap percolation," *Physica A: Statistical Mechanics and its Applications*, vol. 171, no. 3, pp. 453–470, 1991.

[47] J. Adler and U. Lev, "Bootstrap percolation: visualizations and applications," *Brazilian Journal of Physics*, vol. 33, no. 3, pp. 641–644, 2003.

[48] B. B. Mandelbrot, *Fractals: Form, Chance, and Dimension*. Vol. 706. San Francisco, CA: W. H. Freeman, 1977.

[49] D. Weitz and M. Oliveria, "Fractal structures formed by kinetic aggregation of aqueous gold colloids," *Physical Review Letters*, vol. 52, no. 16, p. 1433, 1984.

[50] D. Weitz, M. Lin, and C. Sandroff, "Colloidal aggregation revisited: new insights based on fractal structure and surface-enhanced raman scattering," *Surface Science*, vol. 158, no. 1–3, pp. 147–164, 1985.

[51] H. Hentschel and I. Procaccia, "The infinite number of generalized dimensions of fractals and strange attractors," *Physica D: Nonlinear Phenomena*, vol. 8, no. 3, pp. 435–444, 1983.

[52] T. C. Halsey, M. H. Jensen, L. P. Kadanoff, I. Procaccia, and B. I. Shraiman, "Fractal measures and their singularities: The characterization of strange sets," *Physical Review A*, vol. 33, no. 2, p. 1141, 1986.
[53] A. Chhabra and R. V. Jensen, "Direct determination of the f (α) singularity spectrum," *Physical Review Letters*, vol. 62, no. 12, p. 1327, 1989.
[54] S. Xu, L. Böttcher, and T. Chou, "Diversity in biology: definitions, quantification and models," *Physical Biology*, vol. 17, no. 3, Article 031001, 2020.
[55] A. Kapitulnik, Y. Gefen, and A. Aharony, "On the fractal dimension and correlations in percolation theory," *Journal of Statistical Physics*, vol. 36, no. 5–6, pp. 807–814, 1984.
[56] S. Panyukov, "Correlation functions of an infinite cluster in percolation theory," *Zhurnal Eksperimental'noi i Teoreticheskoi Fiziki*, vol. 93, pp. 1454–1460, 1987.
[57] E. Sander, L. M. Sander, and R. M. Ziff, "Fractals and fractal correlations," *Computers in Physics*, vol. 8, no. 4, pp. 420–425, 1994.
[58] A. Einstein, "Über die von der molekularkinetischen Theorie der Wärme geforderte Bewegung von in ruhenden Flüssigkeiten suspendierten Teilchen," *Annalen der Physik*, vol. 322, no. 8, pp. 549–560, 1905.
[59] M. Kac, "Random walk and the theory of Brownian motion," *The American Mathematical Monthly*, vol. 54, no. 7P1, pp. 369–391, 1947.
[60] B. B. Mandelbrot, "A multifractal walk down wall street," *Scientific American*, vol. 280, no. 2, pp. 70–73, 1999.
[61] B. G. Malkiel, *A Random Walk down Wall Street: Including a Life-Cycle Guide to Personal Investing*. New York, NY: W. W. Norton, 1999.
[62] F. Bartumeus, M. G. E. da Luz, G. M. Viswanathan, and J. Catalan, "Animal search strategies: a quantitative random-walk analysis," *Ecology*, vol. 86, no. 11, pp. 3078–3087, 2005.
[63] M. Rubinstein and R. H. Colby, *Polymer Physics*. Vol. 23. Oxford, UK: Oxford University Press, 2003.
[64] P. D. Gujrati and A. I. Leonov, *Modeling and Simulation in Polymers*. Hoboken, NJ: John Wiley, 2010.
[65] L. Page, S. Brin, R. Motwani, and T. Winograd, "The PageRank citation ranking: bringing order to the web." Stanford InfoLab, Tech. Rep., 1999.
[66] S. Wald and L. Böttcher, "From classical to quantum walks with stochastic resetting on networks," *arXiv preprint arXiv:2008.00510*, 2020.
[67] G. H. Weiss, *Aspects and Applications of the Random Walk*. Amsterdam, Netherlands: North-Holland, 1994.
[68] D. Ben-Avraham and S. Havlin, *Diffusion and Reactions in Fractals and Disordered Systems*. Cambridge, UK: Cambridge University Press, 2000.
[69] M. L. Boas, *Mathematical Methods in the Physical Sciences*. Hoboken, NJ: John Wiley, 2006.
[70] G. F. Lawler, O. Schramm, and W. Werner, "The dimension of the planar Brownian frontier is 4/3," *Mathematical Research Letters*, vol. 8, no. 1, pp. 13–24, 2001.
[71] A. Klenke, *Wahrscheinlichkeitstheorie*. Vol. 1. Berlin, Germany: Springer, 2006.

[72] A. Dvoretzky and P. Erdös, "Some problems on random walk in space," in *Proceedings of the Second Berkeley Symposium on Mathematical Statistics and Probability*, 1951, pp. 353–367.

[73] E. Almaas, R. Kulkarni, and D. Stroud, "Scaling properties of random walks on small-world networks," *Physical Review E*, vol. 68, no. 5, Article 056105, 2003.

[74] M. Shlesinger, B. West, and J. Klafter, "Lévy dynamics of enhanced diffusion: application to turbulence," *Physical Review Letters*, vol. 58, no. 11, p. 1100, 1987.

[75] P. J. Flory and M. Volkenstein, "Statistical mechanics of chain molecules," *Biopolymers: Original Research on Biomolecules*, vol. 8, no. 5, pp. 699–700, 1969.

[76] B. Nienhuis, "Exact critical point and critical exponents of o(n) models in two dimensions," *Physical Review Letters*, vol. 49, no. 15, p. 1062, 1982.

[77] N. Eizenberg and J. Klafter, "Critical exponents of self-avoiding walks in three dimensions," *Physical Review B*, vol. 53, no. 9, p. 5078, 1996.

[78] N. Madras and G. Slade, *The Self-Avoiding Walk*. Berlin, Germany: Springer Science & Business Media, 2013.

[79] J. Millan and P. Prieto, eds., *Some Selected Topics from Critical Phenomena*, 1985, critical Phenomena Phase Transitions Supersymmetry (LASP-85), Cali, Colombia.

[80] A. Guttmann, "On two-dimensional self-avoiding random walks," *Journal of Physics A: Mathematical and General*, vol. 17, no. 2, p. 455, 1984.

[81] I. Majid, N. Jan, A. Coniglio, and H. E. Stanley, "Kinetic growth walk: a new model for linear polymers," *Physical Review Letters*, vol. 52, no. 15, p. 1257, 1984.

[82] F. T. Wall and F. Mandel, "Macromolecular dimensions obtained by an efficient Monte Carlo method without sample attrition," *Journal of Chemical Physics*, vol. 63, no. 11, pp. 4592–4595, 1975.

[83] P. H. Verdier and W. Stockmayer, "Monte Carlo calculations on the dynamics of polymers in dilute solution," *Journal of Chemical Physics*, vol. 36, no. 1, pp. 227–235, 1962.

[84] N. Madras and A. D. Sokal, "The pivot algorithm: a highly efficient Monte Carlo method for the self-avoiding walk," *Journal of Statistical Physics*, vol. 50, no. 1–2, pp. 109–186, 1988.

[85] S. N. Dorogovtsev and J. F. Mendes, *Evolution of Networks: From Biological Nets to the Internet and WWW*. Oxford, UK: Oxford University Press, 2013.

[86] A.-L. Barabási, *Network Science*. Cambridge, UK: Cambridge University Press, 2016.

[87] M. Newman, *Networks*. Oxford, UK: Oxford University Press, 2018.

[88] J. S. Andrade Jr., H. J. Herrmann, R. F. Andrade, and L. R. Da Silva, "Apollonian networks: simultaneously scale-free, small world, euclidean, space filling, and with matching graphs." *Physical Review Letters*, vol. 94, no. 1, Article 018702, 2005.

[89] M. Bastian, S. Heymann, and M. Jacomy, "Gephi: an open source software for exploring and manipulating networks," in *Third international AAAI Conference on Weblogs and Social Media*, 2009.

[90] R. Albert and A.-L. Barabási, "Statistical mechanics of complex networks," *Reviews of Modern Physics*, vol. 74, no. 1, p. 47, 2002.
[91] A.-L. Barabási and R. Albert, "Emergence of scaling in random networks," *Science*, vol. 286, no. 5439, pp. 509–512, 1999.
[92] S. N. Dorogovtsev and J. F. Mendes, "Evolution of networks," *Advances in Physics*, vol. 51, no. 4, pp. 1079–1187, 2002.
[93] A.-L. Barabâsi, H. Jeong, Z. Néda, E. Ravasz, A. Schubert, and T. Vicsek, "Evolution of the social network of scientific collaborations," *Physica A: Statistical Mechanics and Its Applications*, vol. 311, no. 3–4, pp. 590–614, 2002.
[94] S. Nauer, L. Böttcher, and M. A. Porter, "Random-graph models and characterization of granular networks," *Journal of Complex Networks*, vol. 8, no. 5, cnz037, 2019.
[95] L. Böttcher, "Extending buffon's needle problem to 'random-line graphs'," *arXiv preprint arXiv:1911.10679*, 2019.
[96] M. E. Newman, S. H. Strogatz, and D. J. Watts, "Random graphs with arbitrary degree distributions and their applications," *Physical Review E*, vol. 64, no. 2, Article 026118, 2001.
[97] D. J. Watts and S. H. Strogatz, "Collective dynamics of 'small-world' networks," *Nature*, vol. 393, no. 6684, p. 440, 1998.
[98] A. Barrat and M. Weigt, "On the properties of small-world network models," *European Physics Journal B*, vol. 13, no. 3, pp. 547–560, 2000.
[99] J. Saramäki, M. Kivelä, J.-P. Onnela, K. Kaski, and J. Kertesz, "Generalizations of the clustering coefficient to weighted complex networks," *Physical Review E*, vol. 75, no. 2, Article 027105, 2007.
[100] R. Cohen and S. Havlin, "Scale-free networks are ultrasmall," *Physical Review Letters*, vol. 90, no. 5, Article 058701, 2003.
[101] A. Smart, P. Umbanhowar, and J. Ottino, "Effects of self-organization on transport in granular matter: a network-based approach," *Europhysics Letters*, vol. 79, no. 2, p. 24002, 2007.
[102] A. Smart and J. M. Ottino, "Granular matter and networks: three related examples," *Soft Matter*, vol. 4, no. 11, pp. 2125–2131, 2008.
[103] L. Papadopoulos, M. A. Porter, K. E. Daniels, and D. S. Bassett, "Network analysis of particles and grains," *Journal of Complex Networks*, vol. 6, no. 4, pp. 485–565, 2018.
[104] J. Bang-Jensen and G. Z. Gutin, *Digraphs: Theory, Algorithms and Applications*. Berlin, Germany: Springer Science & Business Media, 2008.
[105] W. Ebeling and I. M. Sokolov, *Statistical Thermodynamics and Stochastic Theory of Nonequilibrium Systems*. Vol. 8. Singapore: World Scientific, 2005.
[106] J. W. Gibbs, "On the fundamental formula of statistical mechanics, with applications to astronomy and thermodynamics," in *Proceedings of the American Association for the Advancement of Science*, 1884, pp. 57–58.
[107] R. V. Solé, *Phase Transitions*. Princeton, NJ: Princeton University Press, 2011.
[108] G. Jaeger, "The ehrenfest classification of phase transitions: introduction and evolution," *Archive for History of Exact Sciences*, vol. 53, no. 1, pp. 51–81, 1998.

[109] H. Stanley, *Introduction to Phase Transitions and Critical Phenomena*. Oxford, UK: Clarendon, 1971.

[110] T. Ising, R. Folk, R. Kenna, B. Berche, and Y. Holovatch, "The fate of Ernst Ising and the fate of his model," *Journal of Physical Studies*, vol. 21, p. 3002, 2017.

[111] L. Onsager, "Crystal statistics. I. A two-dimensional model with an order-disorder transition," *Physical Review*, vol. 65, no. 3–4, p. 117, 1944.

[112] A. Böttcher and B. Silbermann, *Introduction to Large Truncated Toeplitz Matrices*. Berlin, Germany: Springer Science & Business Media, 2012.

[113] C. N. Yang, "The spontaneous magnetization of a two-dimensional ising model," *Physical Review*, vol. 85, no. 5, p. 808, 1952.

[114] W. P. Wolf, "The Ising model and real magnetic materials," *Brazilian Journal of Physics*, vol. 30, no. 4, pp. 794–810, 2000.

[115] A. Pelissetto and E. Vicari, "Critical phenomena and renormalization-group theory," *Physics Reports*, vol. 368, no. 6, pp. 549–727, 2002.

[116] S. El-Showk, M. F. Paulos, D. Poland, S. Rychkov, D. Simmons-Duffin, and A. Vichi, "Solving the 3D Ising model with the conformal bootstrap," *Physical Review D*, vol. 86, no. 2, Article 025022, 2012.

[117] A. M. Ferrenberg, J. Xu, and D. P. Landau, "Pushing the limits of Monte Carlo simulations for the three-dimensional Ising model," *Physical Review E*, vol. 97, no. 4, Article 043301, 2018.

[118] F. Kos, D. Poland, D. Simmons-Duffin, and A. Vichi, "Precision islands in the Ising and O(N) models," *Journal of High Energy Physics*, vol. 2016, no. 8, p. 36, 2016.

[119] Z. Komargodski and D. Simmons-Duffin, "The random-bond Ising model in 2.01 and 3 dimensions," *Journal of Physics A: Mathematical and Theoretical*, vol. 50, no. 15, Article 154001, 2017.

[120] H. A. Kramers and G. H. Wannier, "Statistics of the two-dimensional ferromagnet. Part I," *Physical Review*, vol. 60, no. 3, p. 252, 1941.

[121] M. Henkel, H. Hinrichsen, and S. Lübeck, *Non-Equilibrium Phase Transitions, Vol. 1. Absorbing Phase Transitions*. Berlin, Germany: Springer Science & Business Media, 2008.

[122] B. M. McCoy and T. T. Wu, *The Two-Dimensional Ising Model*. North Chelmsford, MA: Courier Corporation, 2014.

[123] H. Stanley, "Scaling, universality, and renormalization: Three pillars of modern critical phenomena," 1999. [Online]. Available: http://link.springer.com/chapter/10.1007%2F978-1-4612-1512-7_39

[124] L. Kadanoff, "Critical behaviour, universality and scaling," in M. S. Green, editor, *Proceedings of the 1970 Varenna Summer School on Critical Phenomena*, 1971.

[125] J. Sengers, J. L. Sengers, and C. Croxton, *Progress in Liquid Physics*. Chichester, UK: Wiley, 1978.

[126] W. Krauth, *Statistical Mechanics: Algorithms and Computations*. Vol. 13. Oxford, UK: Oxford University Press, 2006.

[127] W. H. Holtzman, "The unbiased estimate of the population variance and standard deviation," *American Journal of Psychology*, vol. 63, no. 4, pp. 615–617, 1950.

[128] J. H. Conway and N. J. A. Sloane, *Sphere Packings, Lattices and Groups*, 3rd ed., vol. 290 Berlin, Germany: Springer Science & Business Media, 1998.

[129] W. Hastings, "Monte Carlo sampling methods using Markov chains and their applications," *BioMetrika*, 1969. [Online]. Available: http://biomet .oxfordjournals.org/content/57/1/97

[130] N. Metropolis, A. W. Rosenbluth, M. Rosenbluth, A. Teller, and E. Teller, "Equation of state calculations by fast computing machines," *Journal of Chemical Physics*, 1953. [Online]. Available: http://en.wikipedia.org/wiki/Equation_of_State_Calculations_by_Fast_Computing_Machines

[131] K. Barth, *Oral History Transcript – Dr. Marshall Rosenbluth*, 2003. [Online]. Available: www.aip.org/history/ohilist/28636_1.html

[132] J. Gubernatis, "Marshall Rosenbluth and the Metropolis algorithm," *AIP*, 2005. [Online]. Available: http://scitation.aip.org/content/aip/journal/pop/12/5/10.1063/1.1887186

[133] E. Segré, *Enrico Fermi, Physicist*. Chicago, IL: University of Chicago Press, 1970.

[134] M. Creutz, "Microcanonical monte carlo simulation," *Physical Review Letters*, vol. 50, no. 19, p. 1411, 1983.

[135] S. Manna, H. Herrmann, and D. Landau, "A stochastic method to determine the shape of a drop on a wall," *Journal of Statistical Physics*, vol. 66, no. 3–4, pp. 1155–1163, 1992.

[136] K. Binder and D. Heermann, *Monte Carlo Simulation in Statistical Physics*. Berlin, Germany: Springer, 1997.

[137] P. Grassberger, "Damage spreading and critical exponents for 'model A' Ising dynamics," *Physica A*, vol. 214, no. 4, pp. 547–559, 1995.

[138] M. Nightingale and H. Blöte, "Dynamic exponent of the two-dimensional Ising model and Monte Carlo computation of the subdominant eigenvalue of the stochastic matrix," *Physical Review Letters*, vol. 76, no. 24, p. 4548, 1996.

[139] F.-G. Wang and C.-K. Hu, "Universality in dynamic critical phenomena," *Physical Review E*, vol. 56, no. 2, p. 2310, 1997.

[140] N. Ito, "Nonequilibrium relaxation method – an alternative simulation strategy," *Pramana*, vol. 64, no. 6, pp. 871–880, 2005.

[141] Z. Rácz, "Nonlinear relaxation near the critical point: molecular-field and scaling theory," *Physical Review B*, vol. 13, no. 1, p. 263, 1976.

[142] K. Binder, "Finite size scaling analysis of Ising model block distribution functions," *Zeitschrift für Physik B Condensed Matter*, vol. 43, no. 2, pp. 119–140, 1981.

[143] K. Binder and D. Landau, "Finite-size scaling at first-order phase transitions," *Physical Review B*, vol. 30, no. 3, p. 1477, 1984.

[144] R. B. Potts, "Some generalized order-disorder transformations," in *Mathematical Proceedings of the Cambridge Philosophical Society*, vol. 48, no. 1. Cambridge, UK: Cambridge University Press, 1952, pp. 106–109.

[145] F.-Y. Wu, "The Potts model," *Reviews of Modern Physics*, vol. 54, no. 1, p. 235, 1982.

[146] F. Gliozzi, "Simulation of Potts models with real q and no critical slowing down," *Physical Review E*, vol. 66, no. 1, Article 016115, 2002.

[147] P. Kasteleyn and C. Fortuin, "Phase transitions in lattice systems with random local properties," *Journal of the Physical Society of Japan*, vol. 26, p. 11, 1969.

[148] F. Wu, "Percolation and the Potts model," *Journal of Statistical Physics*, vol. 18, no. 2, pp. 115–123, 1978.

[149] K. Binder, "Coniglio–Klein droplets: a fruitful concept to understand phase transitions geometrically," *Journal of Physics A*, vol. 50, no. 6, Article 061001, 2017.

[150] A. Coniglio and W. Klein, "Clusters and Ising critical droplets: a renormalisation group approach," *Journal of Physics A*, vol. 13, no. 8, p. 2775, 1980.

[151] R. H. Swendsen and J.-S. Wang, "Nonuniversal critical dynamics in Monte Carlo simulations," *Physical Review Letters*, vol. 58, no. 2, p. 86, 1987.

[152] J.-S. Wang and R. H. Swendsen, "Cluster Monte Carlo algorithms," *Physica A*, vol. 167, no. 3, pp. 565–579, 1990.

[153] U. Wolff, "Collective Monte Carlo updating for spin systems," *Physical Review Letters*, vol. 62, no. 4, p. 361, 1989.

[154] H. E. Stanley, "Dependence of critical properties on dimensionality of spins," *Physical Review Letters*, vol. 20, no. 12, p. 589, 1968.

[155] J. M. Kosterlitz and D. J. Thouless, "Ordering, metastability and phase transitions in two-dimensional systems," *Journal of Physics C: Solid State Physics*, vol. 6, no. 7, p. 1181, 1973.

[156] N. Mermin and H. Wagner, "Absence of ferromagnetism or antiferromagnetism in one- or two-dimensional isotropic Heisenberg models," *Physical Review Letters*, no. 17, p. 1133, 1966.

[157] Z. Salzburg, J. Jacobson, W. Fickett, and W. Wood, "Excitation of non-radial stellar oscillations by gravitational waves: a first model," *Chemical Physics*, no. 30, p. 65, 1959.

[158] T. Penna and H. Herrmann, "Broad histogram Monte Carlo," *European Physical Journal B–Condensed Matter and Complex Systems*, vol. 1, no. 2, pp. 205–208, 1998.

[159] P. de Oliveira, T. Penna, and H. Herrmann, "Broad histogram method," *Brazilian Journal of Physics*, no. 26, p. 677, 1996.

[160] J.-S. Wang, "Flat histogram Monte Carlo method," *Physica A: Statistical Mechanics and Its Applications*, vol. 281, no. 1–4, pp. 147–150, 2000.

[161] J. Lee, "New Monte Carlo algorithm: entropic sampling," *Physical Review Letters*, vol. 71, no. 2, p. 211, 1993.

[162] B. A. Berg and T. Neuhaus, "Multicanonical ensemble: a new approach to simulate first-order phase transitions," *Physical Review Letters*, vol. 68, no. 1, p. 9, 1992.

[163] B. A. Berg, "The multicanonical ensemble: a new approach to computer simulations," *International Journal of Modern Physics C*, vol. 3, no. FSU-HEP-92-08-17, pp. 1083–1098, 1992.

[164] F. Wang and D. Landau, "Efficient, multiple-range random walk algorithm to calculate the density of states," *Physical Review Letters*, vol. 86, no. 10, p. 2050, 2001.

[165] C. Zhou, T. C. Schulthess, S. Torbrügge, and D. Landau, "Wang-landau algorithm for continuous models and joint density of states," *Physical Review Letters*, vol. 96, no. 12, Article 120201, 2006.

[166] G. Torrie and J. Valleau, "Nonphysical sampling distributions in Monte Carlo free-energy estimation: umbrella sampling," *Journal of Computational Physics*, vol. 23, pp. 187–199, 1977. [Online]. Available: www.sciencedirect.com/science/article/pii/0021999177901218

[167] T. Niemeijer and J. Van Leeuwen, *Phase Transitions and Critical Phenomena*, vol. 6, London: Academic, pp. 425–505, 1976.

[168] H. J. Maris and L. P. Kadanoff, "Teaching the renormalization group," *American Journal of Physics*, vol. 46, no. 6, pp. 652–657, 1978.

[169] K. G. Wilson, "The renormalization group: critical phenomena and the Kondo problem," *Reviews of Modern Physics*, vol. 47, pp. 773–840, 1975. [Online]. Available: http://link.aps.org/doi/10.1103/RevModPhys.47.773

[170] S. Ma, "Monte Carlo renormalization group," *Physical Review Letters*, vol. 37, p. 461, 1976.

[171] R. Swendsen, "Monte Carlo renormalization group," *Physical Review Letters*, vol. 42, p. 859, 1979.

[172] Y. LeCun, Y. Bengio, and G. Hinton, "Deep learning," *Nature*, vol. 521, no. 7553, p. 436, 2015.

[173] I. Goodfellow, Y. Bengio, and A. Courville, *Deep Learning*. Boston, MA: MIT Press, 2016.

[174] P. Mehta and D. J. Schwab, "An exact mapping between the variational renormalization group and deep learning," *arXiv preprint arXiv:1410.3831*, 2014.

[175] G. E. Hinton, "Deterministic Boltzmann learning performs steepest descent in weight-space," *Neural Computation*, vol. 1, no. 1, pp. 143–150, 1989.

[176] D. H. Ackley, G. E. Hinton, and T. J. Sejnowski, "A learning algorithm for Boltzmann machines," *Cognitive Science*, vol. 9, no. 1, pp. 147–169, 1985.

[177] G. E. Hinton and R. R. Salakhutdinov, "Reducing the dimensionality of data with neural networks," *Science*, vol. 313, no. 5786, pp. 504–507, 2006.

[178] V. Nair and G. E. Hinton, "Rectified linear units improve restricted Boltzmann machines," in *Proceedings of the 27th InterNational Conference on Machine Learning (ICML-10)*, 2010, pp. 807–814.

[179] G. E. Hinton, "A practical guide to training restricted Boltzmann machines," in *Neural Networks: Tricks of the Trade*. Berlin, Germany: Springer, 2012, pp. 599–619.

[180] W. S. McCulloch and W. Pitts, "A logical calculus of the ideas immanent in nervous activity," *Bulletin of Mathematical Biophysics*, vol. 5, no. 4, pp. 115–133, 1943.

[181] J. J. Hopfield, "Neural networks and physical systems with emergent collective computational abilities," *Proceedings of the National Academy of Sciences of the United States of America*, vol. 79, no. 8, pp. 2554–2558, 1982.

[182] D. O. Hebb, *The Organization of Behavior: A Neuropsychological Theory*. Oxfordshire, UK: Psychology Press, 2005.

[183] J. Hertz, A. Krogh, R. G. Palmer, and H. Horner, "Introduction to the theory of neural computation," *Physics Today*, vol. 44, p. 70, 1991.

[184] G. E. Hinton, "Training products of experts by minimizing contrastive divergence," *Neural Computation*, vol. 14, no. 8, pp. 1771–1800, 2002.

[185] F. D'Angelo and L. Böttcher, "Learning the Ising model with generative neural networks," *Physical Review Research*, vol. 2, no. 2, Article 023266, 2020.

[186] D. Kim and D.-H. Kim, "Smallest neural network to learn the Ising criticality," *Physical Review E*, vol. 98, no. 2, Article 022138, 2018.

[187] S. Efthymiou, M. J. Beach, and R. G. Melko, "Super-resolving the ising model with convolutional neural networks," *Physical Review B*, vol. 99, no. 7, Article 075113, 2019.

[188] G. Carleo and M. Troyer, "Solving the quantum many-body problem with artificial neural networks," *Science*, vol. 355, no. 6325, pp. 602–606, 2017.

[189] G. Torlai, G. Mazzola, J. Carrasquilla, M. Troyer, R. Melko, and G. Carleo, "Neural-network quantum state tomography," *Nature Physics*, vol. 14, no. 5, p. 447, 2018.

[190] P. Mehta, M. Bukov, C.-H. Wang, A. G. Day, C. Richardson, C. K. Fisher, and D. J. Schwab, "A high-bias, low-variance introduction to machine learning for physicists," *Physics Reports*, vol. 810, pp. 1–124, 2019.

[191] S. Kirkpatrick, C. D. Gelatt, and M. P. Vecchi, "Optimization by simulated annealing," *Science*, vol. 220, no. 4598, pp. 671–680, 1983.

[192] L. Jacobs and C. Rebi, "Multi-spin coding: A very efficient technique for Monte Carlo simulations of spin systems," *Journal of Computational Physics*, vol. 41, no. 1, pp. 203–210, 1981.

[193] A. Baumgärtner, K. Binder, J.-P. Hansen, M. Kalos, K. Kehr, D. Landau, D. Levesque, H. Müller-Krumbhaar, C. Rebbi, Y. Saito et al., *Applications of the Monte Carlo Method in Statistical Physics*. Vol. 36. Berlin, Germany: Springer Science & Business Media, 2013.

[194] J. Marro and R. Dickman, *Nonequilibrium Phase Transitions in Lattice Models*. Cambridge, UK: Cambridge University Press, 2005.

[195] J. L. Cardy and R. Sugar, "Directed percolation and reggeon field theory," *Journal of Physics A*, vol. 13, p. L423, 1980.

[196] P. Grassberger, "Directed percolation in 2+ 1 dimensions," *Journal of Physics A*, vol. 22, no. 17, p. 3673, 1989.

[197] L. Böttcher, M. Luković, J. Nagler, S. Havlin, and H. J. Herrmann, "Failure and recovery in dynamical networks," *Scientific Reports*, vol. 7, p. 41729, 2017.

[198] L. Böttcher, J. Nagler, and H. J. Herrmann, "Critical behaviors in contagion dynamics," *Physical Review Letters*, vol. 118, no. 8, Article 088301, 2017.

[199] L. Böttcher, H. J. Herrmann, and H. Gersbach, "Clout, activists and budget: the road to presidency," *PloS ONE*, vol. 13, no. 3, Article e0193199, 2018.

[200] P. Richter, M. Henkel, and L. Böttcher, "Aging and relaxation in bistable contagion dynamics," *arXiv preprint arXiv:2006.05937*, 2020.

[201] M. J. Keeling and P. Rohani, *Modeling Infectious Diseases in Humans and Animals*. Princeton, NJ: Princeton University Press, 2011.

[202] R. Pastor-Satorras, C. Castellano, P. Van Mieghem, and A. Vespignani, "Epidemic processes in complex networks," *Reviews of Modern Physics*, vol. 87, no. 3, p. 925, 2015.

[203] L. Böttcher, O. Woolley-Meza, N. A. Araújo, H. J. Herrmann, and D. Helbing, "Disease-induced resource constraints can trigger explosive epidemics," *Scientific Reports*, vol. 5, p. 16571, 2015.

[204] L. Böttcher, J. Andrade, and H. J. Herrmann, "Targeted recovery as an effective strategy against epidemic spreading," *Scientific Reports*, vol. 7, no. 1, p. 14356, 2017.

[205] L. Böttcher, O. Woolley-Meza, E. Goles, D. Helbing, and H. J. Herrmann, "Connectivity disruption sparks explosive epidemic spreading," *Physical Review E*, vol. 93, no. 4, Article 042315, 2016.

[206] L. Böttcher, H. J. Herrmann, and M. Henkel, "Dynamical universality of the contact process," *Journal of Physics A: Mathematical and Theoretical*, vol. 51, no. 12, Article 125003, 2018.

[207] S. Broadbent and J. Hammersley, "Percolation processes I. Crystals and mazes," *Mathematical Proceedings of the Cambridge Philosophical Society*, vol. 53, no. 3, pp. 629–641, 1957.

[208] M. M. S. Sabag and M. J. de Oliveira, "Conserved contact process in one to five dimensions," *Physical Review E*, vol. 66, no. 3, 036115, 2002.

[209] J. Wang, Z. Zhou, Q. Liu, T. M. Garoni, and Y. Deng, "High-precision Monte Carlo study of directed percolation in (d+ 1) dimensions," *Physical Review E*, vol. 88, no. 4, Article 042102, 2013.

[210] P. Grassberger, "On phase transitions in Schlögl's second model," *Zeitschrift für Physik B*, vol. 47, no. 4, pp. 365–374, 1982.

[211] H.-K. Janssen, "On the nonequilibrium phase transition in reaction-diffusion systems with an absorbing stationary state," *Zeitschrift für Physik B*, vol. 42, no. 2, pp. 151–154, 1981.

[212] D. T. Gillespie, "A general method for numerically simulating the stochastic time evolution of coupled chemical reactions," *Journal of Computational Physics*, vol. 22, no. 4, pp. 403–434, 1976.

[213] "Exact stochastic simulation of coupled chemical reactions," *Journal of Physical Chemistry*, vol. 81, no. 25, pp. 2340–2361, 1977.

[214] P. Bak, K. Christensen, L. Danon, and T. Scanlon, "Unified scaling law for earthquakes," *Physical Review Letters*, vol. 88, no. 17, Article 178501, 2002.

[215] R. Baddeley, L. F. Abbott, M. C. Booth, F. Sengpiel, T. Freeman, E. A. Wakeman, and E. T. Rolls, "Responses of neurons in primary and inferior temporal visual cortices to natural scenes," *Proceedings of the Royal Society of London, Series B*, vol. 264, no. 1389, pp. 1775–1783, 1997.

[216] A.-L. Barabasi, "The origin of bursts and heavy tails in human dynamics," *Nature*, vol. 435, no. 7039, p. 207, 2005.

[217] L. Böttcher, O. Woolley-Meza, and D. Brockmann, "Temporal dynamics of online petitions," *PloS ONE*, vol. 12, no. 5, Article e0178062, 2017.

[218] K.-I. Goh and A.-L. Barabási, "Burstiness and memory in complex systems," *Europhysics Letters*, vol. 81, no. 4, p. 48002, 2008.

[219] M. Boguná, L. F. Lafuerza, R. Toral, and M. Á. Serrano, "Simulating non-Markovian stochastic processes," *Physical Review E*, vol. 90, no. 4, Article 042108, 2014.

[220] L. Böttcher and N. Antulov-Fantulin, "Unifying continuous, discrete, and hybrid susceptible-infected-recovered processes on networks," *Physical Review Research*, vol. 2, no. 3, Article 033121, 2020.

[221] N. Masuda and L. E. Rocha, "A Gillespie algorithm for non-Markovian stochastic processes," *SIAM Review*, vol. 60, no. 1, pp. 95–115, 2018.

[222] S. Wolfram, "Universality and complexity in cellular automata," *Physica D: Nonlinear Phenomena*, vol. 10, no. 1–2, pp. 1–35, 1984.

[223] H. J. Herrmann, "Cellular automata," in *Nonlinear Phenomena in Complex Systems*, Amsterdam: Elsevier, 1989, p. 151.

[224] A. Ilachinski, *Cellular Automata: A Discrete Universe*. Singaore: World Scientific, 2001.

[225] "Game of life glider gun." [Online]. Available: https://commons.wikimedia.org/wiki/File:Gospers_glider_gun.gif

[226] G. Y. Vichniac, "Simulating physics with cellular automata," *Physica D*, vol. 10, no. 1–2, pp. 96–116, 1984.

[227] J. G. Zabolitzky and H. J. Herrmann, "Multitasking case study on the cray-2: the q2r cellular automaton," *Journal of Computational Physics*, vol. 76, no. 2, pp. 426–447, 1988.

[228] H. J. Herrmann, "Geometrical cluster growth models and kinetic gelation," *Physics Reports*, vol. 136, no. 3, pp. 153–224, 1986.

[229] F. Family, "Dynamic scaling and phase transitions in interface growth," *Physica A: Statistical Mechanics and Its Applications*, vol. 168, no. 1, pp. 561–580, 1990.

[230] F. Family and T. Vicsek, "Scaling of the active zone in the Eden process on percolation networks and the ballistic deposition model," *Journal of Physics A: Mathematical and General*, vol. 18, no. 2, p. L75, 1985.

[231] M. Eden, "A two-dimensional growth process," in *Proceedings of the fourth Berkeley symposium on mathematical statistics and probability*, vol. 4, Berkeley: University of California Press, 1961, pp. 223–239.

[232] S. F. Edwards and D. Wilkinson, "The surface statistics of a granular aggregate," *Proceedings of the Royal Society of London, Series A, Mathematical and Physical Sciences*, vol. 381, no. 1780, pp. 17–31, 1982.

[233] M. Kardar, G. Parisi, and Y.-C. Zhang, "Dynamic scaling of growing interfaces," *Physical Review Letters*, vol. 56, no. 9, p. 889, 1986.

[234] D. J. Higham, "An algorithmic introduction to numerical simulation of stochastic differential equations," *SIAM Review*, vol. 43, no. 3, pp. 525–546, 2001.

[235] T. A. Witten and L. M. Sander, "Diffusion-limited aggregation," *Physical Review B*, vol. 27, no. 9, p. 5686, 1983.

[236] L. Niemeyer, L. Pietronero, and H. Wiesmann, "Fractal dimension of dielectric breakdown," *Physical Review Letters*, vol. 52, no. 12, p. 1033, 1984.

[237] L. de Arcangelis, S. Redner, and H. Herrmann, "A random fuse model for breaking processes," *Journal de Physique Lettres*, vol. 46, no. 13, pp. 585–590, 1985.

[238] D. C. Rapaport and D. C. R. Rapaport, *The Art of Molecular Dynamics Simulation*. Cambridge, UK: Cambridge University Press, 2004.

[239] I. Newton, *Philosophiae naturalis principia mathematica*. Vol. 1. G. Brookman, 1833.

[240] J.-M. Sanz-Serna and M.-P. Calvo, *Numerical Hamiltonian Problems*. Mineola, NY: Courier Dover Publications, 2018.

[241] R. Mani, L. Böttcher, H. J. Herrmann, and D. Helbing, "Extreme power law in a driven many-particle system without threshold dynamics," *Physical Review E*, vol. 90, no. 4, Article 042201, 2014.

[242] R. Car and M. Parrinello, "Unified approach for molecular dynamics and density-functional theory," *Physical Review Letters*, vol. 55, no. 22, p. 2471, 1985.

[243] B. Alder and T. Wainwright, "Phase transition for a hard sphere system," *Journal of Chemical Physics*, vol. 27, pp. 1208–1209, 1957.

[244] R. Vink, G. Barkema, W. Van der Weg, and N. Mousseau, "Fitting the Stillinger–Weber potential to amorphous silicon," *Journal of Non-crystalline Solids*, vol. 282, no. 2–3, pp. 248–255, 2001.

[245] B. Axilrod and E. Teller, "Interaction of the Van der Waals type between three atoms," *Journal of Chemical Physics*, vol. 11, no. 6, pp. 299–300, 1943.

[246] L. Verlet, "Computer experiments on classical fluids," *Physical Review*, vol. 159, pp. 98–103, 1967.

[247] R. W. Hockney, "The potential calculation and some applications," *Methods in Computation Physics*, vol. 9, p. 136, 1970.

[248] B. Quentrec and C. Brot, "New method for searching for neighbors in molecular dynamics computations," *Journal of Computational Physics*, vol. 13, no. 3, pp. 430–432, 1973.

[249] R. Hockney and J. Eastwood, *Computer Simulations Using Particles*. Boca Raton, FL: CRC Press, 1988.

[250] J. Ryckaert, G. Ciccotti, and H. Berendsen, "Numerical integration of the Cartesian equations of motion of a system with constraints: molecular dynamics of n-alkanes," *Journal of Computational Physics*, vol. 239, pp. 327–341, 1977.

[251] D. Evans, "On the representatation of orientation space," *Molecular Physics*, vol. 34, pp. 317–325, 1977.

[252] "Singularity free algorithm for molecular dynamics simulation of rigid polyatomics," *Molecular Physics*, vol. 34, pp. 327–331, 1977.

[253] D. V. Stäger, N. Araújo, and H. J. Herrmann, "Prediction and control of slip-free rotation states in sphere assemblies," *Physical Review Letters*, vol. 116, no. 25, Article 254301, 2016.

[254] C. Weiner, *Interview with Paul Peter Ewald*, 1968. [Online]. Available: www.aip.org/history/ohilist/4596_1.html

[255] M. Sangster and M.Dixon, "Interionic potentials in alkali halides and their use in simulations of the molten salts," *Advances in Physics*, vol. 25, pp. 247–342, 1976.

[256] D. Wolf, P. Keblinski, S. Phillpot, and J. Eggebrecht, "Exact method for the simulation of Coulombic systems by spherically truncated, pairwise r- 1 summation," *Journal of Physical Chemistry*, vol. 110, no. 17, pp. 8254–8282, 1999.

[257] M. Allen and D. Tildesley, *Computer Simulation of Liquids*. Oxford, UK: Clarendon Press, 1987.

[258] J. W. Eastwood, R. W. Hockney, and D. N. Lawrence, "P3M3DP – The three-dimensional periodic particle-particle/particle-mesh program," *Computer Physics Communications*, vol. 19, no. 2, pp. 215–261 (1980).

[259] J. W. Cooley and J. W. Tukey, "An algorithm for the machine calculation of complex Fourier series," *Mathematics of Computation*, vol. 19, no. 90, pp. 297–301, 1965.

[260] J. W. Cooley, "The re-discovery of the fast fourier transform algorithm," *Microchimica Acta*, vol. 93, no. 1–6, pp. 33–45, 1987.

[261] I. N. Sneddon, *Fourier Transforms*. North Chelmsford, MA: Courier Corporation, 1995.

[262] S. Pfalzner and P. Gibbon, *Many-Body Tree Methods in Physics*. Cambridge, UK: Cambridge University Press, 2005.

[263] D. Potter, J. Stadel, and R. Teyssier, "Pkdgrav3: beyond trillion particle cosmological simulations for the next era of galaxy surveys," *Computational Astrophysics and Cosmology*, vol. 4, no. 1, p. 2, 2017.

[264] L. Onsager, "Electric moments of molecules in liquids," *Journal of the American Chemical Society*, no. 58, pp. 1486–1493, 1936. [Online]. Available: http://pubs.acs.org/doi/abs/10.1021/ja01299a050

[265] H. Yoshida, "Construction of higher order symplectic integrators," *Physics Letters A*, vol. 150, 1990. [Online]. Available: http://cacs.usc.edu/education/phys516/yoshida-symplectic-pla00.pdf

[266] R. Watts, "Monte Carlo studies of liquid water," *Molecular Physics*, vol. 26, pp. 1069–1083, 1974. [Online]. Available: www.tandfonline.com/doi/abs/10.1080/00268977400102381#.U_Wxd9bPHVM

[267] E. Braun, S. M. Moosavi, and B. Smit, "Anomalous effects of velocity rescaling algorithms: the flying ice cube effect revisited," *Journal of Chemical Theory and Computation*, vol. 14, no. 10, pp. 5262–5272, 2018.

[268] K. Huang, *Introduction to Statistical Physics*. London, UK: Chapman and Hall/CRC, 2009.

[269] H. J. Berendsen, J. v. Postma, W. F. van Gunsteren, A. DiNola, and J. R. Haak, "Molecular dynamics with coupling to an external bath," *Journal of Chemical Physics*, vol. 81, no. 8, pp. 3684–3690, 1984.

[270] W. G. Hoover, "William G. Hoover's personal website." [Online]. Available: http://williamhoover.info/

[271] W. G. Hoover, A. J. Ladd, and B. Moran, "High-strain-rate plastic flow studied via nonequilibrium molecular dynamics," *Physical Review Letters*, vol. 48, no. 26, p. 1818, 1982.

[272] D. J. Evans, W. G. Hoover, B. H. Failor, B. Moran, and A. J. Ladd, "Nonequilibrium molecular dynamics via Gauss's principle of least constraint," *Physical Review A*, vol. 28, no. 2, p. 1016, 1983.

[273] W. Hoover, "Canonical dynamics: equilibrium phase-space distributions," *Physical Review*, vol. 31, 1985. [Online]. Available: http://journals.aps.org/pra/abstract/10.1103/PhysRevA.31.1695

[274] S. Nosé, "A molecular dynamics method for simulations in the canonical ensemble," *Molecular Physics*, vol. 52, 1984. [Online]. Available: www.tandfonline.com/doi/abs/10.1080/00268978400101201#.U3Ic-3LI9yA

[275] S. Nosé, "A unified formulation of the constant temperature molecular dynamics methods," *Journal of Chemical Physics*, vol. 81, 1984. [Online]. Available: http://scitation.aip.org/content/aip/journal/jcp/81/1/10.1063/1.447334

[276] S. D. Bond, B. J. Leimkuhler, and B. B. Laird, "The Nosé–Poincaré method for constant temperature molecular dynamics," *Journal of Computational Physics*, vol. 151, no. 1, pp. 114–134, 1999.

[277] H. Andersen, "Molecular dynamics simulations at constant pressure and/or temperature," *Journal of Chemical Physics*, vol. 72, 1980. [Online]. Available: http://scitation.aip.org/content/aip/journal/jcp/72/4/10.1063/1.439486

[278] T. Pöschel and T. Schwager, *Computational Granular Dynamics: Models and Algorithms*. Berlin, Germany: Springer Science & Business Media, 2005.

[279] O. Walton and R. Braun, "Viscosity, granular-temperature and stress calculations for shearing assemblies of inelastic frictional disks," *Journal of Rheology*, no. 30, pp. 949–980, 1986. [Online]. Available: http://arxiv.org/abs/cond-mat/9810009

[280] S. Luding, "Collisions and contacts between two particles," in *Physics of Dry Granular Media*. Berlin, Germany: Springer, 1998, pp. 285–304.

[281] E. Popova and V. L. Popov, "The research works of coulomb and amontons and generalized laws of friction," *Friction*, vol. 3, no. 2, pp. 183–190, 2015.

[282] S. Luding, "Particle–particle interactions and models (discrete element method)," 2010. [Online]. Available: www2.msm.ctw.utwente.nl/sluding/TEACHING/ParticleTechnology/Luding_PForcesModels.pdf

[283] P. Cundall and O. Strack, "A discrete numerical model for granular assemblies," 1979. [Online]. Available: www.icevirtuallibrary.com/content/article/10.1680/geot.1979.29.1.47

[284] B. D. Lubachevsky, "How to simulate billiards and similar systems," *Journal of Computational Physics*, vol. 94, no. 2, pp. 255–283, 1991.

[285] S. Luding, "Clustering instabilities, arching, and anomalous interaction probabilities as examples for cooperative phenomena in dry granular media," *T.A.S.K. Quarterly, Scientific Bulletin of Academic Computer Centre of the Technical University of Gdansk*, vol. 2, pp. 417–443, 1998.

[286] S. Luding and S. McNamara, "How to handle inelastic collapse of a dissipative hard-sphere gas with the tc model," *Granular Matter*, 1998. [Online]. Available: http://arxiv.org/abs/cond-mat/9810009

[287] S. McNamara and W. Young, "Inelastic collapse in two dimensions," *Physical Review E*, vol. 50, no. 1, p. R28, 1994.

[288] S. Gonzalez, A. R. Thornton, and S. Luding, "Free cooling phase-diagram of hard-spheres with short-and long-range interactions," *European Physical Journal Special Topics*, vol. 223, no. 11, pp. 2205–2225, 2014.

[289] S. McNamara and W. Young, "Inelastic collapse and dumping in a one-dimensional granular medium," *Physical Review*, 1991. [Online]. Available: www-pord.ucsd.edu/~wryoung/reprintPDFs/InelasticCollapse.pdf

[290] S. Luding, "Freely cooling granular gas." [Online]. Available: www2.msm.ctw.utwente.nl/sluding/pictures/cooling.html

[291] G. Paul, "A complexity O(1) priority queue for event driven molecular dynamics simulations," *Journal of Computational Physics*, vol. 221, no. 2, pp. 615–625, 2007.

[292] J. W. Perram and M. S. Wertheim, "Statistical mechanics of hard ellipsoids: overlap algorithm and the contact function," *Journal of Computational Physics*, no. 58, pp. 409–416, 1985.

[293] G. W. Delaney and P. W. Cleary, "The packing properties of superellipsoids," *Europhysics Letters*, vol. 89, no. 3, p. 34002, 2010.

[294] A. Donev, I. Cisse, D. Sachs, E. Variano, F. Stillinger, R. Connelly, S. Torquato, and P. M. Chaikin, "Improving the density of jammed disordered packings using ellipsoids," *Science*, vol. 303, no. 5660, pp. 990–993, 2004. [Online]. Available: www.sciencemag.org/content/303/5660/990.full

[295] F. Alonso-Marroquin, "Spheropolygons: a new method to simulate conservative and dissipative interactions between 2d complex-shaped rigid bodies," *Europhysics Letters*, vol. 83, p. 14001, 2008. [Online]. Available: http://iopscience.iop.org/0295-5075/83/1/14001

[296] G. Matheron and J. Serra, *The Birth of Mathematical Morphology*, 2000. [Online]. Available: http://cmm.ensmp.fr/~serra/pdf/birth_of_mm.pdf

[297] R. I. Leine, C. Glocker, and D. H. Van Campen, "Nonlinear dynamics of the woodpecker toy," in *Proceedings of DETC 2001 ASME Symposium on Nonlinear Dynamics and Control In Engineering Systems*, 2001.

[298] O. Maisonneuve, *Commemoration of Jean-Jacques Moreau*, 2014. [Online]. Available: www.isimmforum.tu-darmstadt.de/fileadmin/home/users/186/ISIMM/JJMoreau-ISIMM-forum.pdf

[299] F. Radjai and V. Richefeu, "Contact dynamics as a nonsmooth discrete element method," *Mechanics of Materials*, vol. 41, no. 6, pp. 715–728, 2009.

[300] L. Brendel, T. Unger, and D. E. Wolf, "Contact dynamics for beginners," in *The Physics of Granular Media*, Weinheim: Wiley, 2004, pp. 325–343.

[301] T. Unger, L. Brendel, D. E. Wolf, and J. Kertész, "Elastic behavior in contact dynamics of rigid particles," *Physical Review E*, vol. 65, no. 6, Article 061305, 2002.

[302] D. H. Rothman and S. Zaleski, *Lattice-gas cellular automata: simple models of complex hydrodynamics*. Vol. 5. Cambridge, UK: Cambridge University Press, 2004.

[303] S. Succi, *The lattice Boltzmann equation: for fluid dynamics and beyond*. Oxford, UK: Oxford University Press, 2001.

[304] P. L. Bhatnagar, E. P. Gross, and M. Krook, "A model for collision processes in gases. I. Small amplitude processes in charged and neutral one-component systems," *Physical Review*, vol. 94, no. 3, p. 511, 1954.

[305] J. Rabault, U. Reglade, N. Cerardi, M. Kuchta, and A. Jensen, "Deep reinforcement learning achieves flow control of the 2d karman vortex street," *arXiv preprint arXiv:1808.10754*, 2018.

[306] A. Malevanets and R. Kapral, "Mesoscopic model for solvent dynamics," *Journal of Chemical Physics*, vol. 110, no. 17, pp. 8605–8613, 1999.

[307] G. Bird, "Approach to translational equilibrium in a rigid sphere gas," *Physical Fluids*, vol. 6, p. 1518, 1963.

[308] G. A. Bird, *Molecular gas dynamics and the direct simulation of gas flows*. Vol. 5. Oxford, UK: Clarendon Press, 1994.

[309] P. Hoogerbrugge and J. Koelman, "Simulating microscopic hydrodynamic phenomena with dissipative particle dynamics," *Europhysics Letters*, 1992. [Online]. Available: http://iopscience.iop.org/0295-5075/19/3/001

[310] P. Espanol and P. B. Warren, "Perspective: dissipative particle dynamics," *Journal of Chemical Physics*, vol. 146, no. 15, Article 150901, 2017.

[311] C. Marsh and J. Yeomans, "Dissipative particle dynamics: the equilibrium for finite time steps," *Europhysics Letters*, vol. 37, no. 8, p. 511, 1997.

[312] J. Yeomans, "Mesoscale simulations: Lattice Boltzmann and particle algorithms," *Physica A: Statistical Mechanics and its Applications*, vol. 369, no. 1, pp. 159–184, 2006.

[313] J. J. Monaghan, "Smoothed particle hydrodynamics," *Reports on Progress in Physics*, vol. 68, no. 8, p. 1703, 2005.

[314] R. Gingold and J. Monaghan, "Smoothed particle hydrodynamics – theory and application to non-spherical stars," *Monthly Notices of the Royal Astronomical Society*, vol. 181, pp. 375–389, 1977. [Online]. Available: http://adsabs.harvard.edu/abs/1977MNRAS.181..375G

[315] L. Lucy, "Smoothed particle hydrodynamics – theory and application to non-spherical stars," *Astronomy Journal*, vol. 82, pp. 1013–1024, 1977. [Online]. Available: http://adsabs.harvard.edu/abs/1977AJ.....82.1013L

[316] J. J. Monaghan, "Smoothed particle hydrodynamics," *Annual Review of Astronomy and Astrophysics*, vol. 30, no. 1, pp. 543–574, 1992.

[317] T. Harada, S. Koshizuka, and Y. Kawaguchi, "Smoothed particle hydrodynamics on gpus," 2007. [Online]. Available: http://inf.ufrgs.br/cgi2007/cd_cgi/papers/harada.pdf

[318] NIST, "Rydberg constant times hc in J." [Online]. Available: https://physics.nist.gov/cgi-bin/cuu/Value?rydhcj

[319] P. Pulay, "Ab initio calculation of force constants and equilibrium geometries in polyatomic molecules: I. theory," *Molecular Physics*, vol. 17, no. 2, pp. 197–204, 1969.

[320] P. Hohenberg and W. Kohn, "Inhomogeneous electron gas," *Physical Review*, vol. 136, no. 3B, p. B864, 1964.

[321] W. Kohn and L. J. Sham, "Self-consistent equations including exchange and correlation effects," *Physical Review*, vol. 140, no. 4A, Article A1133, 1965.

[322] C. Lee, W. Yang, and R. G. Parr, "Development of the colle-salvetti correlation-energy formula into a functional of the electron density," *Physical Review B*, vol. 37, no. 2, p. 785, 1988.

[323] F. Herman, J. P. Van Dyke, and I. B. Ortenburger, "Improved statistical exchange approximation for inhomogeneous many-electron systems," *Physical Review Letters*, vol. 22, no. 16, p. 807, 1969.

[324] A. Becke, "Density-functional exchange-energy approximation with correct asymptotic behavior," *Physical Review*, no. 38, p. 3098, 1988. [Online]. Available: http://journals.aps.org/pra/abstract/10.1103/PhysRevA.38.3098

[325] S. Grimme, "Semiempirical GGA-type density functional constructed with a long-range dispersion correction," *Journal of Computational Chemistry*, vol. 27, pp. 1787–1799, 2006. [Online]. Available: http://onlinelibrary.wiley.com/doi/10.1002/jcc.20495/abstract

[326] H. Hellmann, *Hans Hellmann: Einführung in Die Quantenchemie: Mit Biografischen Notizen von Hans Hellmann Jr.* Berlin, Germany: Springer, 2015.

[327] R. P. Feynman, "Forces in molecules," *Physical Review*, vol. 56, no. 4, p. 340, 1939.

[328] R. Car and M. Parrinello, "Unified Approach for Molecular Dynamics and Density-Functional Theory," *Physical Review Letters*, no. 55, p. 2471, 1985. [Online]. Available: http://journals.aps.org/prl/abstract/10.1103/PhysRevLett.55.2471

[329] M. Tuckerman, E. J. Bohm, L. V. Kalé, and G. Martyna, *Car–Parrinello Method*. Boston, MA: Springer US, 2011, pp. 220–227. [Online]. Available: https://doi.org/10.1007/978-0-387-09766-4_200

[330] T. D. Kühne, "Second generation Car–Parrinello molecular dynamics," *Wiley Interdisciplinary Reviews: Computational Molecular Science*, vol. 4, no. 4, pp. 391–406, 2014.

[331] Parrinello, M. and Rahman, A. (1980). Crystal structure and pair potentials: A molecular-dynamics study. Physical Review Letters, 45(14), 1196.

Index